Pulping and Papermaking of Nonwood Plant Fibers

Pulping and Papermaking of Nonwood Plant Fibers

Edited by

Z.M.A. Ainun
Program of Pulp and Paper & Pollution Control, Institute of Tropical Forestry and Forest Products (INTROP), Universiti Putra Malaysia, Serdang, Selangor, Malaysia

S.M. Sapuan
Advanced Engineering Materials and Composites Research Centre (AEMC), Department of Mechanical and Manufacturing Engineering, Universiti Putra Malaysia, Serdang, Selangor, Malaysia

R.A. Ilyas
Department of Chemical Engineering, Faculty of Chemical and Energy Engineering, Universiti Teknologi Malaysia, Johor Bahru, Malaysia

ELSEVIER

ACADEMIC PRESS
An imprint of Elsevier

Academic Press is an imprint of Elsevier
125 London Wall, London EC2Y 5AS, United Kingdom
525 B Street, Suite 1650, San Diego, CA 92101, United States
50 Hampshire Street, 5th Floor, Cambridge, MA 02139, United States
The Boulevard, Langford Lane, Kidlington, Oxford OX5 1GB, United Kingdom

ISBN: 978-0-323-91625-7

For Information on all Academic Press publications
visit our website at https://www.elsevier.com/books-and-journals

Publisher: Nikki P. Levy
Acquisitions Editor: Nancy J. Maragioglio
Editorial Project Manager: Catherine Costello
Production Project Manager: Sajana Devasi P. K.
Cover Designer: Miles Hitchen

Typeset by MPS Limited, Chennai, India

Contents

List of contributors

N. Abdullah Faculty of Plantation and Agrotechnology, Universiti Teknologi MARA Sarawak, Samarahan Campus, Kota Samarahan, Sarawak, Malaysia

Z.M.A. Ainun Program of Pulp and Paper & Pollution Control, Institute of Tropical Forestry and Forest Products (INTROP), Universiti Putra Malaysia, Serdang, Selangor, Malaysia

S. Amran Department of Wood Industry, Faculty of Applied Sciences, Universiti Teknologi MARA, Shah Alam, Selangor, Malaysia

F. Areej Department of Forest Products and Industries, Faculty of Forestry, University of Khartoum, North Khartoum, Sudan

Nishi Kant Bhardwaj Avantha Centre for Industrial Research & Development (ACIRD), Paper Mill Campus, Yamuna Nagar, Haryana, India

J.G. Boon Department of Bio and Natural Resources Technology, Faculty of Bioengineering and Technology, Universiti Malaysia Kelantan (UMK) Kampus Jeli, Jeli, Kelantan, Malaysia

Sunita Boruah Department of Textiles and Apparel Designing, College of Community Science, Assam Agricultural University, Jorhat, Assam, India

J.S. Bujang Department of Biology, Faculty of Science, Universiti Putra Malaysia, Serdang, Selangor Darul Ehsan, Malaysia

Widya Fatriasari Research Center for Biomass and Bioproducts, National Research and Innovation Agency (BRIN), Cibinong, Bogor, West Java, Indonesia

N.M.H. Hassan Wood Industry, Faculty of Applied Sciences, Universiti Teknologi MARA Pahang, Jengka Campus, Bandar Tun Abdul Razak, Pahang, Malaysia

R.A. Ilyas Department of Chemical Engineering, Faculty of Chemical and Energy Engineering, Universiti Teknologi Malaysia, Johor Bahru, Malaysia

M. Sarwar Jahan Pulp and Paper Research Division, Bangladesh Council of Scientific and Industrial Research (BCSIR), Dhanmondi, Dhaka, Bangladesh

Rafidah Jalil Forest Products Division, Forest Research Institute Malaysia (FRIM), Kepong, Selangor, Malaysia

Nurul Ain Nadirah Jamaluddin Forest Products Division, Forest Research Institute Malaysia (FRIM), Kepong, Selangor, Malaysia

Latifah Jasmani Biomass Technology Programme, Forest Products Division, Forest Research Institute Malaysia (FRIM), Kepong Selangor, Selangor, Malaysia; Forest Products Division, Forest Research Institute Malaysia (FRIM), Kepong, Selangor, Malaysia

Binita Baishya Kalita Department of Textiles and Apparel Designing, College of Community Science, Assam Agricultural University, Jorhat, Assam, India

Azizatul Karimah Research Center for Biomass and Bioproducts, National Research and Innovation Agency (BRIN), Cibinong, Bogor, West Java, Indonesia; Department of Forest Products, Faculty of Forestry, IPB University, Dramaga, Bogor, West Java, Indonesia

P. Kaur Avantha Centre for Industrial Research & Development (ACIRD), Paper Mill Campus, Yamuna Nagar, Haryana, India

Abdelhak Maghchiche Department of Pharmacy, Faculty of Medicine, University of Batna 2, Batna, Algeria

K. Mohd Ashadie Program of Pulp and Paper & Pollution Control, Institute of Tropical Forestry and Forest Products (INTROP), Universiti Putra Malaysia, Serdang, Selangor, Malaysia

Z. Muta Harah Department of Aquaculture, Faculty of Agriculture, Universiti Putra Malaysia, Serdang, Selangor Darul Ehsan, Malaysia; International Institute of Aquaculture and Aquatic Sciences (I-AQUAS), Universiti Putra Malaysia (UPM), Port Dickson, Negeri Sembilan, Malaysia

Vinod Kumar Nathan School of Chemical and Biotechnology, SASTRA Deemed to be University, Thanjavur, Tamil Nadu, India

Jannatun Nayeem Pulp and Paper Research Division, Bangladesh Council of Scientific and Industrial Research (BCSIR), Dhanmondi, Dhaka, Bangladesh

A. Nazrin Advanced Engineering Materials and Composites Research Centre (AEMC), Department of Mechanical and Manufacturing Engineering, Universiti Putra Malaysia, Serdang, Selangor, Malaysia

Yonghao Ni Department of Chemical Engineering, University of New Brunswick, Fredericton, NB, Canada; Department of Chemical and Biomedical Engineering, University of Maine, Orono, ME, United States

Razia Sultana Popy Pulp and Paper Research Division, Bangladesh Council of Scientific and Industrial Research (BCSIR), Dhanmondi, Dhaka, Bangladesh

M. Mostafizur Rahman Pulp and Paper Research Division, Bangladesh Council of Scientific and Industrial Research (BCSIR), Dhanmondi, Dhaka, Bangladesh

S.D. Ramaiya Department of Crop Science, Faculty of Agricultural Science and Forestry, Universiti Putra Malaysia Bintulu Campus, Bintulu, Sarawak, Malaysia

Nur Areisman Mohd Salleh Forest Products Division, Forest Research Institute Malaysia (FRIM), Kepong, Selangor, Malaysia

S.M. Sapuan Advanced Engineering Materials and Composites Research Centre (AEMC), Department of Mechanical and Manufacturing Engineering, Universiti Putra Malaysia, Serdang, Selangor, Malaysia

N. Sharma Avantha Centre for Industrial Research & Development (ACIRD), Paper Mill Campus, Yamuna Nagar, Haryana, India

N.M. Suhaimi Faculty of Applied Sciences, Universiti Teknologi MARA, Shah Alam, Selangor, Malaysia

N.N.F. Syed Department of Aquaculture, Faculty of Agriculture, Universiti Putra Malaysia, Serdang, Selangor Darul Ehsan, Malaysia

Sandeep Kumar Tripathi Avantha Centre for Industrial Research & Development (ACIRD), Paper Mill Campus, Yamuna Nagar, Haryana, India

Nyoman J. Wistara Department of Forest Products, Faculty of Forestry, IPB University, Dramaga, Bogor, West Java, Indonesia

S. Zakiah Program of Pulp and Paper & Pollution Control, Institute of Tropical Forestry and Forest Products (INTROP), Universiti Putra Malaysia, Serdang, Selangor, Malaysia; Advanced Engineering Materials and Composites Research Centre (AEMC), Department of Mechanical and Manufacturing Engineering, Universiti Putra Malaysia, Serdang, Selangor, Malaysia

Preface

Pulping and Papermaking of Nonwood Plant Fibers explains the meaning of nonwood fibers that are practical for pulp and paper production due to their chemical content of cellulose, hemicellulose, and lignin percentages. The nonwood fibers can be derived from many types of raw materials provided they are lignocellulosic-based materials. This book covers various types of raw materials from plant or agricultural waste that have proved their scientifical findings suitable for pulp and paper throughout many years worldwide. In pulp and paper technology, the processes involved are listed as plantation, harvesting, chipping, pretreatment, pulping, bleaching, stock preparation, papermaking, finishing, and paper testing as well. Hundreds of approaches are applied since centuries ago comprising mechanical, chemical, thermal, or combination methods in obtaining the pulp. End product of paper should be first identified before determining the pulp process methods. This is very crucial to optimize the performance of fibers produced via different techniques. Pulp properties need to be characterized before entering the papermaking process. This book also exhibits the differences of pulp properties derived from different types of nonwood fibers as well as paper properties. Once the paper is produced, paper testing needs to be carried out in terms of many aspects such as physical, mechanical, thermal, morphological, chemical, and more that are described in this book.

Z.M.A. Ainun
S.M. Sapuan
R.A. Ilyas

Chapter 1

Introduction to nonwood plant fibers for pulp and papermaking production

S.M. Sapuan[1], Z.M.A Ainun[2], S. Zakiah[1], A. Nazrin[1] and R.A. Ilyas[3]
[1]Advanced Engineering Materials and Composites Research Centre (AEMC), Department of Mechanical and Manufacturing Engineering, Universiti Putra Malaysia, Serdang, Selangor, Malaysia, [2]Program of Pulp and Paper & Pollution Control, Institute of Tropical Forestry and Forest Products (INTROP), Universiti Putra Malaysia, Serdang, Selangor, Malaysia, [3]Department of Chemical Engineering, Faculty of Chemical and Energy Engineering, Universiti Teknologi Malaysia, Johor Bahru, Malaysia

1.1 History of pulp and papermaking

Based on reports, the Chinese papermakers started the papermaking by using nonwood material called hemp. This was followed by other nonwoody plants like mulberry in the early second century CE, rattan, bamboo and then rice and wheat straw, and bark of sandalwood (Tsien, 1973). Paper has been applied by humans that began with the use of nonwood materials. At this moment, humans reached for substances that are situated close to them to be processed like nonwoody plants. Papermaking from the nonwood pulp of mulberry trees was reported as early as CE 105 in China (Kamoga Omar Lwako et al., 2013). Mulberry trees or papyrus are some plants that are easily accessible materials. They can be easily cut down and importantly easy to handle for the manufacture of paper pieces. To thin the tree trunk into slices or pieces, humans practice cooking, beating, screening, washing, and drying the sheets from the trunk until human can write on them. At first, this type of paper was a bit rough but the paper from nonwood material was very useful for the purpose of storing information. As time passed until the Industrial Revolution, paper production became more advanced as humans already had the technology of logging wood. Wood logging allowed humans to explore more on the usage of wood in papermaking to be better in paper quality in terms of strength because wood has a good fiber anatomy

Pulping and Papermaking of Nonwood Plant Fibers. DOI: https://doi.org/10.1016/B978-0-323-91625-7.00015-1

structure. The work on modern papermaking from wood pulps started in 1844 in Canada by Charles Fenerty, where he developed papers from woods, replacing traditional papers made from rags and cotton (John Crane, 2019). To date, humans have been relying on the use of wood in producing all categories of paper. However, in the last 30 years, the world is experiencing a decline in timber due to the growth of timber that takes 10–20 years to be logged causing humans to start turning back to nonwood trees. Thus many studies were carried out, and the appearance of industries that began to seriously use nonwood materials to be used as pulp for papermaking.

1.2 Why nonwood plant fibers for pulp and paper recently?

The reason for selecting nonwood plant fibers mainly due to the shortage of wood materials. Besides that, there are few other reasons as stated in Fig. 1.1.

1.2.1 Shortage of wood plant fibers

The world is facing a shortage of wood plant fibers due to the deforestation. The consumption of raw material sources comprises roundwood, chips and residues, and recovered paper. Virgin pulp from roundwood, chips, and residues is one of the important sources in imparting the strength of paper quality derived from wood plant. Most pulp and paper mills still need wood greatly to maintain the quality of paper products, besides blending with nonwood fibers or imparting with other additives. Looking at the high demand of wood, pulp and paper industry is an industry that gives high impact on the global forest and footprint to the environment. An amount of 400 million tonnes of paper and paperboard are produced each year worldwide, which involves 13%–15% of wood consumption. It is predicted that the world production will reach double by 2050 due to the high demand year by year (World Wide Fund For Nature, 2020) meaning that world is facing the declination of forest at about 13.0 million ha per year (Ashori, 2006). The rise of

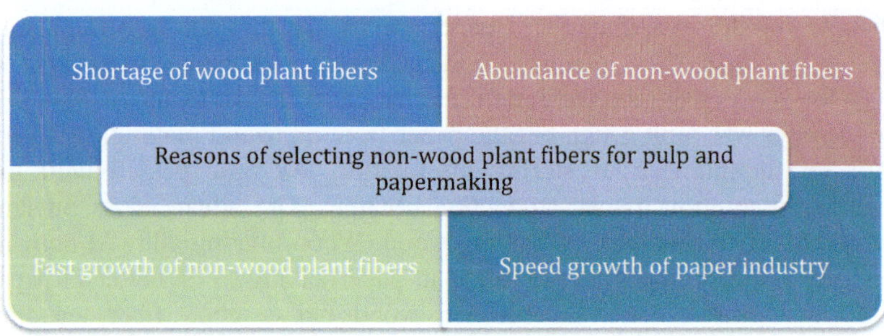

FIGURE 1.1 Reasons of using nonwood plant.

global population, the need for education, and industrialization are expected to increase the demand for paper. In other words, the products of pulp, paper, and packaging contributed to high impact on the social, economic, and infrastructure developments. In response to the concern of environmentalists, governmental, and nongovernmental administrations, new legislative regulations are enacted to supervise the logging activities to maintain the supply and price of the wood to the international market. Shortage of wood-based materials caused the need to seek for alternative materials, and the use of nonwood fibers seemed to be the obvious choice. The forest still must be sustainable, which can be achieved by the responsible management of plantations and reducing the usage of wood by replacing it with other lignocellulosic materials.

1.2.2 Abundance of nonwood plant fibers

The size of nonwood pulp market globally will increase year by year, which involved stalk, bast, leaf, cotton fiber, or more. The market from North America, Europe, Asia Pacific and South America, and the Middle East and Africa, which to date, 133 companies have applied nonwood pulps in their wood-based pulp slurries for commercial papermaking purposes as shown in Table 1.1.

Based on the record by the Food and Agriculture Organization of the United Nations (2022), there are 43 countries that are contributing to pulp production of nonwood fibers as shown in Table 1.2. Two main countries that recently led the production are China and India with more than 5 million and 3 million tonnes, respectively.

The substitution of wood with low-cost alternative raw materials was proposed to utilize nonwood fibers such as agro-wastes and annual plants for

TABLE 1.1 Nonwood pulp producers by region captured in 2021 (Fisher International, 2022).

Region	Number of companies
1. Asia Pacific	93
2. Europe	17
3. North America	8
4. Africa	7
5. Latin America	5
6. Middle East	3
Total	133

TABLE 1.2 Pulp production from other than wood (Food and Agriculture Organization of the United Nations, 2022).

Country	Pulp production (tonnes)		
	2018	**2019**	**2020**
1. Algeria	3000	2000	2000
2. Argentina	129,000	130,000	113,000
3. Bangladesh	18,000	18,000	18,000
4. Belgium	1790	1790	1790
5. Brazil	63,000	63,000	63,000
6. Canada	40,000	40,000	40,000
7. China	6,104,000	5,877,000	5,225,000
8. Colombia	196,486	188,681	177,000
9. Costa Rica	171,000	171,000	171,000
10. Cuba	100	100	100
11. Czechia	428,926	441,278	441,278
12. Republic of Korea	50,000	50,000	50,000
13. Ecuador	16,000	16,000	16,000
14. Egypt	120,000	120,000	120,000
15. Ethiopia	9400	9400	9400
16. France	8100	8100	8100
17. Germany	84,384	66,181	65,285
18. India	3,020,000	3,020,000	3,020,000
19. Indonesia	105,000	105,000	105,000
20. Iran	51,000	51,000	51,000
21. Iraq	2000	2000	2000
22. Italy	202,890	202,890	202,890
23. Japan	6000	5000	5000
24. Jordan	8000	8000	8000
25. Madagascar	2000	2000	2000
26. Mexico	17,000	17,000	17,000
27. Morocco	5000	5000	5000
28. Myanmar	386,000	1000	1000

(*Continued*)

TABLE 1.2 (Continued)

Country	Pulp production (tonnes)		
	2018	2019	2020
29. Nepal	15,000	15,000	15,000
30. Pakistan	370,000	370,000	370,000
31. Philippines	26,000	25,000	33,000
32. Poland	56,716	116,900	120,000
33. Russian Federation	100,000	100,000	100,000
34. Slovakia	2000	1000	1900
35. South Africa	41000	41000	41000
36. Sri Lanka	18,500	18,500	18,500
37. Thailand	118,000	118,000	209,000
38. Tunisia	2000	2000	2000
39. Turkey	53,000	53,000	53,000
40. United Kingdom and Ireland	7000	7000	7000
41. United States of America	168,205	146,000	149,347
42. Uzbekistan	17,300	17,890	21,358
43. Vietnam	60,000	60,000	60,000

reliable resources as well as to compensate the lack of forest resources. Through 1900 years, paper was made from grass and other nonwood fibers instead of wood (Mohieldin, 2015). In comparison to wood, nonwood fibers are higher in hemicelluloses and lower in lignin. Most nonwood fibers were derived from agriculture by-products, which came at lower price, require less refining, have fast annual growth, make excellent filler, and have decent smoothness. The processing of nonwood fiber requires less energy and lower chemical expenses thus providing economically viable operation for the smaller millers. Furthermore, the exploitation of these agro-wastes potentially generates additional income to the smallholders without compromising the yield of the main products.

1.2.3 Speed growth of paper industry

Pulp and paper industry cannot be hindered from human life, although the world is drastically changing into digital era. Last 25 years, the world expected that the consumption of paper will tremendously drop due to the emergence of digital technology (Beckline et al., 2016); however, it is reported that the need is getting greater each year. Paper plays a vital role in the development of global social, economic, and environmental growth in many areas, such as packaging, communication, education, sanitation, security, and more (Ezeudu et al., 2019). In fact, the reality that is happening shows that the industry is thriving and exploding the worldwide industry with interesting trending, especially in packaging paper. It is known that one of the largest industries in the world is pulp and paper industry that is led by countries of the United States of America and China, followed by North American, Northern European, and East Asian. The world production of paper and paperboard is around 390 million tonnes and is expected to reach 490 million tonnes by 2022 (Linchpin et al., 2022). Per capita, paper consumption can be one of the good indicator or a measure of the growth of a country. More than 200 kg per capita consumption is usually obtained by highly developed countries. Paper Export (2022) predicted that there will be an increase in paper packaging in terms of corrugated and containerboard to alternate with plastic packaging. These papers have skyrocketed since 2020 reported for 18% of all global retail sales and will keep on growing via e-commerce business. For example, e-commerce sales in the United States booming with an increase of $80 billion reaching $438 billion in the first half of 2021 (Paper Export, 2022).

1.2.4 Fast growth of nonwood plant fibers

Based on the studies of Hurter (1997), nonwood plants can be categorized into common and specialty nonwoods. It is also called hardwood or softwood substitutes. The common nonwood plant comprises sugarcane bagasse, bamboo, reeds, grasses, and esparto, while specialty nonwood involves cotton linters, flax, hemp, sisal, abaca, kenaf bast, and more. Liu et al. (2018) divided nonwood plants according to plant's components listed as gramineous, bast, seed hull, and leaf fiber materials.

1.3 Nonwood plant fibers

Nonwood plant fibers are lignocellulosic plants that contain mainly cellulose, hemicellulose, aromatic polymer like lignin and other small portion of chemical substances (Tarique et al., 2021; Ilyas et al., 2019, 2021). These substances are considered chemical composition in a plant. Chemical composition differs among lignocellulosic plants based on its ages, climate, soil, and geographical location (Ilyas et al., 2017). In addition, the variation is also found in different parts of the same plant. Lignocellulosic plants can be generally divided into wood and nonwood plants (Hammett et al., 2001). The nonwood plant usually has higher hemicellulose and lower lignin content compared to wood plant that has greater cellulose content yielding higher pulp content after cooking process (Abd El-Sayed et al., 2020). Lignin is responsible for binding the cellulose fiber together. In papermaking, lignin is undesirable as it might lower the performance of the paper (Małachowska et al., 2020). The low content of lignin in nonwood plant is favorable since it eases the delignification process with less chemicals and ultimately reducing hazard contamination to the environment (Daud et al., 2013). Nonwood plants can be classified into three main types according to their origin, which are agricultural by-products, industrial crops, and naturally growing plants (Table 1.3). Agricultural by-products are defined as the secondary products of the principal crops such as rice straw and wheat straw, which possess moderate quality and low cost (Ilyas et al., 2022a,b,c; Azman et al., 2021). Industrial crops such as kenaf, hemp, and jute are cultivated

TABLE 1.3 The category of nonwood plant fibers and their examples.

Categories of nonwood plant fibers		Examples	Ready to be harvested (months)
	Gramineous	Bamboo, bagasse, rice straw, wheat straw, sorghum residue, corn straw, reed	4 years, 12–14 months, 4–6 months, 6 months, 110–140 days, 60–80 days, 12 months
	Bast	Mulberry, cotton stalk bark; hemp, kenaf, jute, hemp, flax	4 years, months, 10 months, 4–6 months, 120–150 days, 10 months, 100 days
	Seed hull	Cotton, cotton linters	2 months.
	Leaf	Banana leaves, sisal, abaca, sugarcane leaves, bagasse, bamboo, cotton, hemp, kenaf	11–12 months, 2–3 years, 12–25 months, 12 months, 4–6 months, 4 years, 2 months, 10 months, 4–6 months

TABLE 1.4 Types of nonwood plant for pulp and paper manufacture.

Types of nonwood plant	Name of plant
Agricultural by-products	Rice straw, wheat straw, bagasse
Industrial crops	Kenaf, hemp, jute
Naturally growing plants	Bamboo, reed grass, sabia grass

particularly to yield fibers, hence, exploiting these crops for pulp production is weighted by the high expense of raw materials. Naturally growing plants, including bamboo, reed, and sabia grass, have also been utilized to produce high-quality pulps (Table 1.4).

Nonwood has some differences compared to wood materials (Table. 1.5).

The problems that rise in managing these resources include collection, transportation, storage, handling, washing, bleaching, papermaking, and chemical recovery.

1.4 Challenges and future of nonwood plants

More countries showing their initiatives to use nonwood fibers for pulp and papermaking besides countries that are having limited wood supply. The future of nonwood plant in pulp and paper industries getting brighter due to the four factors discussed earlier. However, the utilization of nonwood fiber sources in the pulp and paper industry has several disadvantages (Abd El-Sayed et al., 2020). The most common issues are technological (Jahan et al., 2021), and they necessitate research in this field to compete with wood fibers. The following section discusses some of the challenges faced by mills utilizing nonwood plant fibers as source of raw material, as well as some of the solutions proposed to make nonwoods more appealing as papermaking raw material. Currently, a high-tech breakthrough in all areas of papermaking (including raw materials processing, paper machine, and nanotechnology) has made nonwood more economical when compared to wood as a raw material for papermaking. Challenges for nonwood plants are shown in Fig. 1.2 and Table 1.6.

The pulp and paper industry has supplied over 6 billion people worldwide. Paper is used for a variety of applications, such as writing media, printing materials, documentation tools, beauty tools, packaging tools, transaction tools, cleaning tools, and decorations. Pulp and paper mills around the world produce more than 400 million tons of paper annually. These paper-producing activities generated a variety of jobs but led to extreme deforestation (Furszyfer Del Rio et al., 2022). Forest impacts associated with unsustainable pulpwood production can negatively impact climate, such as in

TABLE 1.5 The physicochemical properties of some wood and nonwood plants for papermaking.

Nonwood plant	Bagasse	Bamboo	Hemp	Jute	Kenaf	Reed grass	Rice straw	Wheat straw
Average fiber length (mm)	1.70	1.36–4.03	20	2.5	2.74	1.5	1.41	1.48
Average diameter (μm)	20	8–30	22	18	20	20	8	13
L/D ratio	85:1	135–175:1	1000:1	139:1	135:1	75:1	175:1	110:1
Alpha cellulose (%)	32–44	26–43	55–65	61	31–39	45	28–36	29–35
Ash (%)	1.5–5	1.7–5	5–7	1.6	2–5	3	15–20	4–9
Lignin (%)	19–24	21–31	2–4	11.5	15–18	22	12–16	16–21
Hemicellulose (%)	27–32	15–26	4–7	24	21–23	20	23–28	26–32
Silica (%)	0.7–3	1.5–3	<1	<1	–	2	9–14	3–7
Alcohol benzene solubility (%)	1.7	2.3	2.6	2.4	2.1	6.4	0.56	4.01
Hot water solubility (%)	4.4	4.8	20.5	3.7	5.0	5.4	7.3	12.27
1% sodium hydroxide solubility (%)	33.9	24.9	–	28.5	28.4	34.8	57.7	43.58

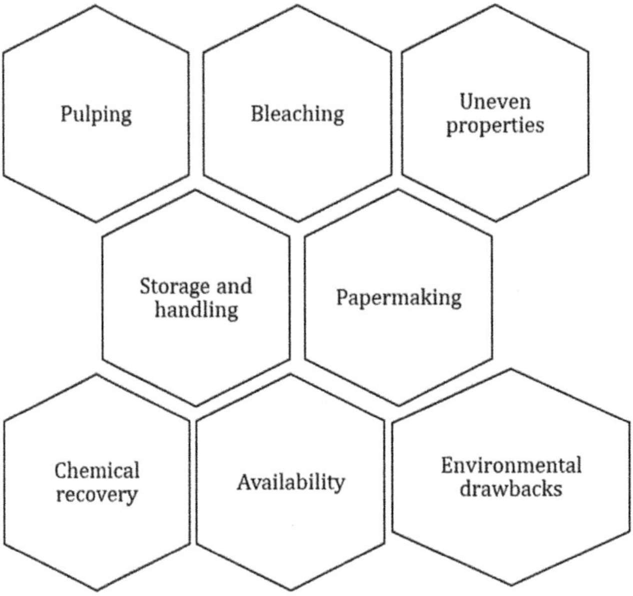

FIGURE 1.2 Challenges for nonwood plants.

TABLE 1.6 Details of challenges for nonwood plants.

Challenges	
Availability	• Collecting and transporting is main cost for these nonwood fibers (Kissinger et al., 2007). • Not easy to handle (Bajpai, 2018a,b). • Manually collected resulted labor costs form a large ratio of total raw material costs. • Volatile raw materials due to agricultural residues that depend on unpredictable weather phenomena (Kissinger et al., 2007).
Storage and handling	• Most nonwood fiber are seasonal availability and have a moderately short harvesting time (González-García et al., 2010; Ashori, 2006). • Bulky and thus requires large storage spaces. • Need protection from deterioration during storage owing to activities of undesirable microorganisms (Abou-Yousef et al., 2005). • Microorganism's activities lead to chemical degradation and discoloration.

(*Continued*)

TABLE 1.6 (Continued)

Challenges	
Pulping	• Most nonwood fibers are high in silica content (Nicacio et al., 2022). • This high content of silica causes many problems. Washing is difficult because of poor drainage of pulp and high viscosity of black liquor (Ashori, 2006). • A lot of water is needed, thus reducing the solid content of the black liquor and increasing the load on the evaporators (Bajpai, 2018a,b). • Low yield during pulping (Jahan et al., 2021).
Bleaching	• During storage, nonwoods tend to deteriorate and the resultant pulp, therefore, has a low initial brightness (Abd El-Sayed et al., 2020). • To make this pulp desirable, it has to be exposed to enzymatic pretreatment or a more severe bleaching action. • The CEH [chlorination (C), alkaline extraction (E), and hypochlorite (H) stage] three-stage bleaching process is mainly adopted in traditional nonwood pulping, but it has heavy waste water pollution (Mboowa, 2021).
Papermaking	• Nonwood raw materials are slow to drain and consequently, cause problems in the paper machine (Jahan et al., 2021). • Paper machine cannot be run at high speeds; thus low the production become low.
Chemical recovery	• Most nonwood paper mills are small in scale as compared to wood-based paper mills. • These mills almost never have a chemical recovery system. • Nonwood paper mills release the effluents directly into the environment. • The present of the silica within the raw material dissolves in the cooking liquor resulting in highly viscous black liquor (Yuan et al., 2016). • Silica causes trouble in evaporator, recovery boilers, causticizing equipment, and soda ovens.
Environmental drawbacks	• Loss of pulping chemicals in black liquor (Abd El-Sayed et al., 2020). • Huge national waste of electricity in manufacturing the chemicals.
Properties	• Uneven chemical properties, such as cellulose content, fiber orientation, crystal structure, and diameter/cross-section area of the fiber. • The fiber cross-section area has a significant impact on fiber strength measurement. • Nonwood biomass fiber is strongly influenced by the strength of its constituent cells, namely, affected by thickness of fiber wall (Dungani et al., 2014).

Sumatra where carbon is released from deep peatlands that are converted to pulp plantations. In addition, pulp and paper manufacturing is among the world's most energy-intensive industries. Although paper mills do utilize some of their own waste products as fuel, emissions and pollution from paper mills can be significant. The largest share of greenhouse gases released in pulp and paper manufacturing comes from the energy production to power the mills. Some companies in the pulp and paper industry leave an unacceptably large ecological footprint on the planet. Irresponsible harvesting from natural forests and an establishment of pulp plantations on converted natural forests can threaten fragile ecosystems and species and cause soil erosion. Some proposed new pulpwood plantations and mills threaten natural habitats in regions with high conservation values and high rates of illegal logging. For example, the remaining natural forests and associated wildlife species in Borneo and Sumatra, Papua New Guinea, the Russian Far East, Southern Chile, and the Atlantic forest region in Brazil are all at risk because of growing demand for pulpwood.

By 2020, global paper consumption is estimated to have reached 500 million tonnes. The challenge of how to meet the growing worldwide demand for paper and paperboard arises. The increasing demand for paper will almost certainly be met by one or more of several potential sources of supply, including:

1. Innovation of environmentally friendly pulp process alternative as bio pulp
2. Increased use of nonwood fiber
3. Increased use of forest waste
4. Increased recycling of waste paper
5. Increased world timber supply
6. Increased yield with better control of pests

1.5 Conclusion

Pulp technology from nonwood sources will continue to develop rapidly due to the significant increase in demand for paper products due to the high cost of pulp imports as well as the lack of timber resources in developing countries. Due to the increasing current rate of investment in pulp and paper mills, the pulp and paper industry in developing countries began to look for alternative raw materials to wood. Furthermore, due to economic and environmental considerations, agricultural waste pulp is becoming increasingly popular. Pulp research from nonwood sources has produced a wide range of products over the years. The use of nonwood fibers in the pulp and paper manufacturing industry is no longer a concern as it once was. However, the nonwood paper industry continues to face some challenges. Through research done over 20 years, bagasse fibers are seen to have good papermaking properties. By-products from the sugar mills such as sugarcane bagasse have provided a source of ancillary income to the industry. In addition, it provides additional benefits to the sugarcane industry. This is due to the reduction in the cost of collecting sugarcane residue, which indirectly the sugarcane mill becomes the center for the collection of sugarcane residue. As a result, in countries where the cost of replacement fuel for sugarcane bagasse in sugar mills is cheap, using sugarcane residue as a raw material for papermaking is an attractive proposition. Besides that, the use of empty fruit bunch as a source of pulp and paper is also seen as a great potential, especially in countries that plant oil palm trees such as in Malaysia and Indonesia. In addition, the utilization of straw as a raw material for papermaking shows several issues, but research has offered solutions to the majority of them. Rice straw is a cheap nonwood fiber material for papermaking. Despite the current concern about the expansion of wood-based pulp mills, the expansion of nonwood pulp mills has not stopped. In fact, for the foreseeable future, it is predicted to grow faster than wood-based paper mills. Taking this factor into account, it is predicted that the use of nonplant wood fibers in the pulp and paper industry will continue to increase at a rapid rate and contribute to a partial percentage of the total raw materials of papermaking.

References

Abd El-Sayed, E.S., El-Sakhawy, M., El-Sakhawy, M.A.-M., 2020. Non-wood fibers as raw material for pulp and paper industry. Nordic Pulp Pap. Res. J. 35 (2), 215–230. Available from: https://doi.org/10.1515/npprj-2019-0064.

Abou-Yousef, H., El-Sakhawy, M., Kamel, S., 2005. Multi-stage bagasse pulping by using alkali/Caro's acid treatment. Ind. Crop. Products 21 (3), 337–341. Available from: https://doi.org/10.1016/j.indcrop.2004.05.001.

Ashori, A., 2006. Nonwood fibers—a potential source of raw material in papermaking. Polymer-Polym. Technol. Eng. 45 (10), 1133–113636. Available from: https://doi.org/10.1080/03602550600728976.

Azman, M.A., Asyraf, M.R.M., Khalina, A., Michal Petrů, C.M., Ruzaidi, S.M., Sapuan, W.B., et al., 2021. Natural fiber reinforced composite material for product design: a short review. Polymers 13 (12), 1917. Available from: https://doi.org/10.3390/polym13121917.

Bajpai, P., 2018a. Non-wood fiber use in pulp and paper, Biermann's Handbook of Pulp and Paper, third ed. Elsevier, Amsterdam, Netherlands, pp. 261–278.

Bajpai, P., 2018b. Brief description of the pulp and papermaking process. Biotechnol. Pulp Pap. Process. 9–26. Available from: https://doi.org/10.1007/978-981-10-7853-8_2.

Beckline, M., Yujun, S., Eric, Z., Kato, M.S., 2016. Paper consumption and environmental impact in an emerging economy. J. Energy Environ. Chem. Eng. 1 (1), 13–18.

Daud, Z., Sari, A., Kassim, M., Aripin, A.M., Awang, H., Hatta, Z.M., 2013. Chemical composition and morphological of cocoa pod husks and cassava peels for pulp and paper production. Austr. J. Basic Appl. Sci. 7 (9), 406–411.

Dungani, R., Abdul Khalil, H.P.S., Sumardi, I., Suhaya, Y., Sulistyawati, E., Nazrul Islam, M., et al., 2014. Non-wood renewable materials: properties improvement and its application.

Biomass and Bioenergy. Springer International Publishing, Cham, pp. 1–29. Available from: https://doi.org/10.1007/978-3-319-07578-5_1.

Ezeudu, O.B., Agunwamba, J.C., Ezeasor, I.C., Madu, C.N., 2019. Sustainable production and consumption of paper and paper products in Nigeria: a review. Resources 8 (1), 53.

Fisher International, 2022. Are More Paper Companies Incorporating Non-Wood Pulps? (fisheri.com).

Food and Agriculture Organization of the United Nations, 2022. Home | Food and Agriculture Organization of the United Nations (fao.org).

Furszyfer Del Rio, D.D., Sovacool, B.K., Griffiths, S., Bazilian, M., Kim, J., Foley, A.M., et al., 2022. Decarbonizing the pulp and paper industry: a critical and systematic review of sociotechnical developments and policy options. Renew. Sustain. Energy Rev. 167 (October), 112706. Available from: https://doi.org/10.1016/j.rser.2022.112706.

González-García, S., Hospido, A., Feijoo, G., Moreira, M.T., 2010. Life cycle assessment of raw materials for non-wood pulp mills: hemp and flax. Resources Conserv. Recycling 54 (11), 923–930. Available from: https://doi.org/10.1016/j.resconrec.2010.01.011.

Hammett, A.L., Youngs, R.L., Sun, X., Chandra, M., 2001. Non-wood fiber as an alternative to wood fiber in Chinas pulp and paper industry. Holzforschung 55 (2), 219–224. Available from: https://doi.org/10.1515/HF.2001.036.

Hurter, R.W., 1997. Nonwood plant fiber characteristics. Agric. Residues 1, 1–4.

Ilyas, R.A., Sapuan, S.M., Ishak, M.R., Zainudin, E.S., 2017. Effect of delignification on the physical, thermal, chemical, and structural properties of sugar palm fibre. BioResources 12 (4), 8734–8754. Available from: https://doi.org/10.15376/biores.12.4.8734-8754.

Ilyas, R.A., Sapuan, S.M., Ishak, M.R., Zainudin, E.S., 2019. Sugar palm nanofibrillated cellulose (*Arenga pinnata* (Wurmb.) Merr): effect of cycles on their yield, physic-chemical, morphological and thermal behavior. Int. J. Biol. Macromol. 123 (February), 379–388. Available from: https://doi.org/10.1016/j.ijbiomac.2018.11.124.

Ilyas, R.A., Sapuan, S.M., Atikah, M.S.N., Asyraf, M.R.M., Ayu Rafiqah, S., Aisyah, H.A., et al., 2021. Effect of hydrolysis time on the morphological, physical, chemical, and thermal behavior of sugar palm nanocrystalline cellulose (*Arenga pinnata* (Wurmb.) Merr). Text. Res. J. 91 (1–2), 152–16767. Available from: https://doi.org/10.1177/0040517520932393.

Ilyas, R.A., Zuhri, M.Y.M., Aisyah, H.A., Asyraf, M.R.M., Hassan, S.A., Zainudin, E.S., et al., 2022a. Natural fiber-reinforced polylactic acid, polylactic acid blends and their composites for advanced applications. Polymers 14 (1), 202. Available from: https://doi.org/10.3390/polym14010202.

Ilyas, R.A., Zuhri, M.Y.M., Norrahim, M.N.F., Mohamad Misenan, M.S., Jenol, M.A., Samsudin, S.A., et al., 2022b. Natural fiber-reinforced polycaprolactone green and hybrid biocomposites for various advanced applications. Polymers 14 (1), 182. Available from: https://doi.org/10.3390/polym14010182.

Ilyas, R.A., Aisyah, H.A., Nordin, A.H., Ngadi, N., Yusoff Mohd Zuhri, M.Y., Asyraf, M.R.M., Sapuan, S.M., et al., 2022c. Natural-fiber-reinforced chitosan, chitosan blends and their nanocomposites for various advanced applications. Polymers 14 (5), 874. Available from: https://doi.org/10.3390/polym14050874.

Jahan, M.S., Rahman, M.M., Ni, Y., 2021. Alternative initiatives for non-wood chemical pulping and integration with the biorefinery concept: a review. Biofuels Bioprod. Bioref. 15 (1), 100–118. Available from: https://doi.org/10.1002/bbb.2143.

John Crane, 2019. Unearthing the History of Pulp and Paper Processing: Celebrating the Father of Modern Papermaking | John Crane.

Kamoga Omar Lwako, M., Byaruhanga, J.K., Kirabira, J.B., 2013. A review on pulp manufacture from non wood plant materials. Int. J. Chem. Eng. Appl. 4 (3), 144–148.

Kissinger, M., Fix, J., Rees, W.E., 2007. Wood and non-wood pulp production: comparative ecological footprinting on the Canadian prairies. Ecol. Econ. 62 (3–4), 552–558. Available from: https://doi.org/10.1016/j.ecolecon.2006.07.019.

Linchpin SEO, 2022. Trends Transforming The Pulp And Paper Industry In 2022 By Linchpin (linchpinseo.com).

Liu, Z., Wang, H., Hui, L., 2018. Pulping and papermaking of non-wood fibers. Pulp Pap. Process. 1, 4–31.

Małachowska, E., Dubowik, M., Boruszewski, P., Łojewska, J., Przybysz, P.P., 2020. Influence of lignin content in cellulose pulp on paper durability. Sci. Rep. 10 (1), 19998. Available from: https://doi.org/10.1038/s41598-020-77101-2.

Mboowa, D., 2021. A review of the traditional pulping methods and the recent improvements in the pulping processes. Biomass Convers. Biorefinery . Available from: https://doi.org/10.1007/s13399-020-01243-6.

Mohieldin, S.D., 2015. Pretreatment approaches in non-wood plants for pulp and paper production: a review. J. For. Prod. Ind. 3 (2), 84–88.

Nicacio, M.A., Paes, J.B., Bufalino, L., Lopez, Y.M., de Sousa, S.F., Verly Lopes, D.J., 2022. Effect of silica bionanoparticles obtained from rice husk for the treatment of wood of *Pinus elliottii*. J. Wood Chem. Technol. 42 (3), 158–17070. Available from: https://doi.org/10.1080/02773813.2022.2055074.

Paper Export, 2022. https://papertoexport.com/predictions-for-the-global-pulp-and-paper-industry-in-2022/.

Tarique, J., Sapuan, S.M., Khalina, A., Sherwani, S.F.K., Yusuf, J., Ilyas, R.A., 2021. Recent developments in sustainable arrowroot (*Maranta arundinacea* Linn) starch biopolymers, fibres, biopolymer composites and their potential industrial applications: a review. J. Mater. Res. Technol. 13 (July), 1191–1219. Available from: https://doi.org/10.1016/j.jmrt.2021.05.047.

Tsien, T.H., 1973. Raw materials for old papermaking in China. J. Am. Orient. Soc. 93 (4), 510–519.

World Wide Fund For Nature, 2020. https://wwf.panda.org/discover/our_focus/forests_practice/forest_sector_transformation_updated/pulp_and_paper/.

Yuan, Z., Kapu, N.S., Beatson, R., Chang, X.F., Martinez, D.M., 2016. Effect of alkaline preextraction of hemicelluloses and silica on kraft pulping of bamboo (*Neosinocalamus affinis* Keng. Ind. Crop. Products 91, 66–75. Available from: https://doi.org/10.1016/j.indcrop.2016.06.019.

Further reading

Abitbol, T., Rivkin, A., Cao, Y., Nevo, Y., Abraham, E., Ben-Shalom, T., et al., 2016. Nanocellulose, a tiny fiber with huge applications. Curr. Opin. Biotechnol. 39 (I), 76–88. Available from: https://doi.org/10.1016/j.copbio.2016.01.002.

Abral, H., Basri, A., Muhammad, F., Fernando, Y., Hafizulhaq, F., Mahardika, M., et al., 2019. "A simple method for improving the properties of the sago starch films prepared by using ultrasonication treatment. Food Hydrocoll. 93, 276–283. Available from: https://doi.org/10.1016/j.foodhyd.2019.02.012.

Chapter 2

Pulping process for nonwoody plants

F. Areej[1], K. Mohd Ashadie[2], S. Zakiah[2] and Z.M.A. Ainun[2]
[1]*Department of Forest Products and Industries, Faculty of Forestry, University of Khartoum, North Khartoum, Sudan,* [2]*Program of Pulp and Paper & Pollution Control, Institute of Tropical Forestry and Forest Products (INTROP), Universiti Putra Malaysia, Serdang, Selangor, Malaysia*

2.1 Introduction

Pulping is a process of converting lignocellulosic materials into pulp. Pulp is an intermediate material applied in papermaking. There are huge numbers of technologies of pulping methods that can be generally divided into three categories: mechanical (MP), chemical (CP), and hybrid pulping (HP).

In comparison, MP can produce higher yields of pulps compared to other pulping methods, but the pulps are having lower strength due to less bonding among the fibers. Hence, the application of HP is found better to prevail over the disadvantages of mechanical pulps. Most previous studies attempted to save energy and reduce chemical usage and, at the same time, optimize pulp yields and quality, especially via bio-based technology.

The objective of pulping is to dislodge and remove as much lignin while maintaining the fiber structure. Selection of pulping method is affected by final paper products, type of lignocellulosic species, etc.

2.2 Type of pulping for nonwoody fibers

Pulps produced from different pulping strategies or techniques have different properties, which are suitable for the specified end products such as paper and fiberboard.

Categorized as gramineous, bast, seed hull, and leaf fiber, nonwoody fibers, such as kenaf, bagasse, bamboo, cotton, wheat straw, hemp, jute, and oil palm fibers, require less chemical, less energy, and less time consumption

Pulping and Papermaking of Nonwood Plant Fibers. DOI: https://doi.org/10.1016/B978-0-323-91625-7.00007-2

at relatively low cost in pulping. The performance of these fiber materials can be estimated by the following factors:

1. economically viability in pulping for the fiber contents and forms;
2. degree of delignification difficulty and fiber dissociation in pulping process;
3. pulping methods adaptability;
4. performance of the pulp (color, degree of bleaching difficulty, and beating performance).

In this chapter, we considered chemical, mechanical, and hybrid as the pulping strategies for nonwoody fibers (Fig. 2.1). Statistical report showed that chemical pulping techniques produce 74.1% of the world pulp, while 21.4% of pulp is attained by mechanical pulping process, and 4.5% of the pulp is produced by using other techniques (Sabharwal et al., 1994). However, some of these pulping techniques had been in use since the ancient times and are ultimately due for improvement to overcome their drawbacks (Azeez, 2018).

2.2.1 Chemical pulping

Chemical methods are mostly utilized when pulping nonwood fibers, and they include alkaline, sulfite, organic solvent (organosolv) and also involve biological and chemi-mechanical processes. In chemical pulping the raw materials are cooked with appropriate chemicals in an aqueous solution such as alkaline (kraft) or acidic (sulfite) at an elevated temperature and pressure

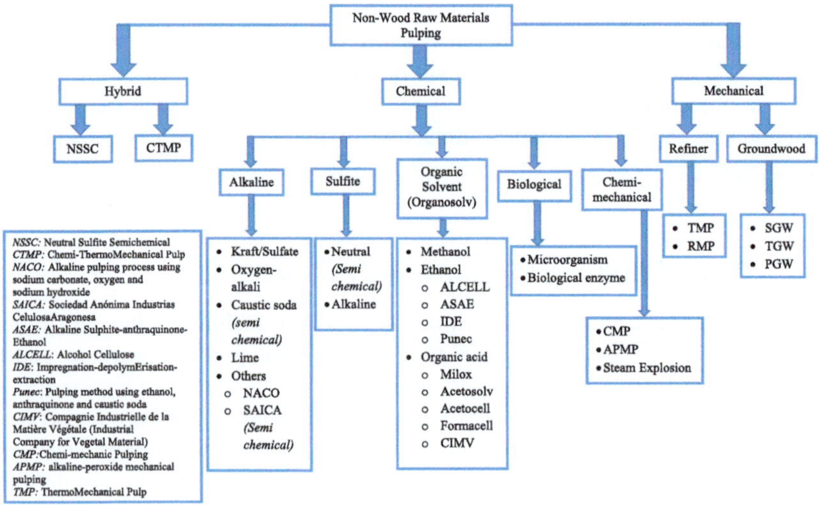

FIGURE 2.1 Pulping for nonwoody fibers strategy.

(Pratima, 2016). This is to degrade lignin and hemicelluloses into small water-soluble molecules that can be washed away from the cellulose fibers without depolymerizing the cellulose fibers (Azeez, 2018) where chemically depolymerizing the cellulose will weaken the fibers.

2.2.1.1 *Alkaline pulping*

Alkaline pulping is a widely used chemical pulping method in which an alkaline solution, such as NaOH and Na_2CO_3, is used to prepare plant fiber raw materials. Kraft or sulfate, oxygen alkali, caustic soda, lime, and others, such as NACO and SAICA, are types of alkaline pulping for nonwood fibers. Generally, alkaline nonwood pulps contain much hemicellulose, while their fibers are short. During alkaline pulping, the aqueous solution of alkaline chemical agent will be utilized to treat fiber materials to dissolve most of the lignin and separate the fibers from the material into pulp.

The main problem with alkaline processes for nonwood fibers is that silicates of nonwood plants dissolve during cooking into the cooking liquor. The presence of silicate ions causes serious problems in recovery such as scaling on the heat transfer surfaces in the evaporator, high viscosity of the concentrated liquor (negative influence on both evaporation and combustion), and also problems in causticizing (Myreen, 2001).

Kraft pulping or known as sulfate pulping is not common for nonwood raw materials. The most immediate factor is that the strength of nonwood kraft pulping is inferior to that of alkaline pulping. However, this process can be considered for pulping of leaf (banana, sisal, etc.) and bast fibrous raw materials (cotton stalk, kenaf, jute, etc.) where it produced bleachable and high-grade unbleached pulps. The main components of kraft cooking liquor are NaOH and Na_2S. However, Na_2S causes environmental problems by producing malodorous gases (mercaptans, ether alcohols, and H_2S) during cooking process. The use of mild lime or sodium carbonate during pulping process results in the production of pulp with durable, strong, and hard properties. For instance, when jute is pulped using this process, long fibers are obtained with increased strength similar or greater than those of softwood pulps.

Oxygen−alkali pulping is a newly developed nonpolluting pulping method by using oxygen as a cooking agent. It can eliminate waste gas pollution generated during cooking. The mechanism of oxygen−alkali cooking is the dissolution of lignin and the breakage of chains. In alkaline condition, through ionization, the free phenolic hydroxyl groups turn into negative ions that can react with oxygen to generate peroxide intermediates of cyclohexadienone, followed by oxidative degradation which can change lignin structure and generate the side chain cleavage. This series of changes in lignin structure can raise hydrophilicity of lignin molecular, which would contribute to the degradation and dissolution of lignin.

However, when lignin is oxidized, the cellulose is also subjected to alkaline hydrolysis and oxidation reaction. In alkaline–oxygen pulping process, the reducing end group of cellulose is easy to reach by oxygen and then produces uronic acid end groups. These terminal groups are unstable under oxygen and high-temperature condition, which leads to rapid decomposition. Result from bagasse pulping showed that, when the oxygen pressure is $9.81-1.47 \times 105\ N/m^2$, the highest pulp yield is higher than that of the same condition.

Caustic soda pulping is a semichemical pulping process that is widely used in pulping of nonwood fiber materials. The main component of the cooking liquor is NaOH, or Na_2CO_3, where the dosage depends on the fiber properties. This process significantly involves cooking temperature, cooking time, and alkali charge, which refers to the amount of the active alkali dosage. Generally, the alkali charge is around 16%, and the cooking temperature is 140°C–170°C.

Lime or lime-soda pulping methods are usually used for the production of lower grade unbleached pulps. This process is applicable in pulping of rags and jute bast fiber in other to obtain high-grade pulps.

Other pulping methods are such as **NACO** and **SAICA**. NACO method, using Na_2CO_3 as chemical agent, is developed from alkaline–oxygen pulping. Some pulp and papermaking mills used NACO to produce straw pulp. The main principle of NACO method is to remove the lignin in Na_2CO_3 solution, with NaOH as a supplementary chemical to reduce the kappa number. NACO process involves the pretreatment of raw materials, delignification, pulp bleaching, combustion of waste liquid, and recovery of chemicals. As for SAICA, it is a semichemical pulping process, using NaOH as a chemical agent. The chemical consumption of this process is lower than in chemical processes. However, the spent liquor must be incinerated or treated biologically.

2.2.1.2 Sulfite pulping

In sulfite pulping process the raw material is cooked with sulfite as cooking liquid, and most of the lignin in the raw material is dissolved, along with the separation of fibers. Sulfite pulping can be divided into neutral sulfite pulping and alkaline sulfite pulping.

Neutral sulfite pulping is a semichemical pulping method where the active chemical Na_2SO_3 is prepared by dissolving SO_2 in Na_2CO_3 solution. With buffer function the Na_2CO_3 left in cooking liquor can let the pH value of the cooking liquor keep at 7–8. For instance, the pulping conditions of bleached wheat straw pulp are 10%–15% of Na_2SO_3 and cooking temperature 165°C. Compared to alkaline pulping, the yield of neutral sulfite pulping is 8%–10% higher, and the pulp is easier to bleach.

Alkaline sulfite pulping method involves cooking raw materials in cooking liquid consisting of NaOH and Na_2SO_3 at pH 10–13.5. One of the advantages of alkaline sulfite process is that the gas produced is odorless. Many studies have shown that, in the same cooking condition, the selectivity of lignin removal will be increased by increasing the concentration of sulfite. Study reported that the optimum concentration of sulfite in the cooking liquor of wheat straw is 0.3%–0.5%.

2.2.1.3 Organic solvent pulping (organosolv)

Organosolv processes are based on cooking with organic solvents such as alcohols or organic acids to dissolve most of the lignin and separate the fibers from the composite material into pulp. There are wide ranges of cooking agents for nonwood organosolv techniques. Methanol and ethanol are common alcohols used and the organic acids are normally formic acid and acetic acid. Other than that are ester organic sovent (ethyl acetate), compound organic solvents (methanol + acetic acid, ethyl acetate + ethanol + acetic acid, etc.), phenol organic solvents (phenol, cresol, and mixed cresol), and active organic solvents (dimethyl sulfoxide, dioxane, diethanolamine, and so on). High cooking temperatures and associated high pressures are needed when alcohols are used in cooking. However, organic acids require lower temperatures and the pressure is closer to atmospheric pressure. Other more unusual solvents include various phenols, amines, glycols, nitrobenzene, dioxane, dimethylsulfoxide, sulfolane, and liquid carbon dioxide (Sundquist, 2000).

Some of the advantages of organosolv process are as follows: it makes possible the breaking up of the lignocellulosic biomass to obtain cellulosic fibers for pulp and papermaking, high-quality hemicelluloses, and lignin degradation products from generated black liquors, thus avoiding emission and effluents (Aziz and Sarkanen, 1989; Hergert, 1998; Paszner, 1998; Sidiras and Koukios, 2004). Organosolv pulping processes, by replacing much or all of the water with an organic solvent, delignify by chemical breakdown of the lignin prior to dissolving it. The cleavage of ether linkages is primarily responsible for lignin breakdown in organosolv pulping.

Methanol organosolv pulping method has been used in a variety of pulping methods, and the addition of methanol can promote the dissolution of lignin and protect carbohydrates, among which alkaline sulfite. Compared to the kraft process, the addition of methanol in alkaline sulfite cooking liquor can significantly improve the delignification rate, which can also result in obtaining the pulp with excellent strength, fine bleaching properties, and high yield. **The alkaline sulfite anthraquinone and methanol (ASAM)** process is basically alkaline sulfite pulping with the addition of anthraquinone (AQ) and methanol (CH_3OH) to achieve a higher delignification level. ASAM pulping process can be regarded as a new technique modified from the conventional alkaline sulfite process supplemented by anthraquinone and methanol (Kordsachia and Patt, 1988; Yawalata, 2002; Klingberg, 2013).

Azeez (2018) associated the popularity of ASAM process in nonwood pulping with the significant improvements achieved in both the yield and quality of the pulp. The active cooking chemicals of the ASAM process are sodium hydroxide, sodium carbonate, and sodium sulfite; the addition of methanol to the alkaline sulfite cooking liquor considerably improves delignification, and the process produces pulp with better strength properties, higher yields, and better bleaching ability compared to the kraft process. ASAM pulping process has been well demonstrated and widely applied to different types of woody plant fiber, such as kenaf (Khristova et al., 2002), cornstalk (Jahan et al., 2006), pearl millet stalk (Elzaki et al., 2020), cotton stalk (Khider et al., 2012), and bamboo (Paridah et al., 2018).

The **ethanol** organosolv method, the main cooking agent of which is ethanol, can be cooked at lower pressure because of its high boiling point. Ethanol organosolv was originally designed to produce clean pulping and was further developed into ALCELL process for pulp production (Pye and Lora, 1991). The **ALCELL** process is a solvent-pulping process that employs a mixture of water and ethanol (C_2H_5OH) as the cooking medium. The process can be viewed as three separate operations: extraction of lignin to produce pulp; lignin and liquor recovery; and byproduct recovery (Stockburger, 1993). Under the cooking temperature 190°C–200°C, with the ethanol solution as the only delignification agent, the cooking liquor of ALCELL is an aqueous solution of 50% ethanol fraction, of which the pH is about 4 due to the effect of acetyl groups. In addition, to maintain a slight overpressure during cooking, nitrogen is injected into the digester. The whole process can be regarded as three independent processes: lignin removal; ethanol recycle from cooking liquid; the recovery of lignin, furfural, and polysaccharide. The ALCELL process has significant capital and cost advantages compared to those of the kraft process, which is purely free of sulfur emissions, has the potential of high-value applications, generates furfural, which is used as the solvent for lubricating oil production, and produces pulps with a higher yield that bleach more easily. **Alkaline Sulfite AQ–ethanol process (ASAE)** was improved on the basis of alkaline sulfite AQ–methanol pulping. However, the amount of ethanol required for alkaline sulfite AQ–ethanol cooking is much greater than that of methanol needed for alkaline sulfite AQ–methanol cooking. Nevertheless, the pulp produced by ASAE is characterized by low lignin content, favorable physical properties, high yield, and good beatability, which manifests that this method can save a large amount of energy, compared to the sulfate process. **IDE** pulping method is a nonsulfur alkaline pulping technology. Because of the addition of ethanol and AQ, the rate of delignification can be accelerated in IDE pulping process. The IDE method consists of three successive steps: impregnation by Na_2CO_3 solution, lignin degradation, and extraction of pulp with ethanol aqueous solution to remove the degraded lignin. When the IDE method is carried out in the pulping process of wheat straw, the yield of pulp would be more than 50%, and lignin residue would be less than 2.5%.

For **PUNEC** pulping method, the active chemicals are ethanol, AQ, and caustic soda. The raw material is pretreated with ethanol−water solution, which precedes high-pressure cooking to remove lignin. During cooking process, lignin and hemicellulose dissolve into the cooking liquid, followed by the ejection of black liquor into the flash tank. Then the black liquor is treated with acid to separate the lignin. To recycle the residual ethanol solution, the hemicellulose-rich black liquor should be distilled.

Organic acid pulping processes are alternative methods of organosol pulping to delignify lignocellulosic materials to produce pulp for paper (Jiménez et al., 1998; Kham et al., 2005; Lam et al., 2001; Poppius-Levlin et al., 1991). Typical organic acids used in the acid pulping methods are formic acid and acetic acid, which can react with lignocellulose to produce the corresponding esters in delignification process. The process is based on acidic delignification to remove lignin, while silicon remains in the pulp. The pulping operation can be carried out at atmospheric pressure. Acid used in pulping can be easily recovered by distillation and reused in the process (Muurinen, 2000). Cellulose, hemicellulose, and lignin can be effectively separated by degradation in aqueous acetic acid or formic acid. The cooking liquor is washed from the pulp, and both cooking chemicals and water are recovered and recycled completely. Formic acid can also be used to enhance acetic acid pulping. The temperature and pressure can be lower when formic acid is used in pulping compared to those used in alcohol or acetic acid pulping. Organic acid lignin is an optimal feedstock for many value-added products, due to its lower molecular weight and higher reactivity (Cetin and Özmen, 2002; Kubo et al., 1998). Another advantage of organic acid pulping is the retention of silica on the pulp fiber that facilitates the efficient recovery of cooking chemicals (Seisto and Poppius, 1997). The organosolv pulping processes based on organic acid cooking are the Milox, Acetosolv, and Formacell processes. For **Milox** pulping method, the chemicals used for cooking are performic acid or peracetic acid, which are produced by the reactions of hydrogen peroxide with formic acid and acetic acid. Milox, a kind of sulfur-free pulping technology, can realize total chlorine-free bleaching, but there are also problems in chemical recovery, due to the production of acetic acid and formic acid in recovery process. If the mixed solution of formic acid and acetic acid can be used as a solvent, the Milox method will be a more economical way of pulping; however, this is the feasibility of technology, which is worth for further exploring. **Acetosolv** pulping method is one of the acetic acid pulping technologies using hydrochloric acid as a catalyst. This pulping process should be carried out at atmospheric pressure, with cooking temperature of 110°C. The **Acetocell** pulping, developed from the Acetosolv pulping, should be taken place in the presence of acetic acid at high temperature (170°C−190°C) without additional catalyst. As for **Formacell** pulping method, it is developed from Acetocell pulping method. At the temperature of 160°C−180°C the raw materials are cooked in the cooking liquid, the mixed solution of formic acid (75%), acetic acid (10%), and water. **CIMV** pulping process is an organic acid process that developed from Formacell pulping method, which can be used to produce the bleaching pulp of straw and bagasse. The cooking liquid of CIMV pulping is the compound of acetic acid (20%–30%), formic acid (50%–60%), and water (20%).

2.2.1.4 Biological pulping

In biological pulping process, microorganisms or enzymes are adopted to pretreat raw material, which precedes the combination with the corresponding mechanical, chemical, and organic solvent methods. Biological pulping includes biochemical pulping and biomechanical pulping. The basic principle is to selectively decompose lignin by microorganisms or biological enzymes. For **microorganism** biological pulping method, there are many microorganisms that can degrade lignin, the most important of which is the basidiomycete, such as white rot fungus. With the existence of oxygen, after the treatment of microorganisms, the lignin would be degraded, and, at the same time, the carboxyl group and conjugated carbonyl group increased significantly, while the methoxy content reduced. Therefore the biodegradation of lignin is mainly oxidation reaction.

Biological enzyme pulping method is derived from microbiological method. In general, the species of biological enzymes mainly include cellulose enzyme, hemicellulose enzyme, and lignin enzyme. With the pretreatment of these biological enzymes the kappa number would decrease and the whiteness of unbleached pulp would increase. Besides, under the same bleaching condition, the whiteness of the enzyme chemical pulp is higher than that of the chemical pulp. The biological enzyme pulping can also overcome destructive effects of temperature and pH on polymerization degree of cellulose. However, the disadvantage is that the biological enzyme preparations need to be added continuously in the system, so the cost of production is high.

2.2.1.5 Chemi-mechanical pulping

In recent years the use of chemi-mechanical pulping (CMP) in nonwood pulping has attracted a great deal of interest. CMP has the advantages of high pulp yield and no need for chemical recovery systems. Chemical processes based on CMP are CMP, alkaline peroxide mechanical pulping (APMP), and steam explosion. **CMP** pulping method is obtained by grinding at atmospheric pressure. The cooking temperature is 100°C–160°C, and the cooking time is 10–30 minutes. A chemical pretreatment involving Na_2SO_3 and NaOH softened the fibers and the energy consumption is also decreased. Hence, the amount of Na_2SO_3 and the maximum temperature of pretreatment would affect sulfonation degree and swelling of lignin, while the dosage of

NaOH and the pretreatment time may influence the brightness and yield of pulp. For unbleached reed pulps, if the amount of NaOH is increased from 1% to 4%, the degree of sulfonation will rise, but the whiteness and yield of the pulp will reduce. **APMP** method is the pulping technology that adopts H_2O_2 to bleach in the process of dipping and grinding, by which the darkening phenomenon can be eliminated, along with the increase of the pulp brightness. The chemicals used in the APMP process are NaOH and H_2O_2, with the addition of DTPA, magnesium salts, or silicates as inhibitors to reduce the decomposition of H_2O_2. The typical APMP pulping consists of two segments that include steaming of raw materials after impregnation by chelating agent, residual alkali, and H_2O_2 and impregnation by alkali, H_2O_2, and chelating agent to remove the metal impurities in raw materials. The application of this pretreatment has been applied to nonwood material due to their relatively open- and easy-to-disintegrate structure (Latibari et al., 2012). The possibility of APMP of various nonwood species has been investigated, including wheat straw (Zhao et al., 2004), oil palm empty fruit bunch (Owolabi et al., 2016), and bagasse (Xu and Narayama, 2001).

Steam explosion pulping method is a process of short-time cooking of raw material in saturated steam, which is followed by rapid-relief pressure and grinding at atmospheric pressure. The impregnating solutions include Na_2SO_3, NaOH, $MgCl_2$, $NaHCO_3$, and $MgCO_3$, which have positive influences on reducing the refining energy. Steam explosion is suitable for pulping of nonwood raw materials. The performances of the obtained pulp are almost the same as that of the CMP and chemi-thermomechanical pulping (CTMP), even better, except for the lower yield.

2.3 Hybrid

Hybrid pulping method uses a combination of chemical and thermal treatments to begin an abbreviated chemical pulping process, followed immediately by a mechanical treatment to separate the fibers. These hybrid methods include thermomechanical pulping (TMP) and CTMP. The chemical and thermal treatments will reduce the amount of energy subsequently required by the mechanical treatment and also reduce the amount of strength loss suffered by the fibers. Sometimes a CTMP mill is located on the same site as a kraft mill so that the effluent from the CTMP mill can be treated in the kraft recovery process to regenerate the inorganic pulping chemicals.

Some mills pretreat wood chips or other plant material like straw with sodium carbonate, sodium hydroxide, sodium sulfite, and other chemicals prior to refining with equipment similar to a mechanical mill. The conditions of the chemical treatment are much less vigorous (lower temperature, shorter time, less extreme pH) than in a chemical pulping process, since the goal is to make the fibers easier to refine, not to remove lignin as in a fully chemical process.

NSSC Neutral sulfite semichemical process is a two-stage process in which wood chips are previously treated under a mild chemical condition at elevated temperature to soften, and the treatment partially removes the lignocellulosic interfiber bonding material. This is followed by a mechanical refining to complete the fiber separation. The cooking liquor in the early development of the process contained sodium sulfite and was buffered with sodium carbonate to reduce corrosion, and thus the process became known as the NSSC process (Do et al., 2020). NSSC pulping uses a solution of sodium sulfite and carbonate or bicarbonate to impregnate the wood chips (Thang, 2020). The sulfonation mainly introduced at middle lamella lignin causes a partial dissolution so that the resistance of fibers is weakened for the subsequent mechanical defibration. NSSC pulp remains a significant amount of residual lignin (more than 15%) and, therefore, is mostly used for unbleached products where good strength and stiffness are particularly important (Evtuguin, 2016).

Recently, with the rising demand for using nonwood fiber sources, NSSC pulping of wood-alternative material also takes interests. Study of NSSC pulping for bagasse at 165°C with different sodium sulfite charge showed that the influence of pulping time is greater than chemical charge on yield and pulp strength (Khakifirooz et al., 2013). Pulping sunflower stalk for producing fluting paper grade revealed that NSSC delignification rate of sunflower stalk is quite slow even at high chemical charge and cooking temperature, and to obtain good fibers, pulping process should be done in a relatively intensive condition with low yield and high kappa number (Rudi et al., 2016). In addition, anatomical structure of nonwood material, such as silvergrass, with less amount of fibers on unit mass, thin wall of fibrous elements, and high content of parenchyma cells, would make the strength of NSSC pulp become poorer comparing with wood (Bocianowski et al., 2019).

(CTMP) CTMP needs to impregnate the lignocellulosic materials aided with steam and chemicals to soften the lignin prior to refining. Carbonates, hydroxides, sulfite, and peroxides are among the lignin-softening chemicals with pretreatment at high temperatures that break down the structure and hence remove the lignin. This process uses less energy and chemicals to soften and fibrillate lignocellulosic materials but still maintains high yields of 85%–95%. This pulp is beneficial for high-grade printing papers. **CTMP** pulping processes provide excellent bulk properties of paper at a high production capacity. The key subprocesses in CTMP pulping cover all the way from chip preparation, high-consistency refining, steam recovery, fractionation, treatment of long fiber fraction, bleaching, and dewatering to pulp storage prior to papermaking process or drying and baling for market pulp. In the CTMP production process, wood chips are first impregnated with chemicals (sulfite and/or sodium hydroxide) and preheated at 90°C–150°C to help soften the adhesive material (lignin) between the fibers. The chips are then refined mechanically in the pressurized refiner to produce pulp.

Further development of CTMP process is HT-CTMP (high-temperature chemi-thermomechanical pulp). The HT-CTMP process uses higher temperatures of 150°C–180°C to help soften lignin, even more, allowing the fibers to be separated more easily. The result is a pulp with fibers that are more intact and have very low shive content (fewer bundles of fibers). The binding quality of the pulp is slightly reduced, but the bulk is greatly increased. As it is easier to separate the fibers, the total energy consumption of the refining process is reduced.

2.4 Mechanical pulping

Mechanical pulping is a process that converts lignocellulosic materials into pulp by using mechanical energy. Such process involves grinding and refiner methods to break the fiber bonding, hence separating the fibers (Ilyas et al., 2022a,b,c; Norfarhana et al., 2022). There are four types of MP: groundwood pulping, refiner mechanical pulping (RMP), thermomechanical pulping, and chemi-thermomechanical pulping. Mechanical pulps have weaker strength properties, softness, high bulk, and low brightness. An amount of about 90%–98% yield per unit volume can be produced by MP which is considered high pulp yields. This is due to the retention of most lignin content. Such pulp can be utilized in making low-grade products like newsprints, magazines, printing paper, and low-cost-grade packaging paper. Generally, mechanical pulping process physically tears the cellulose fibers one from another. However, much of the lignin remains adhering to the fibers. Strength is impaired because the fibers may be cut. Mechanical pulping process includes refiner and groundwood process.

Refiner pulping process includes TMP and RMP. TMP is a method in which chips are steamed while being refined. Steam treatment significantly reduces the total energy needed to make the pulp and decreases the damage (cutting) to fibers. Mechanical pulp mills use large amounts of energy, mostly electricity to power motors, which turn the grinders. A rough estimate of the electrical energy needed is 10,000 MJ/tonne of pulp (2750 kWh/tonne) TMP pulping processes provide excellent strength-related properties of paper at a high production capacity. The key subprocesses in TMP cover all the way from chip preparation, high-consistency refining, steam recovery, fractionation, treatment of long fiber fraction, bleaching, and dewatering to pulp storage prior to papermaking process or drying and baling for market pulp. TMP process can consist of single stage, two or three refiners in a series in one production line supplying paper machine, and the TMP pulp is dewatered and stored before papermaking.

The **thermomechanical pulping (TMP)** consists of impregnation and preheating of lignocellulosic materials under steam pressure (20–40 psi) and temperature (115°C–155°C) followed by refining. This process has two stages: (1) refining at an elevated temperature and pressure and (2) refining at ambient temperature and atmospheric pressure. The temperature is very useful to soften lignin and increase fiber separation by having less cutting and fine generation. The TMP process is two-staged, with the first stage involving refiners being at an elevated temperature (glass transition temperature of lignin, 140°C) and pressure and the second stage encompassing refiners being at ambient temperature and atmospheric pressure. The high temperature of steam softens lignin and enhances fiber separation with minimal cutting and fine fiber generation. TMP produces stronger pulps compared to stone groundwood pulp (SGWP) and RMP that is applicable for packaging materials, magazines, printing papers, paperboards, tissue paper, and more. In addition, the main effects of TMP or refining were found to be internal fibrillation, external fibrillation, and fiber shortening. Internal fibrillation increases fiber flexibility and collapsibility and improves fiber-to-fiber contact in the paper sheet (Rusu et al., 2011). With the primary wall of the fiber partially removed the secondary wall is in direct contact with water and subsequent swelling consequently improves the flexibility of the fiber (Samariha, 2011). The refining process, equipped with refining pressure, is a very important determinant of the physical and mechanical properties of refined fibers (Xing et al., 2008). In this process the preheating retention time significantly affects bending strength (Xing et al., 2007). However, the strength properties of pulps produced via mechanical pulping are generally inferior to those of pulps produced via soda pulping (Hosseinpour et al., 2014; Zhai and Zhou, 2014). Furthermore, TMP processes have been used for various nonwood materials, including sugarcane bagasse, oil palm EFB, bamboo, kenaf, canola, wheat, and rice straw. A mechanical process with a 0.10 mm refiner gap under atmospheric conditions used for oil palm EFB pretreated with 2% NaOH at 121°C for 2 hours exhibited maximum tensile and tear indices strengths (Mulyantara et al., 2015). Oil palm EFB treated with caustic soda was refined using four stages under atmospheric refining to obtain suitable fibers, and results indicated that the higher caustic concentration gave better properties (Daud et al., 2013). In addition, the TMP process employing a 0.508 mm refiner gap and presteaming at 0.62 MPa for 5 and 10 minutes for rice straw pretreated with oxalic acid resulted in a product with good mechanical strength (Li et al., 2013).

Refiner mechanical pulp (RMP) method involves chips that are just ground up with ridged metal disks called refiner plates. The RMP involves shredding and grinding of lignocellulosic materials between rotating metal disk by using a refiner at atmospheric pressure. RMP may involve thermo and chemical pulping pretreatments. There are two main stages in RMP: (1) defibration into individual fibers and fiber fragments and (2) loosening the fiber structure to achieve fiber flexibility and fibrillation. RMP produces stronger, more flexible, and darker pulps than SGWP. The pulp is bulkier, stronger, and contains fewer fines. (Madhuri and Reddy, 2020).

Groundwood mechanical pulping process includes SGW, thermo groundwood (TGW), and pressurized groundwood (PGW). The basic unit of these processes is a grinding stone that mechanically separates the fibers of round wood logs from each other (Hartmann, 2009). **Stone groundwood pulp (SGW)** is a manufactured stone with embedded silicon carbide or aluminum oxide that can be used to grind small wood logs called "bolts" to make stone pulp SGW (Marella et al., 2014). The stone groundwood pulping (SGWP) is known as the first mechanical pulping method that **ground wood** into pulps. The process involves debarking the wood log in a rotating motion at speeds ranging from 1000 to 1200 rev/min. The wood is torn into thin fibers and fragments that are washed away by a stream of water. Grinding involves friction that creates and raises temperature that simultaneously helps in the softening of wood. The pulp slurry is screened to split accepted and rejected (oversized particles) pulps. In SGW process the groundwood is conducted through the groundwood channel to coarse screening, then the white water is pumped into shower pipes through a pressure screen that has a basket with holes of the diameter 0.8 mm to remove any disturbing shives or fibers, and, further, to thickening. The filtrate from the thickening is returned to the previous process phases. The flow of each shower pipe is measured and adjusted individually. Shower pipe flows are also used as a grinder interlocking to prevent and burn on the grindstones.

Thermo groundwood (TGW) includes grinding under atmospheric pressure, shower water temperature of 80°C, or more. Grinding is a thermomechanical process that is divided into two parts, each of which overlaps one another: (1) softening and breakdown of the fiber structure and (2) peeling of the softened fibers from the wood matrix in the grinding zone (Sixta, 2006). In addition, grinding of the water-showered wood logs produces a lot of low-pressure steam and warm water that can be used as heat source for other process steps. Process waste such as bark from the wood handling plant and rejects from screening and cleaning operations can be combusted for energy generation purposes (Hartmann, 2009).

Pressurized groundwood (PGW) is a process where wood is ground in a pressurized, sealed grinder. PGW pulping method utilizes roundwood as raw material. The main subprocesses of PGW include grinding, screening, reject treatment, bleaching, dewatering, and storage. The PGW process is unique in allowing separate choice of pulping temperature and pressure. As a result, it is possible to customize pulp properties according to specific paper quality requirements. The PGW process is highly automated, featuring automatic log feed, extensive control, and information systems as well as online measurement. The WaterJet pulp-stone conditioning system is used as an improvement on traditional burr sharpening of a pulp stone. For producers of wood-containing printing papers, visual quality improvement is the key to satisfying customer needs. Valmet's pressure grinders and complete PGW processes produce mechanical pulps for products of the highest final quality—and with the lowest energy consumption and environmental load (area)

Acknowledgments

The authors would like to thank the Ministry of Higher Education for the grant of Vot No. 5524855 and 6369111 under Fundamental Research Grant Scheme (FRGS) and Higher Education Center of Excellence (HICoE), respectively. The authors also would like to thank all team members from the Department of Forest Products and Industries, University of Khartoum, and Institute of Tropical Forestry and Forest Products (INTROP), and Universiti Putra Malaysia for their endless support.

References

Azeez, M.A., 2018. Pulping of non-woody biomass. Pulp Pap. Process. 55–86.

Aziz, S., Sarkanen, K., 1989. Organosolv pulping: a review. Tappi J. 72 (3), 169–175.

Bocianowski, J., Fabisiak, E., Joachimiak, K., Wójciak, A., 2019. NSSC pulping of miscanthus giganteus and birch wood Part 2: a comparison of papermaking potential and strength properties. Wood Res. 64 (2), 281–291.

Cetin, N.S., Özmen, N., 2002. Use of organosolv lignin in phenol–formaldehyde resins for particleboard production: I. Organosolv lignin modified resins. Int. J. Adhes. Adhes. 22 (6), 477–480.

Daud, W.R.W., Wahid, K.A., Law, K.N., 2013. Cold soda pulping of oil palm empty fruit bunch (OPEFB). BioResources 8 (4), 6151–6160.

Do, V.T., Tanifuji, K., Jin, G., Ohi, H., 2020. Neutral sulfite semi–chemical pulping of moso–bamboo. Jpn. TAPPI J. 74 (4), 379–387.

Elzaki, O.T., Khider, T.O., Omer, S.H., Mohieldin, S.D., Shomeina, S.K., 2020. Suitability of Sudanese pearl millet stalks *Pennisetum glaucum* (L.) for alkaline pulping with additives. Songklanakarin J. Sci. Technol. 42 (1).

Evtuguin, D.V., 2016. Sulphite Pulping, Lignocellulosic Fibers and Wood Handbook. John Wiley & Sons, Inc., pp. 225–244. Available from: https://doi.org/10.1002/9781118773727.ch8.

Hartmann, M., 2009. Efficient Use of Energy in the European Paper Industry (PhD diss.). University of Leoben.

Hergert, H., 1998. Developments in organosolv pulping-an overview. In: Environmentally Friendly Technologies for Pulp and Paper Industry. John Wiley & Sons Inc., NY, pp. 5–66.

Hosseinpour, R., Latibari, A.J., Fatehi, P., 2014. Hydrogen peroxide bleaching of canola straw chemimechanical pulp. BioResources 9 (1), 1201–1211.

Ilyas, R.A., Aisyah, H.A., Nordin, A.H., Ngadi, N., Mohd Zuhri, M.Y., Muhammad Asyraf, M. R., et al., 2022a. Natural-fiber-reinforced chitosan, chitosan blends and their nanocomposites for various advanced applications. Polymers 14 (5), 1–36.

Ilyas, R.A., Zuhri, M.Y.M., Aisyah, H.A., Asyraf, M.R.M., Hassan, S.A., Zainudin, E.S., Sapuan, S.M., et al., 2022b. Natural fiber-reinforced polylactic acid, polylactic acid blends and their composites for advanced applications. Polymers 14 (1), 1–39.

Ilyas, R.A., Zuhri, M.Y.M., Norrahim, M.N.F., Mohamad Misenan, M.S., Jenol, M.A., Samsudin, S.A., et al., 2022c. Natural fiber-reinforced polycaprolactone green and hybrid biocomposites for various advanced applications. Polymers 14 (1), 1–28.

Jahan, M.Sarwar, Chowdhury, D.A.Nasima, Russel, M.A.N., Phil Mun, Sung, Quaiyyum, M.A., 2006. Alkaline sulfite-anthraquinone-methanol (ASAM) pulping of corn stalks. Cellulose Chem. Technol. 40 (7), 531.

Jiménez, L., Torre, M., Maestre, J., Ferrer, P.F., 1998. Delignification of wheat straw by use of low molecular weight organic acid. Holzforschung 52 (2), 191–196.

Khakifirooz, Alireza, Ravanbakhsh, Foad, Samariha, Ahmad, Kiaei, Majid, 2013. Investigating the possibility of chemi-mechanical pulping of bagasse. BioResources 8 (1), 21–30.

Kham, L., Bigot, Y.L., Mlayah, B.B., Delmas, M., 2005. Bleaching of solvent delignified wheat straw pulp. Appita J. 58 (2), 135–137.

Khider, O.T., Omer, S., Taha, O., Shomeina, S.K., 2012. "Suitability of Sudanese cotton stalks for alkaline pulping with additives. Iran. (Iran.) J. Energy Environ. 3 (2).

Khristova, P., Kordsachia, O., Patt, R., Khider, T., Karrar, I., 2002. Alkaline pulping with additives of kenaf from Sudan. Ind. Crop. Products 15 (3), 229–235.

Klingberg, A., 2013. On the Analysis of Alkaline Sulfite Pulps by means of Pyrolysis: Gas chromatography/Mass spectrometry (PhD diss.). Staats-und Universitätsbibliothek Hambg. Carl von. Ossietzky.

Kordsachia, O., Patt, R., 1988. Full bleaching of ASAM pulps without chlorine compounds. Holzforschung 203–209.

Kubo, S., Uraki, Y., Sano, Y., 1998. Preparation of carbon fibers from softwood lignin by atmospheric acetic acid pulping. Carbon 36 (7–8), 1119–1124.

Lam, H.Q., Bigot, Y.L., Delmas, M., Avignon, G., 2001. Formic acid pulping of rice straw. Ind. Crops Prod. 14 (1), 65–71.

Latibari, A.J., Pourali, K., Roghani, A.F., 2012. Alkaline peroxide mechanical pulping of fast growth paulownia wood. BioResources 7 (1), 0265–0274.

Li, X., Wu, Y., Cai, Z., Winandy, J.E., 2013. Primary properties of MDF using thermomechanical pulp made from oxalic acid pretreated rice straw particles. Ind. Crop. Products 41, 414–418.

Madhuri, P., Reddy, K., 2020. Effect of pulping, bleaching and refining process on fibers for paper making – a review. Int. J. Eng. Res. Technol. (IJERT) 09 (12), 310–316.

Marella, J.B.R., Madireddy, S., Maripi, A.N., 2014. Production of pulp from banana pseudo stem for grease proof paper. Table Content Topics Page no 61.

Mulyantara, L.T., Rizaluddin, A.T., Nakagawa-izumi, A., Ohi, H., Nakamata, K., 2015. Properties of fibers prepared from oil palm empty fruit bunch for use as corrugating medium and fiberboard. Jpn. Tappi J. 69 (12), 1349–1359.

Muurinen, E., 2000. Organosolv pulping: a review and distillation study related toperoxyacid pulping. Department of Process Engineering, University of Oulu, Finland, p. 314.

Myreen, B., 2001. A new approach to the non-wood black liquor problem. Proc. TAPPI Pulping Conf., Seattle, Washington. CD-ROM. p 3.

Norfarhana, A.S., Ilyas, R.A., Ngadi, N., 2022. A review of nanocellulose adsorptive membrane as multifunctional wastewater treatment. Carbohydr. Polym. 291, 1–19.

Owolabi, A.W.T., Arniza, G., Daud, W.W., Alkharkhi, A.F.M., 2016. Effect of alkaline peroxide pre-treatment on microfibrillated cellulose from oil palm fronds rachis amenable for pulp and paper and bio-composite production. BioResources 11 (2), 3013–3026.

Paridah, M.T., Moradbak, A., Mohamed, A.Z., Taiwo Owolabi, F.A., Asniza, M., Abdul Khalil, H.P.S., 2018. Alkaline sulfite anthraquinone and methanol (ASAM) pulping process of tropical bamboo (*Gigantochloa scortechinii*). Bamboo-Curr. Fut. Prospects. IntechOpen.

Paszner, L., 1998. Catalysed alcohol organosolv pulping. In: Environmental Friendly Technologies for the Pulp and Paper Industry. John Wiley & Sons Inc., NY, pp. 69–100.

Poppius-Levlin, K., Mustonen, R., Huovila, T., Sundquist, J., 1991. Milox pulping with acetic-acid. Paperi ja Puu-Paper Timber 73 (2), 154–158.

Pratima, B., 2016. Pulp and paper production processes and energy overview. Pulp and paper industry. 15–49.

Pye, E.K., Lora, J.H., 1991. The Alcell process: a proven alternative to Kraft pulping. Tappi J. 74 (3), 113–118.

Rudi, H., Resalati, H., Eshkiki, R.B., Kermanian, H., 2016. Sunflower stalk neutral sulfite semi-chemical pulp: an alternative fiber source for production of fluting paper. J. Clean. Prod. 127, 562–566.

Rusu, M., Mörseburg, K., Gregersen, Ø.W., Yamakawa, A., Liukkonen, S., 2011. Relation between fibre flexibility and cross-sectional properties. BioResources 6 (1), 641–655.

Sabharwal, H.S., Akhtar, M., Young, R.A. Blanchette, R.A., 1994. Bio-refiner mechanical pulping of bast type fibers. In Tappi Pulping Conference (623–623). Tappi Press.

Samariha, A., 2011. Effect of beating value on bagasse CMP pulp fiber characteristics. World Appl. Sci. J. 12 (11), 1987–1988.

Seisto, A., Poppius, L. 1997. Peroxyformic acid pulping of nonwood plants by MILOX methods—Part-1 Pulping and Bleaching. Tappi J. 80(9), 215–221.

Sidiras, D., Koukios, E., 2004. Simulation of acid-catalysed organosolv fractionation of wheat straw. Bioresour. Technol. 94 (1), 91–98.

Sixta, H., 2006. Handbook of Pulp, 2. Wiley-Vch Verlag Gmbh & Co, Weinheim.

Stockburger, P., 1993. An overview of near commercial and commercial solvent based pulping process. Tappi J. 76 (6), 71–74.

Sundquist, J., 2000. Organosolv pulping, In: Chemical Pulping, Papermaking Science and Technology Book, 6B. Fapet Oy., Finland, pp. 411–427.

Thang, D.V., 2020. Application of Thermomechanical Pulping Method for Fabricating Bamboo Fiber (PhD diss.). University of Tsukuba.

Xing, C., Deng, J., Zhang, S.Y., 2007. Effect of thermo-mechanical refining on properties of MDF made from black spruce bark. Wood Sci. Technol. 41 (4), 329–338.

Xing, C., Wang, S., Pharr, G.M., Groom, L.H., 2008. Effect of thermo-mechanical refining pressure on the properties of wood fibers as measured by nanoindentation and atomic force microscopy. Holzforschung 230–236.

Xu, E.C., Narayama, R., 2001. APMP (alkaline peroxide mechanical pulping) pulps from non-wood fibers, part 3: Bagasse. In: Pulping Conference. Proceedings of the Technical Association of the Pulp and Paper Industry, pp. 951–956.

Yawalata, D., 2002. Catalytic Selectivity in Alcohol Organosolv Pulping of Spruce Wood. National Library of Canada = Bibliothéque nationale du Canada, Ottawa.

Zhai, R., Zhou, X., 2014. Enhanced effect of NaOH/thiourea/urea aqueous solution on paper strength of high yield pulp. BioResources 9 (2), 2154–2166.

Zhao, J., Li, X., Qu, Y., Gao, P., 2004. Alkaline peroxide mechanical pulping of wheat straw with enzyme treatment. Appl. Biochem. Biotechnol. 112 (1), 13–23.

Chapter 3

Bleaching of nonwood pulp

Nurul Ain Nadirah Jamaluddin, Nur Areisman Mohd Salleh, Rafidah Jalil and Latifah Jasmani
Forest Products Division, Forest Research Institute Malaysia (FRIM), Kepong, Selangor, Malaysia

3.1 Introduction

Pulp and paper manufacturing, considered a high-capital investment industry, is one of the important forest-based sectors. Despite digitalization era, the global market for pulp and paper products is projected to grow from USD351 billion in 2021 to USD370 billion in 2028 with a Compound Annual Growth Rate (CAGR) of 0.74% between 2021 and 2028 (Fortune Business Insights, 2021). Pulp and paper have received positive demand across the globe amid the COVID-19 pandemic between 2020 and 2021. The increased demand during pandemic period was mostly on personal care and hygiene products, such as toilet papers, tissue, and the like.

Bleaching is one of the essential processes in papermaking to make the end product whiter or brighter. The term bleaching, when used for pulp, refers to the treatment of pulp with chemicals to enhance its reflection of visible light thus increasing the pulp brightness. The color of unbleached pulp is typically brown, which is due to the residual lignin after pulping. The bleaching is then applied to remove or decolorize the lignin. Bleaching process experiences huge development and changes over the past two centuries. In the early years of bleaching history the aim was just to bleach pulp using one chemical namely hypochlorite. Nevertheless, over the years, more chemicals were used to obtain effective brightness but with little concern on the impact to the environment. Simultaneously, an increased number of mills lead to increased pollution effect to the environment. In addition the use of elemental chlorine or hypochlorite during that period results in the release of carcinogenic dioxins and furans in the wastewater effluent to unexpectedly great extent. Due to this alarming discovery, stricter rules and regulations in terms of effluent release, air emission, and solid waste disposal are enforced in most countries since then. As a result of this requirement, pulp and paper mills need to change and improve their processes toward greener approach.

Pulping and Papermaking of Nonwood Plant Fibers. DOI: https://doi.org/10.1016/B978-0-323-91625-7.00008-4

New technology involves the use of different bleaching chemicals or nontoxic chemicals to meet the environmental requirement that has been adopted. The mill around the globe changes their chemical from elemental chlorine to more environmental chemicals and processes. The chemicals used during this new type of bleaching include chlorine dioxide, oxygen, ozone, hydrogen peroxide, and enzyme. Bleaching technology using this chemical combination is further classified into elemental chlorine-free (ECF) and total chlorine-free (TCF). ECF bleaching is a chlorine-free process wherein the use of elemental chlorine in any bleaching sequence is totally banned. ECF commonly uses chlorine dioxide as its primary bleaching agent. On the other hand, TCF bleaching does not use any chlorine chemicals in the process. TCF involves the use of oxygen, hydrogen peroxide, ozone, and other peroxygens. ECF, however, has become the most typical bleaching technology and is expected to stay long enough as there is no pressure for an alternative yet. ECF is the preferred method over TCF and dominates the global bleached chemical pulp market. The reason for such trend and preference is owing to its capability to maintain good yield, high strength, and high and stable brightness. This is also supported by research findings showing that wastewater discharged from well-managed mill adopting ECF bleaching is free of dioxin and poses a negligible risk to aquatic life (Bajpai, 2012).

3.2 Brief history of bleaching

In principle, bleaching detaches lignin from the pulp to give pulp a stable brightness. Compared to an unbleached pulp, bleached pulp has superior brightness, softness, and cleanliness. Printing paper, tissue, sanitary materials, and adsorbent products all benefit from these qualities (Sharma et al., 2020). However, bleaching process also produces by-products that are highly toxic and nonbiodegradable. Currently, many studies were conducted, and modifications were made to solve the environmental issues caused by bleaching process.

The mass production of paper started in the mid-1800s in Europe. There are few methods to whiten the fiber, either via laundering or cleaning to true chemical bleaching process. In the 1770s, textile bleaching process developed into several stage processes including watering of the fiber with lye, then heat treatment with lye followed by sun bleaching in sunny area (grassing). Finally, the fiber was treated with lactic acid (souring) that is derived from sour milk or soap. Nevertheless, this traditional method could not sustain for the mass production of paper. In 1756 Francis Home suggested the use of dilute sulfuric acid to replace the traditional sour milk and reduced the time of souring by about 4% (Home, 2010). This method led to the revolution in bleaching technology a few years later.

In 1774 C.W. Scheele, a Swedish pharmacist discovered chlorine. He found that red, blue, and yellow flowers and green plants turns to white in color after an exposure to chlorine over a period of time. Later, F.L. Berthollet, a French chemist suggested the use of chlorine for commercial bleaching. However, this chlorine solution is not stable until Charles Tennant in 1799, produced a bleaching powder-chlorine on slaked lime (hypochlorite). This innovation solved transportation issues, and it became a prominent bleaching agent for textiles and later for wood pulp industries in the 1900s. In the 1920s, multistage bleaching process was reintroduced at pulp mills using chlorine gas. Although all the bleaching equipment were made from stainless steel and reduce corrosive effects of the gas, Scheele realized that chlorine gas reacted robustly with cellulose and damaged the paper feedstock. Then, chlorine dioxide properties were studied and introduced as a commercial bleaching agent due to its properties that did not damage the cellulose.

The use of peroxide and ozone bleaching has been studied since the 1920s but on a small scale. The main factor that constrains the technology is because it is too costly. However, hydrogen peroxide is still used in some bleaching processes, mainly for mechanical pulp. In the 1950s, oxygen gas was found as a bleaching agent, and in the 1960s, a new bleaching process was developed in the United States. The technology used ozone with peroxides and caustic soda, and in 1972, the first ozone-based bleaching process was commissioned in Muskegon, Michigan. In spite of the technology developed in the United States, most of the technology had only been used in the Scandinavian region. In 1996, 16% of pulp mills in Sweden used ozone based bleaching. More chlorine-free bleaching agents have been studied on a laboratory scale. Today, more approaches on biobleaching methods were studied such as the use of enzymes like xylanase or fungi for more environmental-friendly bleaching process.

3.3 Bleaching agents

3.3.1 Hypochlorite

In the early days, bleaching was a straight forward process. By bubbling chlorine gas into lime slurry, calcium hypochlorite was created. Then, the solution was added to a warm suspension of untreated pulp at low consistency and wait until desired degree of bleaching had been achieved and finally, the pulp was washed from the bleaching solutions. Later, the technology used two-stage (HH) bleaching which removed some dissolve but still oxidizable reaction products after the initial stage and reduced chemical consumption and time-averaged concentrations of the oxidizing agent, resulting in low cellulose degradation. This two stage bleaching technique was a pioneering technique in development of modern's multiple bleaching stage technique.

3.3.2 Chlorine

The use of chlorine gas as a bleaching agent is regarded as a watershed moment in the evolution of current bleaching techniques. In the 1930s, Nekoosa-Edwards Paper Co. reported the first commercial chlorination (C) stage. It has been discovered that lignin is easily chlorinated, and that chlorinated lignin is alkali-soluble. By using extraction (E) stage, the chlorinated lignin can be removed following the C stage. Removal of lignin in C and E stages results in good bleached pulp with low hypochlorite usage. Furthermore, the usage of chlorine and a relatively low amount of hypochlorite produced a better bleached pulp and proved that the good brightness levels can be achieved with the multistage CEH technique.

3.3.3 Chlorine dioxide

In the 1940s, chlorine dioxide had already been used as a commercialized bleaching agent. In comparison to hypochlorite, chlorine dioxide has been demonstrated to be a good bleaching agent. It was possible to reach a higher brightness level by adding a chlorine dioxide (D) stage following the CEH process. In the 1960s, a five-stage bleaching sequence consists of a first stage chlorine treatment, a second stage alkali extraction stage, a third stage chlorine dioxide treatment, a fourth stage alkali extraction treatment, and a final fifth chlorine dioxide treatment (CEDED) had become a standard for the industry. However, for many years, chlorine dioxide was considered too costly for complete chlorine replacement. It is only used at the early stage due to prevent losing pulp strength and to improve delignification efficiency. Only in the late 1980s in Sweden, the total replacement of chlorine with chlorine dioxide was done commercially. Then, there was a new environmental regulation setup. The bleaching effluents that potentially pollute the environment is normally gauged in several different chemical test methods, and one of the methods is adsorbable organic halide, known as AOX. Mainly, the AOX limits have been set at low levels by the authorities. Usage of chlorine dioxide produced less AOX compared to the chlorine usage, and later, the total replacement of chlorine dioxide coined the term "elemental chlorine-free" ECF.

3.3.4 Oxygen

Many research studies had been done on the feasibility of oxygen as a bleaching agent. In the oxygen delignification (O) stage, oxygen was applied with alkali, and the process can remove much residual lignin, especially in kraft pulp. However, due to the low solubility and its tendency to be nonselective in the presence of transition metal ions (iron, manganese, or copper), the process took a slow progress in delignification process. In the 1960s a

French scientist discovered a method to protect cellulose from the degradation of oxygen bleaching. When a little amount of magnesium ion is introduced to the pulp as a pretreatment before the oxygen delignification (O) stage, half of the lignin can be removed without impacting pulp strength. Later, in 1970, the first mill using a commercial oxygen delignification system was set up. The importance of this system is the low environmental impact compare to a system that uses other bleaching agents. The chloride ion is not suppressed in large amounts in oxygen stage (O) effluents, allowing it to be recycled and eventually broken up by burning in the recovery furnace. The use of oxygen stage (O) before conventional bleaching (OCED) can bring down 50% of bleach plant discharges. The technique was improved in 1980 where oxygen is dispersed in the pulp with the presence of alkali, turning E stage into an (EO) stage, resulting in better removal of lignin that has been dissolved by preceding oxidative stage. This technique reduced the cost of bleaching process and reduce AOX. Now, adding both oxygen and hydrogen peroxide during the extraction stage has better effect than only using oxygen, resulting in (EPO) stage.

3.3.5 Ozone

Theoretically, ozone has exceptional potential compared to oxygen as a bleaching agent. It is more reactive and can remove almost all of the lignin and the process is recyclable. However, the ozone reactivity also affects the cellulose degradation and pulp strength if not applied according to suitable parameters (amounts, conditions). Due to this requirement, the technology development of ozone as a bleaching agent took a long time for commercialization. In 1992 Union Camp Corporation established a technique that was able to delignify kraft pulp using ozone (Z) stage. The company process 1000-ton per day of kraft pulp using ozone and chlorine dioxide stages (OZED) and become the pioneer for commercial ozone stages used later.

3.3.6 Biobleaching

Recently, environmental issues and demand for the high quality of pulp and paper products lead to the development of green bleaching techniques. In 1987 Finnish scientists discovered the hemicellulose-degrading enzymes (xylanases) as pretreating agents to ease the bleaching process. The xylanase (X) stage is capable to reduce chlorine dioxide usage to as low as 5 kg/ton of pulp, and this technology has been commercialized in many bleached kraft mills. Today, there are so many studies on biobleaching agents, such as pectinases, mannanases, ligninolytic, lignin peroxidase, manganese peroxidases, and laccases (Kumar, 2021), to develop more environmental friendly bleaching technique that can produce high-quality pulp and paper products.

3.4 Bleaching of chemical pulp

Bleaching of chemical pulp is a treatment of cellulosic fiber that involves chemical processing to increase brightness, whiten the pulp and brighten its color where the coloring materials (lignin) of the pulp are removed (delignification) or modified (Douglas, 1989; Mc Donough, 2004). Even after chemical pulping and bleaching, lignin continues to be removed from the pulp in a process that is never finished. Nevertheless, lignin remains the major component of pulp. Bleaching helps to clean the pulp of dirt and foreign materials that escaped the digestion process, in addition, to remove the lignin to give higher brightness to the paper (Douglas, 1989). It is also removing hemicellulose and extractives. Other than that, bleaching of chemical pulp also functions to remove the chromophores inside the pulp without affecting the strength of the pulp (Castellan and Grelier, 2016).

In the papermaking process, bleaching chemical pulp is a critical stage, used to enhance the brightness of paper pulp as well as provide high-quality pulp with a certain whiteness, cleanliness, purity, and excellent physical and chemical qualities. The bleaching of nonwood pulps made from oil palm, bamboo, or kenaf may also be conducted using same techniques and chemistry. Utilizing bleached pulp in the production of excellent paper and refined pulp is particularly advantageous. Chemical pulps produced using kraft process or sulfite pulping contain less lignin ($<5\%$) as compared to mechanical pulps ($\sim 40\%$). Initially, sodium hypochlorite was used to bleach chemical pulps but was replaced with chlorine in the 1930s, and in the 1970s, the new bleaching agents were introduced (oxygen, hydrogen peroxide, ozone, and peracids) due to the stricter environmental regulations (Castellan and Grelier, 2016; Mc Donough, 2004). These concerns have directed the development of new processes of bleaching such as TCF and ECF processes. In chemical pulp bleaching the process is often divided into four stages, each of which is represented in Table 3.1 by a different letter (Table. 3.2).

Generally, a mixture of bleaching agents is applied to bleach chemical pulps. It is used to create the "bleaching sequence that contains two phases which are delignification stage (to remove the lignin) and brightening stage (to increase the brightness of the pulp) until the target brightness is achieved" (Loureiro et al., 2011; Bajpai, 2015). Bleaching sequences are identified by describing each treatment sequentially. CEDED is a common five-stage bleaching sequence consisting of chlorine treatment (first stage), alkali extraction stage (second stage), chlorine dioxide treatment (third stage), alkali extraction treatment (fourth stage), and chlorine dioxide treatment (final stage) where washing is conducted between each stage.

One of the most common bleaching sequences for TCF is OQP with brightness ranging from 70% to 75% ISO that could be achieved. Increasing the brightness to about 90% ISO might be achievable by including additional stages involving the use of ozone and/or peracetic acid (Castellan and Grelier, 2016).

TABLE 3.1 History of fiber bleaching technology (Torén and Blanc, 1997; Chaurasia and Bhardwaj, 2019).

Year	Technologies used
1700s	Lactic acid, lye, and sunlight
1756	Sulfuric acid speeds up the bleaching process
1774	Discovery of chlorine gas by Scheele
1785	Chlorine as bleaching agent by Berthollet
1799	Development of bleaching powder
1815	Synthesis of chlorine dioxide by Davy
1920s	Reused of chlorine gas as bleaching agent
1946	First commercial chlorine dioxide plant
1970	First commercial oxygen plant
1972	First plan equipped with ozone bleaching
1990s	TCF method was introduced
2000s	Xylanase as pretreatment (bio bleaching)

TABLE 3.2 Bleaching process of chemical pulp (Douglas, 1989).

Chemical or process used	Letter designation	Function
Chlorine	C	Oxidize and chlorinate lignin
Sodium hypochlorite	H	Oxidize, brighten, and solubilize lignin
Chlorine dioxide	D	Oxidize, brighten, and solubilize lignin. In tiny amounts with chlorine gas to prevent pulp deterioration
Oxygen	O	Oxidize and solubilize lignin
Hydrogen peroxide	P	Oxidize and brighten lignin in chemical and high-yield pulps
Extraction with sodium hydroxide	E	Hydrolyze chloro-lignin and solubilize lignin
Ozone	Z	Solubilize lignin
Peracetic acid	P_a	Pulp delignifying and brightening agents
Chelating agent	Q	Removal of transition metals (copper, manganese, and iron)

Sequence DED requires less chemicals and is capable to achieve higher brightness. As for DEDED, this sequence also requires less chemicals and is capable to reach even higher brightness as compared to DED (Mc Donough, 2004). During the brightening stage, chlorine dioxide is very selective to produce high-brightness pulp that attacks lignin without degrading its cellulose as well as to reduce chromophores and normally apply at the end of bleaching sequences (Bajpai, 2015). Application of chlorine dioxide as a bleaching agent can produce strong, stable, clean, and high-brightness kraft pulp (Bajpai, 2012).

Chemical pulp bleaching has the potential to cause serious environmental destruction in its early stages, particularly due to the release of organic compounds into the water stream. Conventional bleaching with elemental chlorine releases chlorinated organic compounds, including dioxins. ECF and TCF (completely chlorine-free) bleaching methods replaced elemental chlorine. In 2005 kraft pulp manufacturing utilized 19%–20% elemental chlorine, compared to 90% in 1990. It was also found that there is no environmental impact for pulp derived from ECF and TCF processes.

3.5 Stability of brightness

Bleached chemical pulps' brightening stability is an essential property for many paper grades and is primarily measured by species or types of wood and bleaching process conditions (Milanez et al., 2008; Castellan and Grelier, 2016). It is also the reflecting properties of a sheet of pulp, and it is one of the most important quality parameters of commercial pulps that should be kept after transportation and storage. The brightness stability of bleached chemical pulp in storage may be poor, and brightness reversion or aging can happen during the pulp drying process, during shipping in a pulp bale, or in a pile of paper, as well as during storage and transportation (Sêco et al., 2008; Suess, 2008). In general, the brightness stability depends on the amount of residual lignin, where the more the lignin is removed, the brighter the pulp becomes. In general, pulps that have low levels of residual lignin may age more quickly than pulps that have higher levels of the residual lignin component.

Besides that, the bleaching process affects the brightness stability. The brightness reversion tendency of pulp bleached with hypochlorite in the final stage is higher than that of pulp bleached with chlorine dioxide. It has the potential to oxidize cellulose and become a new source of chromophores. In chlorine dioxide bleaching, cellulose oxidation is significantly reduced. The brightness and bleachability of pulp are also affected by the pulping conditions. Pine kraft pulp with a high amount of quinones was produced via low-sulfidity cooking, making it less reactive to chlorine dioxide. The bleaching sequence has an impact on the brightness stability.

The TAPPI (Technical Association of the Pulp and Paper Industry) technique T-452 has been used to analyze pulp brightness, and the brightness is expressed in %ISO. The sample is illuminated at 45 degrees, and the reflected light is measured perpendicularly (90 degrees).

During transportation, storage, and usage, light and/or heat are often used to cause pulp brightness reversion. Since bleached chemical pulps are subjected to long-term storage, heat promoted considerable brightness reversion. Heat reversion in chemical pulps is not well recognized, despite the fact that the environmental and chemical elements that influence reversion are widely understood. Traces of residual lignin, extractives, transition metals, carboxyl groups, carbonyl groups, pH, and other chemical properties of pulp are among them. Apart from that, the pulp brightness stability is affected by wood species, tree growing region, pulping, and bleaching processes (Oliveira et al., 2006). The brightness stability also influences by pH, where increasing of pH from 2.5 to 3.5 will increase the brightness with a kappa number of 2.0 was achieved at pH 3.0 and a kappa factor of 0.25 (Milanez et al., 2008; Suess, 2008).

Brightness reversion is needed to measure brightness reduction in the aging test. It is calculated by comparing brightness before and after reversion (Home, 2010). Measurements of brightness loss (Dr457 in % ISO), postcolor number (PCN), and increase in b* coordinate (Db*) across aging are the methods that are used most frequently. Very small PCN value is equivalent to very high brightness stability, and it is a useful tool to detect a small difference as it increases the visibility of these differences (Suess, 2008).

There are also a variety of techniques for aging the pulp to measure the brightness reversion, and these techniques varies with time, temperature, and relative humidity. In general, heat reversion has a positive correlation with the quantity of bleached pulp carbonyl, carboxyl, acid soluble lignin, and permanganate and is unaffected by the amount of metals or DCM extractive content (Oliveira et al., 2006).

Sodium hypochlorite oxidation was used to add carbonyl (CO) groups to entirely bleached pulp. The CO groups were found to be responsible for the loss of brightness stability. An alkaline extraction step helped to increase the brightness stability by partially removing these groups. An alkaline peroxide step seemed more effective in improving brightness stability while reducing CO loss. Furthermore, an unbleached dissolving pulp was bleached to the same brightness by ECF and TCF sequences, and CO content was extremely low in both cases. When compared to TCF bleached pulp, ECF-bleached pulp had a lower brightness stability (Perrin et al., 2017).

According to Oliveira et al. (2006), the findings of a research that was conducted on 13 different samples of eucalyptus, the brightness reversion, as evaluated by the Dr457 values, greatly differed across the various types of wood, depending on the bleaching sequence. For the ODEDD bleached pulps, fluctuations in the range of 2.1%–3.6% ISO were identified, while for the ODEDP bleached pulps, variations in the range of 0.8%–1.7% ISO were observed. In addition, the chemical consumption and bleachability of the pulp samples bleached to 90%–91% ISO brightness using the sequences ODEDD and ODEDP were only slightly different. Also, unbleached kraft pulp is brown and has a brightness of 20–30% ISO, while sulfite and fully bleached chemical pulps have the brightness of 60%–70% ISO and 88%–92% ISO, respectively.

3.6 Bleaching of mechanical pulp

The composition of lignin, cellulose, and polyoses is unaffected by mechanical defiberization. Only some compounds (arabinose and some acetyl group) are cleaved hydrolytically from polyoses releasing acetic acid. Chemical bleaching employed strong electrophiles (chlorine dioxide or ozone) to oxidize and solubilize lignin and polyoses. However, this cannot be applied in bleaching of mechanical pulp as it would alter its opacity qualities, reduce yield, and necessitate the use of considerable amount of chemicals. Bleaching of mechanical pulp involves the use of less harsh chemicals under ideal conditions to keep extraction of lower molecular weight carbohydrate and lignin to a minimum level. Mechanical pulps are bleached with lignin-retaining chemicals, for example, sodium hydrosulfite, whereas chemical pulps undergo bleaching with lignin-degrading chemicals.

The high temperature of the defiberization procedure dissolves the polyoses and initiates acetyl group hydrolysis. As the temperature and intensity of defiberization rise, so does the amount of dissolved organic material. At moderate pH (pH 5 and pH 6), the resins and lignans are partially dispersed and dissolves, thus improving the bleaching process. To preserve yield, bleaching should be carried out under moderate alkaline or acidic conditions.

The brightness of softwood pulp from forests or wood chips from sawmills is sufficient enough for use in newsprint manufacture without the need for bleaching. Unbleached pulp brightness is influenced by the bark removal method, wood species, and storage period between harvest and refining. When sawmill waste is used, the unbleached brightness is usually reduced. Apart from that, longer time during storage in humid piles can also lead to poor brightness. Most applications involving mechanical pulp in higher quality paper grades, such as SC or LWC, necessitate bleaching.

Some hardwood species on the other hand are very bright. Aspen and poplar primarily have high unbleached brightness (more than 60% ISO). Others, however, could be much darker but relatively easy to bleach due to their low lignin content. The color is due to the presence of aromatic ring structures with conjugated side chains, quinine, and metal catechol complexes. Despite the fact that lignin has few chromophoric structures, it is recognized to be a source of chromophores in mechanical pulp. Destruction of the chromophoric compounds or mechanical bleaching is accomplished either by reductive or oxidative process. However, for pulp to obtain maximum brightness, two bleaching procedures are normally required: oxidative bleaching with peroxide and reductive treatment with dithionite (hydrosulfite). Another method is by conducting two stages of hydrogen peroxide bleaching.

3.6.1 Reductive bleaching

3.6.1.1 *Bleaching with bisulfate*

After screening, reductive treatment with sulfur dioxide at a low consistency can be used to increase the brightness. The reduction of transition metals decolorizes the catechol metal complexes as well as solubilizes the metal and permits the removal by washing through dewatering. However, removing metal ions does not automatically improve brightness as air oxidation usually regenerates the chromophores. It is best to conduct all reductive bleaching treatments in a closed system to limit the excess of atmospheric oxygen.

3.6.1.2 *Bleaching with dithionite*

Dithionite helps the efficiency of reduction process. It removes not only catechol complexes but also quinones and quinine methods. Today, sodium dithionite has been one of the most frequently used in bleaching process. Bleaching with dithionite is usually conducted at acidic pH between 5 and 6. It is critical to keep the pH below 5. The pulp is combined with dithionite solution at 60°C–80°C temperature. While a higher temperature speeds up the process, it also increases the chance of compounds being lost in disproportionation reactions.

3.6.2 Oxidative bleaching

3.6.2.1 *Bleaching with hydrogen peroxide*

Hydrogen peroxide is known to be the most common oxidative bleaching agent in mechanical bleaching. Hydrogen peroxide is very unique as it can be used either as lignin preserving or lignin degrading depending on the process condition. For example, hydrogen peroxide is known as "lignin preserving" because during the bleaching process the side chains are cleaved and quinones are oxidized to more-soluble carboxylic acids. However, reaction between hydrogen peroxide and lignin's aromatic structure is quite complex, which limits the elimination process.

Hydrogen peroxide bleaching occurs by the addition of alkali or increased in temperature (more than 100°C) if it is used as the primary bleaching and delignification reagent. In aqueous solution, it releases per hydroxyl ion (HO^{2-}) and chemically behaves like a weak dibasic acid. The per-hydroxyl ion is very unstable and with the presence of oxidizable substance, it decomposes leading to bleaching process. Sodium hydroxide activates hydrogen peroxide because H ion is neutralized by alkali. However, it is important to maintain the pH (not above 10.8) to avoid the rapid liberation of HO^{2-} which may become unstable due to the formation of oxygen gas. Moreover, if the rate of decomposition keeps on increasing, it may damage the fiber (Saravanan et al., 2010). The delignification technique, which uses a moderate amount of hydrogen peroxide, improves the yield percentage and opacity features of mechanical pulp bleaching. Several essential parameters to control the bleaching process are: temperature, consistency, chemical charge as well as residence time. In bleaching, it is important to use optimum temperature to maintain the properties of the pulp and control the final brightness. Consistency is very important to gain the high concentration of active chemicals and avoid side reactions with dissolved compounds.

Reductive bleaching has been commonly used because it only requires one chemical to be added to the pulp and can be conducted in any consistency. Peroxide bleaching on the other hand is more complicated as high-consistency pulp is very essential to ensure bleaching effectiveness. According to standard, it also requires the help of four different chemicals (caustic soda, sodium silicate, hydrogen peroxide, and water) to mix with the pulp. High consistency is very crucial to ensure all the chemicals are evenly distributed.

Besides dithionite and bisulfate, tetrakis (hydroxyl methyl) phosphonium chloride is also utilized as a new bleaching agent for mechanical pulping. However, it is still far from being scaled up due to its unpleasant smell from the resulting products and the chemicals are very expensive and difficult to recycle. The most ideal for achieving maximum brightness is hydrogen peroxide; however, the amount of organic matter dissolved by the alkaline bleaching condition is the key concern. Alternatively, alkalization with magnesium or calcium hydroxide can be used to dissolve lignanes and polyoses. Another arising problem might be due to the application of chelants (EDTA/DTPA) as they are not biodegradable. Acidic treatment can be used to remove the transition metal, but dilution with water and thickening needs to be conducted. This increases the water demand, and press needs to be installed.

3.7 Brightening of secondary fiber

3.7.1 Recycling of paper and board

Due to the rising population and great paper demand, recycling has increased dramatically in recent years. When compared to the manufacture of new fiber, recovering fibers from recycled fiber requires less energy. As a result, recycled waste paper has emerged as a significant fiber source for paper products such as packaging paper and board.

The consumption of fiber raw material over the last few decades has been increased. Within those years, the use of virgin fibers climbed by 2.6%, while secondary fiber consumption increased by more than twice as much (Sixta, 2008).

As for weak fiber or high filler content, starch or fresh fiber is added in an adequate amount to overcome the loss of fillers, fiber fines, and printing ink. Bleaching of recycled paper requires the same type of chemical as chemical or mechanical bleaching. Before bleaching process, fillers, printing ink, plastic, metal staples, and other material are usually separated mechanically. The ink removal is necessary to ensure the success of bleaching process.

However, recycling wastepaper through bleaching process may cause a number of problems. Majority of wastepaper contains a high amount of fillers that lead to longer process, are time-consuming, and need a large amount of chemical as it consists of mechanical fiber. It requires complete de-ashing and a sufficient amount of acid for the dissolution of calcium carbonate. A huge amount of calcium carbonate may eventually lead to serious scaling problems.

3.7.2 Recycling of printing paper

To remove big particles, filler, and other waste, a mechanical technique is required. In the case of printing ink, it is either floated or washed away. Printing ink is easy to remove, and an alkaline solution is usually recommended. Separating water-based printing, ink, and residual ink has recently been accomplished using a colloidal gas aphron (Voss and Schabel, 2009). Hydrogen peroxide is introduced to the repulper together with other deinking agents, such as caustic soda, surfactant, and sodium silicate during pulping. Hydrogen peroxide is added to focus on color correction and to avoid yellowing. The process was conducted at low temperature (40°C–45°C). Printing ink is a combination of finely powdered and dispersed pigments. Thus the oxidation of printing ink pigments requires very aggressive compounds which may harm the fibers' structure. Although ozone or oxygen bleaching may be used for the bleaching of printing ink, it may result in low yield.

3.7.3 Recycling for the production of board

Board is produced by combining several layers of paper. To suit for advertising or printing, higher brightness on the top layer is necessary. Nowadays, reductive bleaching with dithionite is said as a better option to remove the dyestuff rather than hydrogen peroxide. This is due to the stability of the dyestuff against ambient oxygen and higher sensitivity toward reductive bleaching.

3.8 Bleaching of other materials

Bleaching of other materials is very similar to the bleaching of wood pulp. The level of contamination with contaminants differs between flex, linen, hemp, ramie, and wood pulp. Before bleaching, purification methods are conducted dependent upon the level of contaminants. The knowledge from bleaching of mechanical pulp can be applied during the brightening of wood. Bleaching agents such as hydrogen peroxide and mild alkalinity cleave chromophores without dissolving the substance. The more buffered the solution, the less aggressive the bleaching process. As for clay, bleaching involved the elimination of colored metal complexes, which applies the concept of dithionite bleaching of mechanical pulp. Metal reduction causes the destabilization of the complexes, washing away the metal ion thus removing the color. Bleaching clay is favored over other adsorbents due to its low cost and high adsorption capacity.

3.8.1 Textile fibers

3.8.1.1 Cotton, cotton linters and linen

Natural fibers have been utilized widely for textile production, with cotton being the most crucial product of these. Usually, after scouring process, natural fibers still contain yellowish and brown discoloration (naturally coloring matter) due to the pigments of the cotton flower. Climate, soil frost, and drought may also lead to a certain degree of yellowness. Dirt, dust, and insect could be another reason of discoloration. The fiber only contains wax and pectin. Bleaching is performed to produce white fabrics with the aid of bleaching agents to destroy the colorings while minimizing the degradation of the fiber (Narendra and Abdul, 2013). However, any contamination with dust and dirt could interfere with the bleaching process, which makes an acidic washing step (with the presence of chelant) ahead of bleaching is necessary. The bleaching agents could either oxidize or reduce the coloring. Alkaline peroxide bleaching has become one of the most frequently used methods for bleaching of cotton. It resulted in less cellulose chain oxidation leading to higher brightness stability and longer fabric's life. As for buffer, sodium silicate is added to replace or combined with an organic stabilizer. Cellulose linters have a higher amount of contamination due to lignin traces from seed capsules. To remove it, usually alkaline cooking step will be conducted at temperature above 100°C. Later, washing and acidic delignification stage with chlorine dioxide continued by alkaline peroxide will eventually lead to brightness.

Bleaching of flex is the same as another type of plants such as hemp or ramie. Mechanical action is conducted to remove any impurities in the flax stem. It was later subjected to retting, which uses bacteria to decompose pectin that binds the fibers tightly. Scutching (crushing them between metal rollers) helps to remove the woody portion and separates the bast fiber from the inner layer. Sodium carbonate buffer was used to remove the excess lignin (90°C at 1 hour). Some would also use caustic soda at ambient temperature for more than 8 hours. Later, the fiber was washed and acidified with sulfuric acid followed by 2–3 hours of treatment with 4% $NaClO_2$ at 80°C. Once completed, peroxide is mixed with about 1.5% H_2O_2, organic stabilizer as well as caustic soda to maintain the pH. These three stages were conducted to produce high brightness and high cleanliness.

3.8.1.2 *Wool*

Wool is a protein fiber, the structure of which consists of amino acid and is very sensitive toward alkali. In order to conduct alkali washing and bleaching, mild pH acid is preferable along with a buffer solution. Usually, tetrasodium diphosphate is used for alkaline bleaching. After passing through shearing process, wool is thoroughly washed and scouring (with the presence of surfactant). Peracetic acid is used as a disinfectant for moderate brightness and bleaching is conducted using sodium dithionite at pH 5 or 6 (Cegarra et al., 1994). In order to obtain better brightness, hydrogen peroxide bleaching was conducted at low temperature (less than 45°C). These two stages of bleaching will give higher brightness.

3.8.1.3 *Synthetic fiber*

At a mild acidic pH, cotton with polyamide or polyester can be bleached using chlorite. Hot hydrogen peroxide is not recommended in this situation since it may cause the polyamide or polyester to lose its strength. Polyurethane fiber, on the other hand, is very sensitive to chlorine chemicals so, hydrogen peroxide is used as a bleaching agent.

3.8.1.4 *Solid organic material*

Other materials require bleaching processes such as straw made from pandan or bamboo grass. It undergoes delignification with sodium chlorite, which this type of product is typically used by florists. At room temperature, rattan reeds for furniture and wicker baskets are bleached with hydrogen peroxide, sodium carbonate, and silicate, followed by dithionite treatment. Solid wood is brightened with hydrogen peroxide and ammonium carbamate for activation. For furniture, bleaching is necessary for the correction of off-color. Overall, 5%–15% of H_2O_2 is applied and later dried. After 24 hours, the remaining peroxide is removed by sodium bisulfite solution. The bleaching process may only take place at the surface of the furniture (less than 1 mm thick). However, this depends on the type of wood, some hardwood may allow deeper bleaching effect.

3.8.1.5 Vegetable and other oil

Some vegetable oil used in food production or chemical production may also need purification. The aim of bleaching is to remove the excess residue traces such as phosphatide, soap lipid peroxidation, and other impurities (Gharby, 2022). It also helps to minimize the levels of colored pigments (carotene) that resulted in dark oil (Gotor and Rhazi, 2016). In this case, the oil will undergo oxidation with hydrogen peroxide and acetic acid at elevated temperature. Unfortunately, an epoxidation of the double bond will occur as a result of the chromophore breakdown, resulting in the increase of oxidation number. This situation might not be favorable in food-grade vegetable oil. Fish oil also undergoes bleaching and is deodorized with air at 80°C, and the finished product may be essential for leather industry.

3.9 Conclusion

The aim of pulp bleaching technology nowadays is always to minimize the negative impact to the environment and pose zero health risks to human. The technology must include among others chlorine-free, zero effluent, no emissions, no dioxins, and other relevant risk factors. While taking into consideration bleaching efficiency and risk assessment, bleaching technology adopted will not influence to much extent the pulp yield, strength, and brightness stability.

Reference

Bajpai, P., 2012. Environmentally benign approaches for pulp bleaching. Environmentally Benign Approaches Pulp Bleaching . Available from: https://doi.org/10.1016/B978-0-444-51724-1.X5000-X. *Second Edition.*

Bajpai, P., 2015. Pulp and Paper Industry: Chemicals.

Castellan, A., Grelier, S., 2016. Color and color reversion of cellulosic and lignocellulosic fibers. Lignocellulosic Fibers and Wood Handbook: Renewable Materials for Today's Environment 531–551. Available from: https://doi.org/10.1002/9781118773727.ch21.

Cegarra, J., Gacén, J., Cayuela, D., Riva, M.C., 1994. Comparison of stabilisers for wool bleaching with hydrogen peroxide. J. Soc. Dyers Colour. 110 (9), 308–310. Available from: https://doi.org/10.1111/j.1478-4408.1994.tb01664.x.

Chaurasia, S.K., Bhardwaj, N.K., 2019. Biobleaching – an ecofriendly and environmental benign pulp bleaching technique: a review. J. Carbohydr. Chem. 38 (2), 87–108. Available from: https://doi.org/10.1080/07328303.2019.1581888.

Douglas, W.R., 1989. Pulp bleaching technology. Technologies for Reducing Dioxin in the Manufacture Bleached Wood Pulp. Congress of the U.S., Office of Technology Assessment, 1989, pp. 39–52.

Fortune Business Insights, 2021. Pulp and Paper Market. 2021. Available from https://www.fortunebusinessinsights.com/pulp-and-paper-market-103447.

Gharby, S., 2022. Refining vegetable oils: chemical and physical refining. Sci. World J. 2022. Available from: https://doi.org/10.1155/2022/6627013 (Table 1).

Gotor, A.A., Rhazi, L., 2016. Effects of refining process on sunflower oil minor components: a review. OCL – Oilseeds Fats, Crop. Lipids 23 (2), 1–10. Available from: https://doi.org/10.1051/ocl/2016007.

Home, F., 2010. Experiments on Bleaching.

Kumar, A., 2021. Biobleaching: an eco-friendly approach to reduce chemical consumption and pollutants generation. Phys. Sci. Rev. 6 (4). Available from: https://doi.org/10.1515/psr-2019-0044.

Loureiro, P.E.G., Evtuguin, D.V., Carvalho, M.G.V.S., 2011. The final bleaching of eucalypt kraft pulps with hydrogen peroxide: relationship with industrial ECF bleaching history and cellulose degradation. J. Chem. Technol. Biotechnol. 86 (3), 381–390. Available from: https://doi.org/10.1002/jctb.2527.

Mc Donough, T.J., 2004. Encylopedia of forest sciences. Pulping; Bleaching of Pulp. Elsevier Academic Press, pp. 910–918.

Milanez, A.F., Colodette, J.L., Carvalho, A.M.M.L., 2008. Optimal conditions for bleaching eucalyptus kraft pulp with a three stage sequence. Sci. Forestalis/Forest Sci. 78, 105–113.

Narendra, G., Abdul, S.B., 2013. Accelerated bleaching of cotton material with hydrogen peroxide. J. Text. Sci. Eng. 03 (04). Available from: https://doi.org/10.4172/2165-8064.1000140.

Oliveira, R.L., Colodette, J.L., Morais Eiras, K.M., Ventorim, G., 2006. The effect of wood supply and bleaching process on pulp brightness stability. Rev. Arvore 30 (3), 439–450. Available from: https://doi.org/10.1590/s0100-67622006000300014.

Perrin, J., Lachenal, D., Chirat, C., 2017. Brightness stability of eucalyptus-dissolving pulps: effect of the bleaching sequence. Holzforschung 71 (7–8), 625–631. Available from: https://doi.org/10.1515/hf-2016-0208.

Saravanan, D., Vasanthi, N.S., Senthil Raja, K., Das, A., Ramachandran, T., 2010. Bleaching of cotton fabrics using hydrogen peroxide produced by glucose oxidase. Indian J. Fibre Text. Res. 35 (3), 281–283.

Sêco, I.M.C.L., Neto, C.P., Silvestre, A.J.D., 2008. Strategies to reduce the brightness reversion of industrial ECF bleached eucalyptus globulus kraft pulp. J. Chem. Technol. Biotechnol. 83 (3), 218–226. Available from: https://doi.org/10.1002/jctb.1791.

Sharma, N., Godiyal, R.D., Thapliyal, B.P., 2020. A review on pulping, bleaching and papermaking processes. J. Graphic Era Univ. 8 (2), 95–112.

Sixta, H., (Ed.), 2008. Raw material for pulp. Handbook of Pulp. Wiley.

Suess, H.U., 2008. Bleaching of chemical pulp. Pulp Bleaching Today. Walter De Gruytr.

Torén, K., Blanc, P.D., 1997. The history of pulp and paper bleaching: respiratory-health effects. Lancet 349 (9061), 1316–1318. Available from: https://doi.org/10.1016/S0140-6736(96)10141-0.

Voss, D., Schabel, S., 2009. Use of colloidal gas aphrons for seperating of water based printing inks and impurities from paper stock suspensions. Zellcheming Annual Meeting. Wiesbaden.

Chapter 4

Pulping and paper mechanical properties of *Bambusa vulgaris*, *Gigantochloa levis*, and *Gigantochloa scortechinii* bamboo

N.M.H. Hassan[1], N.M. Suhaimi[2] and N. Abdullah[3]

[1]*Wood Industry, Faculty of Applied Sciences, Universiti Teknologi MARA Pahang, Jengka Campus, Bandar Tun Abdul Razak, Pahang, Malaysia,* [2]*Faculty of Applied Sciences, Universiti Teknologi MARA, Shah Alam, Selangor, Malaysia,* [3]*Faculty of Plantation and Agrotechnology, Universiti Teknologi MARA Sarawak, Samarahan Campus, Kota Samarahan, Sarawak, Malaysia*

4.1 Introduction

Bamboo is increasingly being used as a substitute for Mengkulang, Cengal Merbau, Balau, Red Balau, Keruing, Kempas, Kapur, Tualang (Medium Hardwood), Dark Red Meranti, Red Meranti, Mixed Heavy Hardwood (Heavy Hardwood), White Meranti, Yellow Meranti, Mersawa, Nyatoh, Sepetir, Jelutong (Light Hardwood), and Blended Light Harwood (Light Hardwood) as an alternative for furniture and other industries. This versatile plant is utilized not only for furniture, but also for structural purposes, textiles, pulp and paper, and as a food source. Bamboo has long been utilized as a building material in the East, with its toughness being put to good use in China's oldest suspension bridges and different designs. Bamboo, which is naturally sturdy and visually attractive, is gaining popularity as a renewable alternative in various regions of the world (Anonymous, 2021). This includes sources for papermaking and pulp.

Bamboo is a nontimber fiber that has been proven to be a possible pulp and paper production substitute. Significant versions withinside the overall performance of bamboo paper were diagnosed because of the extensive version in organisms, anatomical structure, and chemical composition (Tripathi et al., 2016). Bamboo paper, like softwood paper, had great stiffness, bending consistency,

Pulping and Papermaking of Nonwood Plant Fibers. DOI: https://doi.org/10.1016/B978-0-323-91625-7.00002-3

tear strength, and crush resistance, all of which might be ideal for paperboard (Moradbak et al., 2016).

Moradbak et al. (2016) stated that, due to its tight bundle sheaths, impermeable epidermis, total lack of ray cells, and limited area of conducting tissues, bamboo requires a more intensive pulping routine than wood. They pulped bamboo (*Gigantochloa scortechinii*) using an alkaline sulfite anthraquinone (AQ) method to develop high-performance packing paper.

The effects of chemical pretreatment (14% and 18% NaOH) before pulping on the pulp and paper properties of bamboo *G. scortechinii* were investigated (Ainun et al., 2018; Deniz et al., 2017; Naithani et al., 2015). The pulp that had not been bleached was obtained and turned into paper. The pulp and paper were tested for pulp yield, pulp freeness, paper optical, paper physical, and paper mechanical characteristics. The use of treated bamboo chips enhanced pulp production, freeness, and paper brightness, according to the findings (Naithani et al., 2015). Tensile, tear, and burst intensity of mechanical paper all exhibited a small reduction. As a consequence, mild kraft pulping may be used to create bamboo unbleached pulps. Additionally, the use of bleaching repetitions can be minimized (Ainun et al., 2018).

Gigantochloa thoii, Gigantochloa scortechinii, Gigantochloa ligulata, Gigantochloa wrayi, Gigantochloa brang, Schizostachyum brachycladum, Schizostachyum grande, Schizostachyum zollingeri, Bambusa vulgaris, Bambusa blumeana, Bambusa heterostachya, Bambusa vulgaris cv Vittata, and *Dendrocalamus asper* were among the thirteen species of Malaysian bamboo. The typical fiber length of the 13 bamboo species ranged from 2330 to 4070 μm, according to the data (Gominho et al., 2014). Bamboo fiber is longer on average than *Eucalyptus* spp. (Gominho et al., 2014; Carrillo et al., 2015), and *Pinus* spp. (Atac and Eroğlu, 2013; Gülsoy and Ozturk, 2015).

Bamboo fibers are categorized as long fibers by the International Society for Wood Anatomy (Wheeler et al., 1989). The greatest mean Runkel ratio of 4.53 was *G. thoii*, then *B. vulgaris* (4.25) and *G. ligulata* (4.30), and *S. zollingeri* (1.01) had the lowest value (Abdullah Siam et al., 2019). Fine raw material has a Runkel ratio of less than one, whereas inferior raw material has a Runkel ratio of more than one (Zhan et al., 2015). Secondary fiber raw material is described as fiber with a Runkel ratio of one (Wu, 1997; Zhan et al., 2015). The Runkel ratios of all the bamboo species in this sample were greater than 1, indicating that they are inferior raw materials for papermaking. According to Kiaei et al. (2014), a high Runkel ratio implies that the fiber is rigid and less flexible.

The highest mean aspect ratio was found in *B. vulgaris* (177.15), second by *B. vulgaris cv. Vittata* was 176.70, and *G. ligulata* was 174.13, with *G. wrayi* having the least value of 114.10. The aspect ratio of pulp fibers should be substantially greater than 100 for papermaking, and a higher ratio leads to superior fiber quality (Wang et al., 2008; Yang et al., 2008; Wang et al., 2016).

The aspect ratios for all of the bamboo types studied were greater than 100, indicating that they could still be used as pulp raw materials, although the Runkel ratio was higher. Identical findings were stated by Chandra and Susi (2018), Wang et al. (2016), Zhan et al. (2015), and Sharma et al. (2011); impedance values were over 100, longer fibers were discovered, yet the Runkel ratio seemed larger than 1 (Zhan et al., 2015). Nevertheless, Chandra and Susi (2018) indicated that to maximize the utilization of this fiber, additional chemicals should be employed during the cooking process. The objective of this study is to evaluate the paper mechanical properties of paper made from different bamboo species.

4.2 Materials and methods

4.2.1 Raw materials preparation

B. vulgaris (*Aur* bamboo), *Gigantochloa levis* (*Beting* bamboo), and *G. scortechinii* (*Semantan* bamboo) were harvested at 1 ft above the ground and up to 16 ft in height. All of the specimens were taken at the Forest Research Institute Malaysia (FRIM) in Kepong, Selangor, on a bamboo plantation. The bamboo culm was chopped into bamboo chips, which were then screened using a chip classifier to produce uniform chips with a length of 10–30 mm and a thickness of 2–5 mm. Fig. 4.1A shows the illustration of the harvested area from the bamboo clump, Fig. 4.1B shows the *Aur* bamboo, Fig. 4.1C shows the *Beting* bamboo, and Fig. 4.1D shows the *Semantan* bamboo.

4.2.2 Bamboo pulping

The bamboo was digested into pulp by using the soda–anthraquinone (AQ) pulping technique. A sample of 1000 g was weighed and placed into the rotary digester. The pulping parameter used for pulping was 20% NaOH, the

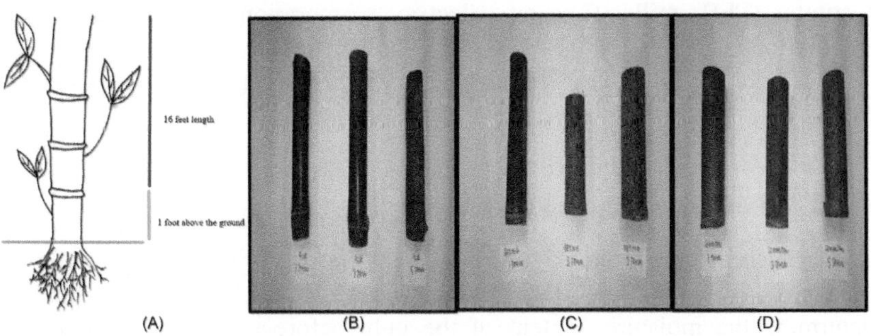

(A) (B) (C) (D)

FIGURE 4.1 (A) Bamboo harvesting illustration and (B) the *Aur* bamboo, (C) the *Beting* bamboo, and (D) the *Semantan* bamboo.

(A) (B)

FIGURE 4.2 (A) Rotary digester and (B) the Somerville screener.

pulping temperature at 170°C, the time to reach pulping temperature at 90 minutes, the time at pulping temperature at 90 minutes, and the percentage used for AQ is 0.1% based on oven-dried weight. After the pulping process was completed, the pulp was washed under running water to remove the black liquor. Then the pulp was screened using the Somerville screener, and only fiber that can go through the 0.15 mm screener plate size was used for papermaking, as indicated in Technical Association of Pulp and Paper Institute (TAPPI) standard T 275 sp-02. Fig. 4.2A shows the rotary digester and Fig. 4.2B shows the Somerville screener used in the pulping and screening processes.

4.2.3 Beating

The primary goal of the beating process is to improve fiber-to-fiber interaction during the papermaking process by improving fiber flexibility, which is enhanced by surface fibrillation. The beating level enhances the surface area of pulp fibers, which is important for improving paper physical properties (Suphat, 2007). The beating was done by using Paper and Fiber Research Institute (PFI) mills (Leeson Electric Corporation, Wisconsin, USA). Revolutions of 1000, 2000, 4000, and 8000 per minute were used to beat the pulp samples (rpm). The PFI beater used in the beating process is shown in Fig. 4.3.

4.2.4 Handsheet making and testing

The handsheet or papermaking process was based on TAPPI standard practice (sp) T 205 sp-02 (TAPPI, 1995). The Moisture Analyzer was used to determine the moisture content of the pulp before papermaking. For each batch of handsheets, 24 g of oven-dried pulp was weighed, and the distilled water was added up to 2000 mL, and the sample was disintegrated at

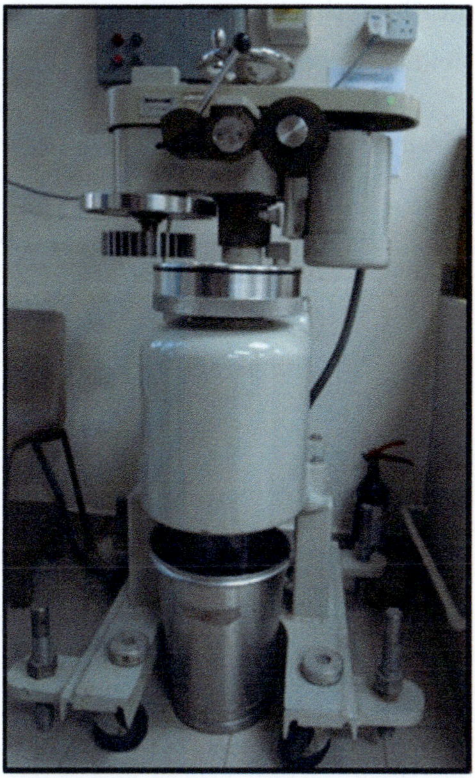

FIGURE 4.3 PFI beater. *PFI*, Paper and Fiber Research Institute.

TABLE 4.1 Technical Association of Pulp and Paper Institute (TAPPI) official methods for paper testing.

Mechanical testing	TAPPI official methods
Tearing index	T414 om-98 (TAPPI, 1998)
Folding endurance	T511 om-96 (TAPPI, 1996)
Tensile index	T494 om-01 (TAPPI, 2001)
Bursting index	T403 om-97 (TAPPI, 1997)

3000 rpm in a disintegrator. The freeness calculation was carried out using 1000 mL of paper stock. The paper stock was thoroughly stirred before 1000 mL of paper stock was taken for papermaking of each handsheet. The mechanical properties of bamboo paper testing were conducted according to the TAPPI official method procedures as listed in Table 4.1.

4.3 Results and discussion

4.3.1 Pulping yield and paper mechanical properties

Pulp yield is the recovery percentage of bamboo chips after the pulping process. Fig. 4.4 shows the pulping total yield of *Aur*, *Beting*, and *Semantan* bamboo at the age of 1, 3, and 5. The highest pulp yield was *Semantan* (SMT), which is 51.7% at age of 1, followed by SMT at age of 5 (46%) and BTG at the age of 3 (45.2%). The lowest pulp yield was BTG at age of 5 (35.7%). Based on the total pulping yield, all the bamboo species at age of 1 were chosen for the next papermaking process that involved the beating process to enhance the paper's mechanical properties.

Moradbak (2016) used an alkaline sulfite AQ with methanol (ASAM) process with an alkali ratio of (80/20:0.1%:15%) and an alkali ratio of (Na_2SO_3/NaOH: AQ: methanol) to pulp *G. scortechinii* at 170°C. He observed that using 14% of sodium hydroxide and 90 minutes of cooking time yielded the highest yield (52.36%) while using 18% and 120 minutes yielded the lowest Kappa number (10.38). The Kappa number and pulp output were significantly reduced as alkali and boiling time were increased with a Kappa number of 14.17 and a pulp yield of 49.06%, the optimum pulping conditions were 16% alkali and 90 minutes of boiling time, which was within this study's pulping yield ranged of 35.7%–51.7%.

Ainun et al. (2018) investigated the effect of sodium hydroxide pretreatment before bamboo chemical pulping on *G. scortechinii* in primary research. At 60°C, the bamboo chips were steeped in a 2% sodium hydroxide solution for 6 hours. Using the kraft pulping process, these treated chips

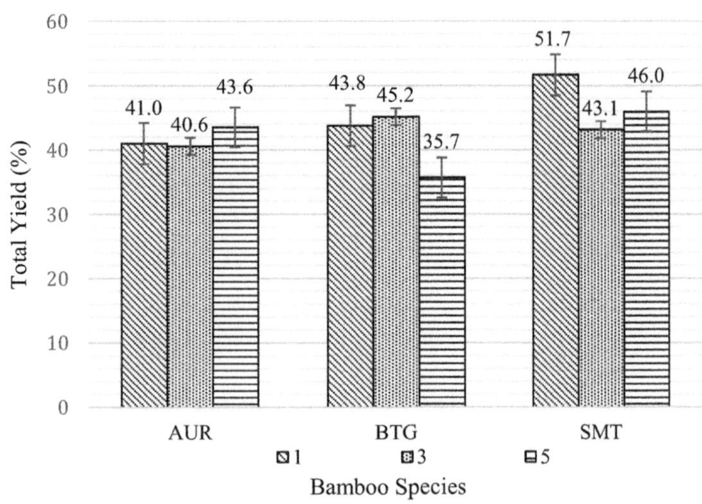

FIGURE 4.4 Bamboo pulp total yield at different species.

TABLE 4.2 The effect of beating revolution on the paper stock freeness and drainage.

	Aur		*Beting*		*Semantan*	
Beating (Rev.)	Freeness (mL)	Drainage (s)	Freeness (mL)	Drainage (s)	Freeness (mL)	Drainage (s)
0	660.5	4.1	667.5	4.1	690.5	3.9
1000	651	4.1	536.5	4.3	594	4.1
2000	570.5	4.3	519	4.6	563.5	4.5
4000	380	5.3	339.5	6.0	367.5	5.4
8000	124	21.0	113	48.0	151	24.0

mL, Milliliter; *Rev.*, revolutions; *s*, second.

were heated at 14% and 18% alkali active. They discovered that the screened yield varied from 34.85% to 46.35%, which is nearly identical to the pulping yield obtained in this study (35.7%–51.7%).

Table 4.2 shows the effect of beating revolution on bamboo paper stock freeness and drainage during the paper formation. The pulp freeness reduced as the beating revolutions rose, and the drainage time rose as the beating revolutions increased. When the beating process occurred, the fiber cell wall fibrillation allowed water to enter the cell wall thus holding the water from draining out of the fiber. The further the beating process was performed to the fiber, hence more water is absorbed, contributing to the pulp freeness diminishing as the beating revolution increased. On the other hand, the drainage time becomes slower as the beating revolutions increase, due to the fiber holding up more water, and the water was released at a slower rate. This can contribute to the better formation of hydrogen bonding during the paper formation on the wire mesh and thus also contributes to higher paper mechanical properties, especially in bursting index, folding endurance, and the tensile index.

Ainun et al. (2018) investigated the effect of 14% and 18% NaOH pretreatment before kraft pulping, finding that pulp freeness rose and was estimated to be lower than that of untreated samples. There were no changes in the pulp freeness of the 18% NaOH samples. This condition may be described as a pulp state. Although the pulps were swollen, they have a good probability of sticking together and moving through the refining process. This is due to the addition of a large number of chemicals to pulps during the pretreatment or pulping operations.

Table 4.3 shows the effect of beating revolutions on bamboo paper mechanical properties. Grammage is the mass per area of paper measured in

TABLE 4.3 The effect of beating revolution on bamboo paper mechanical properties.

Sample	Grammage (g/m^2)	Burst index (kPa (m^2/g))	Folding endurance (double folds)	Tearing index (mN (m^2/g))	Tensile index (Nm/g)
A0	61.98	1.44 (0.1)	3 (0.4)	9.02 (0.0)	18.88 (0.3)
A1000	59.22	2.37 (0.2)	11 (1.1)	10.25 (0.0)	30.73 (0.7)
A2000	58.95	4.12 (0.4)	77 (0.8)	10.96 (0.0)	46.99 (0.4)
A4000	60.28	4.75 (0.4)	285 (0.8)	9.76 (0.0)	55.41 (0.2)
A8000	58.66	6.64 (0.7)	579 (0.5)	9.70 (0.0)	64.44 (1.1)
B0	61.55	2.36 (0.2)	28 (0.9)	14.90 (0.1)	29.76 (0.6)
B1000	59.95	4.41 (0.3)	219 (0.5)	17.03 (0.0)	46.88 (0.1)
B2000	57.69	4.93 (0.7)	438 (1.0)	19.68 (1.0)	56.51 (1.0)
B4000	60.16	5.82 (0.3)	708 (0.9)	16.26 (0.3)	67.32 (0.9)
B8000	59.38	7.20 (0.4)	1231 (1.1)	14.70 (0.1)	69.97 (0.5)
S0	62.81	1.32 (0.1)	2 (0.0)	7.48 9 (0.0)	16.02 (0.4)
S1000	66.21	2.12 (0.2)	12 (1.1)	13.34 (0.1)	26.66 (0.9)
S2000	61.21	3.75 (0.6)	42 (1.1)	16.83 (0.3)	38.4 (0.8)
S4000	60.91	4.74 (0.4)	222 (1.2)	16.63 (0.5)	50.73 (0.8)
S8000	58.67	5.64 (0.2)	413 (0.8)	14.44 (1.0)	56.93 (0.72)

A, Aur bamboo; *B, Beting* bamboo; *S, Semantan* bamboo.

grams per square meter. The burst, tearing, and tensile strength are divided with the paper grammage to get the burst, tearing, and tensile index. This is important to compensate the strength of the paper achieved with the grammage, to avoid bias to the paper strength, as the paper strength increased with the increasing of the paper grammage.

As can be seen in Table 4.3, the burst index, folding endurance, and tensile index ranged from 1.32 to 7.20 kPa (m^2/g), 2 to 1231 double folds, 16.02 to 69.97 Nm/g increased with the increase of the beating revolutions from 1000 to 8000, for all bamboo species, except for tearing index. The beating revolution was halted at 8000 revolutions based on Nurul Husna et al. (2011), where they found that there was no significant increase in paper mechanical characteristics with extending the beating revolution from 8000 to 24,000 revolutions. Beating reacts as the mechanical treatment on the fiber. As the fiber was beaten, the fibrillation process occurred, and more water was absorbed into the fiber cell wall, thus increasing the fiber flexibility and the fiber becoming closer to each other. This contributed to the enhancement of the fiber-to-fiber bonding during the paper formation and thus increased the resistance of the paper sheet during the paper testing, and this increased the paper's mechanical strength.

On the other hand, for tearing strength, it can be seen that the tearing index increased with the increase of the beating revolutions from 1000 to 2000, but at the beating revolutions of 4000–8000, the tearing index started to drop. The tearing strength involved the pulling of the fiber from the paper; the longer the fiber, the more force is needed to tear the paper sheet. As the beating revolution increased from 2000 to 4000 revolution, the fiber might become shorter and thus decrease the tearing index. Sharma et al. (2011) confirmed this by stating that increasing fiber length contributes to increased paper tearing resistance.

ASAM unbleached bamboo paper has tensile, break, and break failure indexes of 26 3318.64 mN (m^2/g), 24 817.87 N m/g, and 10 839.27 kPa, respectively, according to Amin (2016), making it excellent for packaging paper. Bamboo tends to need a great amount of beating (as high as 9500 revolutions) to accomplish promising applications for packaging material, based on the paper characteristics. Since the lumen width of bamboo pulp fibers is so small, swelling of the cell wall is difficult. As a result, the swelling of bamboo pulp fibers is more difficult, necessitating a greater number of beatings. CNC was implemented to improve the paper with minimal beating because beating has a big impact on its intensity and requires a lot of energy.

4.3.2 Number in bracket indicate the standard deviation for the testing results

A total of 8000 beating rotations of *G. scortechinii* pulp using the soda–AQ pulping technique resulted in paper mechanical characteristics ranging from 42.04 to 91.09 Nm/g, 2.68 to 7.10 kPa (m^2/g), 11.03 to 26.64 mN (m^2/g), and 30 to 1127 double folds, as reported by Nurul Husna et al. (2013).

Table 4.4 shows the statistical analysis on the effect of beating for each species on paper mechanical properties, and Table 4.5 shows the statistical analysis on the effect of beating on paper mechanical properties. The data shows that all species are highly significant differences for all the testing except for the tearing index for *Aur* and *Beting*. Beating revolutions for 1000 on folding endurance were not significantly different from 4000 and 8000, but highly significant between both 4000 and 8000 revolutions. Each revolution shows highly significant differences to each testing, which are burst index, tearing index, and tensile.

TABLE 4.4 Statistical analysis on the effect of beating for each species on paper mechanical properties of *Aur, Beting,* and *Semantan.*

Species	Burst index [kPa (m^2/g)]	Folding endurance (double folds)	Tearing index [mN (m^2/g)]	Tensile index (Nm/g)
Aur	3.87[b]	191.00[b]	15.86[b]	43.18[b]
Beting	4.89[c]	525.28[c]	16.51[b]	53.87[c]
Semantan	3.55[a]	138.24[a]	14.01[a]	38.24[a]

Note: Common letter(s) in the same column are not significantly different.

TABLE 4.5 Statistical analysis on the effect of beating revolutions on paper mechanical properties of *Aur, Beting,* and *Semantan.*

Sample	Burst index [kPa (m^2/g)]	Folding endurance (double folds)	Tearing index [mN (m^2/g)]	Tensile index (Nm/g)
0	1.71[a]	11.20[a]	10.65[a]	21.39[a]
1000	2.99[b]	80.80[b]	15.09[b]	35.59[b]
2000	4.21[c]	186.07[c]	18.85[d]	47.45[c]
4000	5.11[d]	405.20[d]	17.11[c]	56.68[d]
8000	6.49[e]	740.93[e]	15.61[b]	64.38[e]

Note: Common letter(s) in the same column are not significantly different.

4.4 Conclusions

Pulping and papermaking from Malaysian bamboo are very promising as they showed good pulp yield and mechanical properties, regardless of types of pulping used for the pulping process, such as soda–AQ pulping, alkaline ASAM, or pretreatment of kraft pulping. The mechanical paper testing was significantly affected by the beating process applied to the bamboo pulp. The optimum beating revolutions are suggested up to 8000 as higher beating revolutions do not have a significant effect in increasing the paper's mechanical properties. Furthermore, it is believed that at severe beating revolution, the fiber is being cut and thus contributes to lowering paper's mechanical properties.

Acknowledgments

The authors acknowledge Ministry of Higher Education Malaysia through Fundamental Research Grant Scheme (FRGS/1/2018/TK10/UITM/02/4) as a fund provider. The authors also express their appreciation to Forest Research Institute Malaysia (FRIM) for providing the raw materials for this study.

References

Abdullah Siam, N., Uyup, M.K.A., Husain, H., Mohmod, A.L., Awalludin, M.F., 2019. Anatomical, physical, and mechanical properties of thirteen Malaysian bamboo species. Bioresource 14 (2), 3925–3943.

Ainun, Z.M.A., Muhammad, K.I., Rasmina, H., Hazwani, H.A., Sharmiza, A., Naziratulasikin, A.K., et al., 2018. Effect of chemical pretreatment on pulp and paper characteristics of bamboo *Gigantochloa scortechinii* Kraft fibers. IOP Conf. Series: Mater. Sci. Eng. 368, 12–44. Available from: https://doi.org/10.1088/1757-899x/368/1/012044.

Anonymous, 2021. What Is Bamboo, Camelback Display Inc. From https://www.camelbackdisplays.com/what-is-bamboo/ (retrieved 10.03.21.).

Atac, Y., Eroğlu, H., 2013. The effects of heartwood and sapwood on kraft pulp properties of *Pinus nigra* J.F. Arnold and *Abies bornmuelleriana* Mattf. Turkish J. Agric. For. 37, 243–248. Available from: https://doi.org/10.3906/tar-1205-20.

Carrillo, I., Graciela, M.A., Valenzuela, S., Mendoca, R.T., Elissetche, J.P., 2015. Variations in wood anatomy and fiber biometry of *Eucalyptus globulus* genotypes with different wood density. Wood Res. 60 (1), 1–10.

Chandra, A.P., Susi, S., 2018. Pembuatan dan karakterisasi dissolving pulp serat panjang dari bambu duri (*Bambusa blumeana*). J. Selulosa 8 (1), 21–32.

Deniz, I., Okan, O.T., Serdar, B., Şahin, H.I., 2017. Kraft and Modified Kraft Pulping of Bamboo (*Phyllostachys bambusoides*). Available from: https://doi.org/10.12841/wood.1644-3985.224.05.

Gominho, J., Lopes, C., Lourenço, A., Simões, R., Pereira, H., 2014. *Eucalyptus globulus* Stumpwood as a raw material for pulping. BioResources 9 (3). Available from: https://doi.org/10.15376/biores.9.3.4038-4049.

Gülsoy, S.K., Ozturk, F., 2015. Kraft pulping properties of European black pine cone. Maderas. Cienc. y. Tecnología 17 (4), 875–882. Available from: https://doi.org/10.4067/S0718221X2015005000076.

Kiaei, M., Tajik, M., Vaysi, R., 2014. Chemical and biometrical properties of plum wood and its application in pulp and paper production. Maderas. Cienc. y. Tecnología 16 (3), 313–322.

Moradbak, A., Tahir, P.M., Mohamed, A.Z., Halis, R., 2016. Alkaline sulfite anthraquinone and methanol pulping of bamboo (*Gigantochloa scortechinii*). BioResources 11 (1), 235–248. 10.15376/biores.11.1.235-248.

Naithani, V., Phillips, C., Jameel, H., 2015. Technical and economic analysis of bamboo fiber in pulp and paper manufacture. In: PEERS Conference 2015: Sustainable Solutions for Our Future, vol. 2, pp. 850–864.

Nurul Husna, M.H., Mansur, A., Rushdan, I., 2011. Enhancement on mechanical properties of Semantan bamboo paper by beating. In: Gan, K.S., Mahmudin, S., Mohd Nor, M.Y. (Eds.), Proceedings of the International Symposium on Forestry and Forest Products 2010: Addressing Global Concerns and Changing Societal Needs, 5–7 October 2010, Kuala Lumpur 91–100.

Nurul Husna, M.H., Suhaimi, M., Rushdan, I., 2013. Effect of soda-anthraquinone pulping conditions and beating revolutions on the mechanical properties of paper made from *Gigantochloa scortechinii* (*Semantan* bamboo). Malaysian J. Anal. Sci. 17 (1), 75–84.

Sharma, A.K., Dutt, D., Upadhyaya, J.S., Roy, T.K., 2011. Anatomical, morphological and chemical characterization of *Bambusa tulda, Dendrocalamus hamiltonii, Bambusa balcooa, Malocana baccifera, Bambusa arundinacea* and *Eucalyptus tereticornis*. BioResources 6 (4), 5062–5073. Available from: https://doi.org/10.15376/biores.6.4.5062-5073.

Suphat, K., 2007. Preliminary study of anthraquinone in sweet bamboo (*Dendrocalamus asper* Backer) alkaline sulfite pulping. Chiang Mai J. Sci. 34 (2), 235–247.

TAPPI, 1995. Test Method T 205 sp-95. Forming Handsheets for Physical Tests of Pulp. Technical Association of Pulp and Paper Institute (TAPPI), 1–9.

TAPPI, 1996. Test Method T511 om-96. Folding Endurance of Paper. Technical Association of Pulp and Paper Institute (TAPPI), 1–5.

TAPPI, 1997. Test Method T 403 om-97. Bursting Strength of Paper. Technical Association of Pulp and Paper Institute (TAPPI), 1–6.

TAPPI, 1998. Test Method T 414 om-98. Internal Tearing Resistance of Paper (Elmendorf-Type Method). Technical Association of Pulp and Paper Institute (TAPPI), 1–7.

TAPPI, 2001. Test Method T494 om-01. Tensile Properties of Towel and Tissue Products (Using Constant Rate of Elongation Apparatus). Technical Association of Pulp and Paper Institute (TAPPI), 1–14.

Tripathi, S.K., Bhardwaj, N.K., Varadhan, R., 2016. Pulp and papermaking properties of bamboo species *Melocanna baccifera*. Cellilose Chem. Technol. 8.

Wang, C.M., Wang, J., Wang, W.J., Mu, Q.Y., Deng, Q.P., 2008. The property and papermaking performance of the major bamboo species in Yunnan Province. China Pulp Paper 27 (8), 10–12.

Wang, Y., Hui, Z., Yulong, D., Shuguang, W., Shuyan, L., 2016. Variability of anatomical and chemical properties with age and height in *Dendrocalamus brandisii*. BioResources 11 (1), 1202–1213. Available from: https://doi.org/10.15376/biores.11.1.1202-1213.

Wheeler, E.A., Baas, P., Gasson, P.E., 1989. IAWA list of microscopic features for hardwood identification. IAWA Bull. 10 (3), 219–332.

Wu, Y.M., 1997. Plant Fibre Chemistry. *China Light Industry Press, Beijing, China*, pp. 70–74.

Yang, Q., Su, G.R., Duan, Z.B., Wang, Z.L., Hang, L., Sun, Q.X., et al., 2008. Fibre characteristics and papermaking feasibility of major sympodial bamboos in Xishuangbanna. Trans. China Pulp Paper 23 (4), 1–7.

Zhan, H., Tang, G., Wang, C., Wang, S., 2015. Chemical properties and fibre morphology of *Fargesia fungosa* at different culm ages and heights. BioResources 10 (3), 5666–5676. Available from: https://doi.org/10.15376/biores.10.3.5666-5676.

Chapter 5

Pulping and papermaking of cornstalk

Sandeep Kumar Tripathi and Nishi Kant Bhardwaj
Avantha Centre for Industrial Research & Development (ACIRD), Paper Mill Campus, Yamuna Nagar, Haryana, India

5.1 Introduction

The pulp and paper industry is considered to be one of the highest consumers of forest-based raw materials as these woody raw materials are the main source of papermaking fibers. The major raw materials for pulp and papermaking can be categorized in three groups: wood, nonwood, and nonplant (mainly wastepapers). In the last several decades, the rise in world paper production along with increasing world population, improved literacy, and improved standard of living has created opportunities for renewable nonwood raw materials as an alternative fiber source. Nonwood fibers are largely used for papermaking in countries with a lack of forest resources, such as India, China, and others. The regions where forest-based raw materials are limited, the nonwood raw materials are considered an important source of fibers for pulp and papermaking (Liu et al., 2018). Maize crop is widely grown throughout the world, 40% of the world's harvest being produced in the United States. The other maize producing countries include Brazil, China, Argentina, Ukraine, Mexico, Canada, India, and Indonesia. Cornstalks are available in abundant quantities as one of the nonwood fiber sources in many maize producing countries.

Presently about 1148 million MT (metric tonne) of corn or maize is produced worldwide from 194 million hectare area (FAO, 2020). The estimated global availability of corn stalk as waste is 750,000,000 bone dry metric tonne (El-Sayed et al., 2020). Following nomenclatures are generally used for different portions of the whole maize plant:

Stalk: Represents mainly the stems and leaves.
Husk: The leafy outer covering over the corn seeds.
Cobs: The central core on which corn seeds are grown.

Pulping and Papermaking of Nonwood Plant Fibers. DOI: https://doi.org/10.1016/B978-0-323-91625-7.00009-6

Stover: The residue left after harvesting the corn seed consisting of stalks, leaves, husk, sheath and cobs. The yield of stover is about 50% of corn production.

The whole maize plant contains about 31.9% grain, 52.1% stalk residue, 11.5% cob, and 4.5% husk (Singh et al., 2015). Corn stalk residue is considered potential fiber source for making different grades of papers. Fiber morphological properties, specifically fiber length and width of cornstalks pulp, are similar to the sugarcane bagasse (Pfaffli and Sisko, 1995). The chemical properties of the cornstalks varied with the species and the location of cultivation.

The use of cornstalks (mainly the stems and leaves) as papermaking material is not new. As early as the 17th century, a small paper mill in Rimini, Italy produced paper from corn husk (leafy outer covering over the corn seeds). In the 18th century, Schaeffer in Germany produced paper from the husks, leaves, and stalks as well as from whole stalks. In 1602, B. Allison and J. Hawkins received a patent in the United States on a process for the manufacture of paper from com husk. In 1828, paper was made from cornstalks in the mill of one, Mr. Rowland and Guilford, England. The stalks and leaves were made into an olive-colored paper, while the husk produced a white paper. The weekly Political Register for January 10, 1829 was printed on corn husk paper (Arnold, 1930). Afterward, the utilization of cornstalks for producing paperboard and packaging paper pulp was initiated in Europe, especially in Austria, in 1837. This was followed by the work of Weber in 1929, and research was completed soon as the results were not favorable.

Presently, the paper industry is facing problems with a shortage of raw materials, stringent requirements to reduce effluent volume and contaminants, and high capital cost of technically superior equipment. Shortage of raw material resources is the most dominating factor (El-Sakhawy et al., 1996). Cornstalk may fulfill part of the raw material requirement for paper industry as it has low lignin and high carbohydrate contents, fiber characteristics of cornstalk are also similar to the hardwood and sugarcane bagasse. The chemical composition and morphological characteristics of cornstalk are also reported to be suitable as raw material for producing pulp in combination with other raw materials like hardwood, softwood, Napier grass, and pineapple leaf pulps (Won and Ahmed, 2004; Daud et al., 2013, 2014; Aremu et al., 2015a). Researchers have also reported on the utilization of cornstalk for making particleboard along with industrial wood mixture, as well as urea-formaldehyde as adhesive (Kadam and McMillan 2003; Guler et al., 2016).

Some nonwood fibers have very long fiber lengths and are considered similar to or better than the softwood materials in terms of the pulp and papermaking properties like sisal, jute, hemp, flax, kenaf, abaca, and cotton (Liu et al., 2018). Pulp obtained from banana stems was also reported to be similar or superior to the hardwood pulp. Fractionated long fiber pulp from banana stems was found suitable for making high-strength papers like currency notes and for making teabag papers (Tripathi et al., 2017, 2019a).

Like many other nonwoods, such as bagasse, sarkanda, and wheat straw, cornstalks are considered a promising source of raw material for making paper of different grades (Lal et al., 2004, 2005; Bajpai et al., 2008; Tripathi et al., 2010, 2018, 2021; Sharma et al., 2020). Cornstalk, specifically the bottom 3–4 ft of corn stalk, is reported to be as good as hardwood material for making pulp and paper (Won and Ahmed, 2004). As compared to other nonwood fibers, the research works on the utilization of cornstalks fibrous source for pulp and paper industry are scanty.

Chemical constituents, morphological properties, raw material preparation for pulping, suitability of different pulping processes like soda, soda–anthraquinone (AQ), alkaline sulfite, organosolv, different bleaching sequences, and papermaking properties for cornstalk are reviewed in this chapter.

5.2 Chemical and morphological properties

5.2.1 Chemical properties

The structural framework of wood or nonwood plant fibers is made from cellulose, hemicellulose, lignin, solvent extractives, and minerals. Suitability of a nonwood plant for pulping and papermaking is governed by the proportionate presence of these chemical constituents. Presence of higher amounts of cellulose and hemicellulose and lower amounts of lignin, extractives, and minerals are desirable for a raw material to be good for pulping and papermaking. Higher amounts of cellulose and hemicellulose in raw material give higher pulp yield after pulping. Presence of higher amounts of lignin in raw material consumes higher chemicals during pulping and results in low pulp yield. Presence of higher amounts of minerals makes the material unsuitable for pulp and papermaking. The chemical properties of the cornstalks varied with the species and the location of cultivation. The average values of chemical constituents of cornstalk reported by different research studies are shown in Fig. 5.1.

The chemical analysis of an Egyptian corn stalk has holocellulose 77.0%, α-cellulose 53.6%, hemicellulose 16.2%, lignin 22.2%, and ash 3.0%. Cornstalk produced in the Moldova region of Romania was reported to have significantly lower holocellulose 62.5%–67.7%, cellulose 34.6%–42.1%, pentosans 17.6%–20.7%, lignin 19.1%–20.9%, and ash 4.5%–5.7% (Chesca et al., 2018). Cornstalk grown in northern India has cellulose 53.6%, hemicellulose 26.4%, lignin 20.0%, ash 4.6%, silica 0.65%, and acetone extractives 3.5%. The waxes, resins, fats, tannins, and gums are mainly extracted in the acetone (El-Sayed et al., 2020; Mishra et al., 2020). The mineral content in cornstalk is comparatively lower than the other nonwood raw materials except sugarcane bagasse (Shakhes et al., 2011; Tripathi et al., 2013, 2019b; Sridach 2010). Detailed values reported in different research articles for cornstalk are given in Table 5.1.

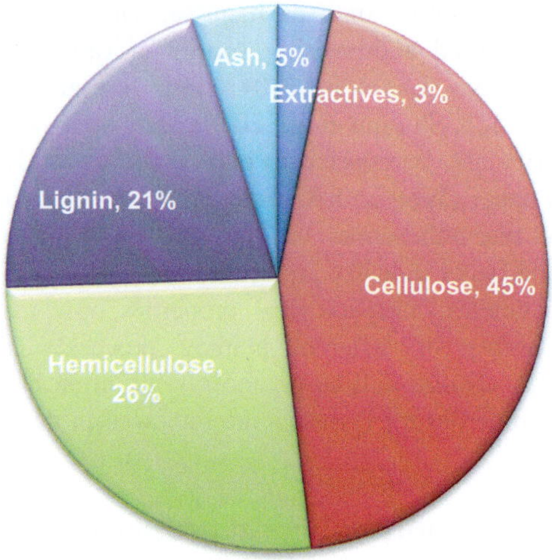

FIGURE 5.1 Chemical constituents of cornstalk.

In the studies carried out earlier in 2004 and 2006, researchers compared the chemical constituents and morphological properties of cornstalk pulp with those of hardwood and softwood pulps. It was reported that cornstalk contains the higher hemicellulose content (about 27%) as compared to that in other nonwood raw materials and most of the hardwoods. Cornstalks also contain low lignin compared to wood making it a prospective source of papermaking fiber. The protection of hemicelluloses in papermaking fiber is very important for increased fiber–fiber bonding (Ahmed and Zhu, 2006). The average fiber length of cornstalk pulp is better than or similar to hardwood pulp. The characteristics of fiber for softwood and hardwood compared to cornstalk are shown in Table 5.2.

Cornstalk contains the highest amount of glucan followed by xylan and arabinan among the different sugar monomers. Table 5.3 contains the detailed values of different sugar monomers in the cornstalk (Ahmed and Zhu, 2006; Li et al., 2012). Percentage of monomer sugars in cornstalk was 55.6% only, and the remaining fraction may consist of lignin, extractives, and minerals.

Cornstalk rind and pith are reported to have a significant amount of potassium, chlorides, and silica along with a trivial amount of other metals and minerals. A range of these metals and minerals present in cornstalk rind and pith are given in Table 5.4.

5.2.2 Morphological properties

The morphological properties of any plant fiber are the most important from a pulp and papermaking point of view. Fiber length and width are the most important properties among different morphological properties.

TABLE 5.1 Chemical constituents of cornstalk.

Particulars	Values (%)						
	Mishra et al. (2020)	Ferdous et al. (2020)	Alagbe et al. (2019)	Chesca et al. (2018)	Daud et al. (2016)	Barbash et al. (2012)	Arnold (1930)
Hot water solubility	11.4	18.1	15.2	22.9	14.88	12.6	18.89
1% NaOH solubility	41.3	38.9	18.4	49.3	39.62	69.6	53.82
Acetone extractives	3.52	0.87	2.80[a]	2.6	3.57[a]	–	–
Cellulose	53.6	–	53.3	38.7	42.64	39.0	43.68
α-Cellulose	–	35.1	48.2	–	–	–	53.34
β Cellulose	–	–	–	–	–	–	26.33
γ Cellulose	–	–	–	–	–	–	16.57
Pentosans	25.4	17.8	–	19.5	25.66	–	27.11
Klason lignin	20.0	19.7	13.7	20.2	17.98	7.3	30.36
Holocellulose	77.5	59.5	62.3	64.4	–	–	–
Ash	4.65	4.48	1.27	5.1	4.74	4.7	6.20

[a] *Alcohol benzene extractives.*

TABLE 5.2 Fiber characteristics for cornstalk and wood (Ahmed and Zhu, 2006).

Particulars	Cornstalk	Softwood	Hardwood
Cellulose (%)	46–50	40–52	38–50
Lignin (%)	16–17	26–32	18–28
Pentosans (%)	27–28	8–12	15–25

TABLE 5.3 Sugar monomers in cornstalk (Ahmed and Zhu, 2006).

Particulars	Value (%)	
	Ahmed and Zhu (2006)	Li et al. (2012)
Arabinan	1.25	1.4
Galactan	0.57	0.27
Rhamnan	0.04	0.04
Glucan	38.47	–
Xylan	15.13	14.6
Mannan	0.11	0.19

TABLE 5.4 Metals and minerals in cornstalk (Li et al., 2012).

Particulars	Value
Potassium as K (%)	4.03–4.17
Chloride as Cl (%)	2.35–2.52
Silica as Si (%)	0.37–0.38
Magnesium as Mg (%)	0.14–0.26
Calcium as Ca (%)	0.22–0.25
Sulfur as S (%)	0.10–0.11
Phosphorus as P (ppm)	394–485
Sodium as Na (ppm)	242–326
Iron as Fe (ppm)	52–108
Aluminum as Al (ppm)	65–86
Zinc as Zn (ppm)	19–32
Copper as Cu (ppm)	8.1–8.6
Manganese as Mn (ppm)	6.5–6.9

Generally, the fibers having a length-to-width ratio of about 100:1 are considered good for papermaking and a length-to-width ratio of more than 1000:1 as textile fiber. Softwood fibers are having the length-to-width ratio of 60:1–100:1, whereas hardwood fibers have the said ratio up to 60:1 (Hurter, 1988; Hunsigi, 1989; McDougall et al., 1993; Saijonkari-Pahkala, 2001). The nonwood fibers have average fiber length-to-width ratio of 50:1–1500:1 (Hurter, 1988). Cornstalk pulp fiber is having a length-to-width ratio of 41:1–82:1. Cell wall thickness and lumen size of fibers govern the refining ability of pulp and rigidity, physical strength properties of the paper. Cornstalk pulp fibers have wide lumen and thin cell wall, which makes the fiber easily collapsible. There is a huge variation in the values of various morphological properties reported by different authors for cornstalk pulp fiber. The variation is mainly due to the climate, soil, and the species reported by the authors. The average fiber length of cornstalk pulp fiber ranges from 0.93 to 1.49 mm, which is comparable to the sugarcane bagasse, higher than that of hardwood and lower than the softwood pulp fibers. Similarly, the fiber width of cornstalk pulp fiber is also comparable to the sugarcane bagasse higher than that of hardwood and lower than the softwood pulp fibers. Morphological properties of different cornstalk pulps are given in Table 5.5.

TABLE 5.5 Morphological properties of corn stalk.

Particulars	Value (%)				
	Mishra et al. (2020)	Mishra et al. (2020)	Ferdous et al. (2020)	Alagbe et al. (2019)	Chesca et al. (2018)
Fiber length (mm)	0.96	0.93	0.9	1.49	0.93
Width (μm)	19.3	19	15.7	18.0	22.9
L:W ratio	50	49	57	82	41
Cell wall thickness (μm)	–	–	–	2.3	5.8
Coarseness (μg/m)	95.7	94.1	–	–	–
Length distribution					
0.2–0.5 mm	32.6	34.4	–	–	–
0.5–1.5 mm	47.9	47.2	–	–	–
1.5–3.0 mm	18.2	16.7	–	–	–
>3.0 mm	1.3	1.7	–	–	–

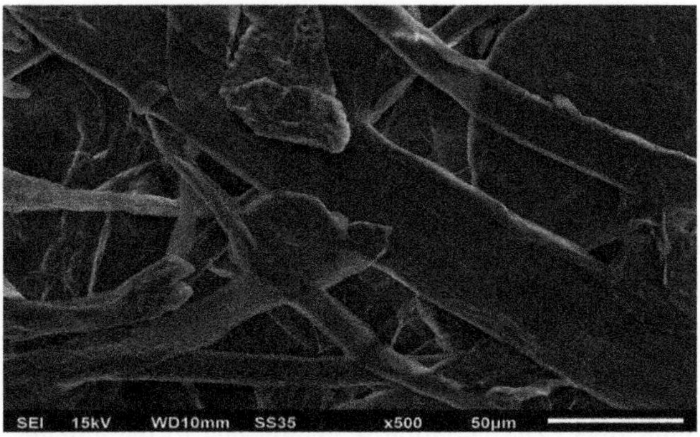

FIGURE 5.2 Scanning electron microscope (SEM) micrograph of unrefined bleached cornstalk pulp (Mishra et al., 2020).

Nonwood fibers are comparatively weaker than hardwood and softwood pulps. The S1 layer of nonwood fiber is mostly deteriorated during pulping process itself. A lot of fibrillation can be seen on nonwood pulp fibers after bleaching process even without refining. Most of the nonwood fibers do not require refining to develop physical strength properties. Similar to the other nonwood raw materials, fibrillation in the cornstalk fiber is also generated during bleaching process even without mechanical refining. Cornstalk fibers easily get flattened during papermaking due to their wide lumen and thin cell wall thickness. An scanning electron microscope (SEM) micrograph of cornstalk bleached unrefined pulp is shown in Fig. 5.2 with visible fibrils on flattened fibers.

5.3 Suitability of different pulping processes

5.3.1 Raw material preparation for pulping

Cornstalk is required to be processed for drying, chopping, and wet cleaning before being used for pulping. Cornstalk contains about 70%–80% moisture at the time of harvesting (Igathinathane et al., 2004; Gould, 2007; Sadawarte, 1995). Initially, the moisture is required to be reduced through air/sun drying to reduce the decomposition of organic matters and transportation costs from farm to the mill. The cornstalk grows about 4–12 ft in height and thus requires cutting or chopping in small pieces prior to pulping. Moreover, cornstalk contains a large amount of sand and silica on its outer surface, which is also needed to be removed for better pulping performance.

5.3.1.1 Chopping

For making cornstalk suitable for pulping, it is chopped in 3–5 cm length pieces for better impregnation of cooking liquor during pulping (Chesca et al.,

2018). Due to the mechanical action of cutter/chipper a lot of nonfibrous material gets separated from the surface of cornstalk. Separation of nonfibrous material improves the pulping and bleaching efficiencies of nonwood materials (Saijonkari-Pahkala, 2001).

5.3.1.2 Wet cleaning and screening

After chopping, cornstalk is washed with hot water or process water to remove nonfibrous materials, specifically the sand and silica. Washing of cornstalk with normal tap water removed ash content by 2.9% and with hot water reduced ash content by 4.8%. It is beneficial to treat the raw material with water before pulping to remove the undesired nonfibrous materials before pulping (Abou-State et al., 1973; Leponiemi, 2008).

5.3.2 Pulping process

The objective of pulping is to separate the fibers from fibrous raw material and convert into the pulp form. Fibrous biomass can be converted into pulp using chemical or mechanical pulping processes. In chemical pulping process, alkaline or acidic chemicals are used at elevated temperature to separate the fibers. In the mechanical pulping process, fibers are separated using mechanical force at mild temperature and sometimes using mild chemicals. The pulps obtained from mechanical pulping process have less strength and durability because severe damage occurs to the fibers during mechanical treatment and presence of the significant amount of lignin with the pulp. The mechanical and chemi-mechanical pulping processes are also known as high yield pulping processes as they yielded with to 88%–97% pulp. Pulps obtained from mechanical pulping processes are generally utilized for making newsprint, book, and magazine papers. Softwood and few species of hardwood are generally used for the production of mechanical pulp. Nonwood materials are sparingly used to produce mechanical pulp.

In the chemical pulping processes, lignin is selectively removed, leaving behind cellulosic fibers in the form of pulp. The pulps obtained from chemical pulping process have more strength and durability therefore used mainly for making writing, printing, and packaging papers (Flandez et al., 2010). Pulp yield from chemical pulping processes is varied from 43% to 57% only as lot of hemicellulose and cellulose are also dissolved along with lignin during pulping. The lignin and carbohydrates dissolved during chemical pulping process are separated, concentrated using evaporators, and burnt in the boilers to generate energy and recover the chemicals. Results obtained from some commonly used chemical pulping processes for cornstalk are given in Table 5.6.

Brief description about most commonly followed pulping process for nonwood materials specifically for cornstalk are given below.

TABLE 5.6 Pulping condition, results obtained from cornstalk.

Pulping process	Chemical used (%)	Cooking temperature (°C)	Pulp yield (%)	Kappa no.	References
Soda	12 16 14	110 140 170	54.1 48.9 41.6	33.2 20.5 15.6	Jahan et al. (2004)
	10	122	–	14.3	Fagbemigun et al. (2014)
	18	170	35.9	19.2	Chesca et al. (2018)
	15 15	135 160	47.3 46.1	8.1[a] 5.2[a]	Fahmy et al. (1992)
Sulfate	15 17	135 135	46.0 44.4	6.7[a] 6.1[a]	
	18	170	37.0	32.4	Chesca et al. (2018)
Soda–AQ	14 16 18	150 150 150	58.6 54.9 51.8	27.4 24.8 21.8	Jahan and Rahman (2012)
	16	160	42.4	10.4	Ates et al. (2015)
	14 16 18 20	150 150 150 150	32.8 31.8 31.8 27.6	9.9 8.3 8.7 7.6	Ferdous et al. (2020)
	13 14.5 16	168 168 168	56.0 53.7 47.7	20.3 16.1 11.9	Mishra et al. (2020)
Sulfite	14 16	145 145	50.6 50.2	33.9 32.3	Latibari et al. (2011)

AQ, Anthraquinone.
[a]*Lignin content, %.*

5.3.2.1 Soda and soda–anthraquinone

Soda and soda–AQ pulping processes are the most commonly used pulping processes for nonwood (Assumpcao, 1992; Byrd et al., 2006; Behin and Zeyghami, 2009, Jahan and Rahman, 2012; Mayowa 2018; Liu et al., 2018; El-Sayed et al., 2020). In this process, only sodium hydroxide is used as a cooking chemical, and in few cases, AQ is also added to fasten the delignification and protect the carbohydrate from degradation. Use of AQ in soda pulping process for cornstalk is reported to improve pulp yield by 6.4% at a similar kappa number of pulp (Jahan et al., 2004). But the use of AQ in pulping is now banned in several countries due to its carcinogenic nature (Hart and Rudie, 2014).

5.3.2.2 Sulfate or kraft

In the sulfate or kraft process, sodium hydroxide and sodium sulfide are the main active chemicals. Pulp/paper produced using this process has higher strength compared to soda process; therefore this process is called "kraft" (in German which means the "strength"). This process is also called "sulfate" because sodium sulfate is added as makeup chemical. Sulfate pulping processes are the most commonly used pulping processes for wood-based materials and are also used for nonwood materials.

5.3.2.3 Alkaline sulfite

In the alkaline sulfite process, sodium sulfite is used along with alkaline chemicals like sodium hydroxide or sodium carbonate. The pulp obtained from sulfite process is comparatively brighter than the pulp obtained from soda or sulfate process. Unbleached pulp of 32.3 kappa number at 50.3% yield was obtained in a study carried out on cornstalk using 16% of sodium sulfite and sodium hydroxide in 50:50 ratio at 145°C for 30 minutes. The pulp has tensile index of 62.4 N (m/g), burst index of 3.80 kPa (m^2/g), and tear index of 10.53 mN (m^2/g) (Latibari et al., 2011).

5.3.2.4 Organosolv

In the organosolv pulping process, organic solvents mainly alcohol like methanol, ethanol and acids like formic acid, acetic acid are used at elevated temperature to dissolve the lignin. Organosolv pulping process is considered more environmentally friendly than soda, kraft, or sulfite pulping process but has higher production cost. Other than the methanol, ethanol, formic acid, and acetic acid, the use of other solvents like liquid carbon dioxide, dimethylsulfoxide, dioxane, phenols, sulfolane, glycols, amines, and nitrobenzene are also reported in lab scale studies (Sundquist, 2000; Sridach, 2010; Weiss et al., 2010). Organosolv pulping of nonwood materials are more suitable for converting biomass in ethanol or methanol as the strength of pulp obtained is comparatively weaker than the pulps obtained from soda or sulfate pulping processes (Snelders et al., 2014; Mesa L et al., 2016; Park et al., 2018).

5.3.2.5 Other pulping processes

Cornstalk was also evaluated for different nonconventional pulping processes like alkaline sulfite AQ methanol pulping process (Jahan et al., 2006), organosolv (Barbash et al., 2012; Balea et al., 2016), thermomechanical (Theng et al., 2017), oxygen–alkali (Fahmy et al., 1992), alkaline sulfite (Latibari et al., 2011), biopulping (Wu et al., 2011), soda–butanol (Chesca et al., 2018), and Natron pulping (Chesca et al., 2018b) for producing pulp, paper, paper board, particle board, nano fibrillated cellulose etc. The suitability of cornstalk has been also evaluated for making dissolving pulp using water and acid prehydrolysis, followed by kraft pulping and bleaching with HEHP sequence (Reddy and Yang, 2005; Behin et al., 2008).

5.3.3 Suitable digester types

Batch or continuous types of the digesters are used for making pulp from wood or nonwood materials. Conventionally, batch-type rotary digesters were used for pulping of nonwood materials using alkaline pulping processes. Nowadays, the use of batch-type digester for nonwood materials is substantially reduced due to productivity and quality issues. Due to the bulkier nature of nonwood materials, batch-type digesters required more time, high energy, and water/liquor to produce pulp. Pandia/tube-type continuous digesters are now been preferred to produce pulp from nonwood fibers as they require less pulping time, less energy, and less space compared to batch digesters. Moreover, in pandia type digesters, the pulp after cooking is blown at low pressure and temperature, which protects the fiber from damage and results in improved physical strength properties.

5.4 Suitability of different bleaching sequences

The pulp bleaching can be attained through chemical action by the removal of lignin in the pulp or by changing the structure of chromophoric group. The oxidizing chemicals, reductive chemicals, sodium hydroxide, chelating agents, and enzymes can be used in different combinations for bleaching of pulp. Traditionally, the three-stage bleaching (CEH) process was largely adopted for bleaching of nonwood pulps. However, due to the usage of chlorine and chlorine-containing compounds, this bleaching method has high wastewater pollution due to the generation of adsorbable organic halogens (AOX) that are harmful toxic and difficult to degrade. So, the emission of AOX in bleaching wastewater has become more and more stringent in last one to two decades. Presently, the main clean bleaching methods are elemental chlorine-free (ECF) bleaching, main chemical being chlorine dioxide with strong oxidation capacity, and total chlorine free (TCF) bleaching, main chemicals being oxygen, hydrogen peroxide, and ozone.

Abou-state et al. (1973) investigated the effects of the amount of active chlorine in the hypochlorite step on the optical and the physical strength properties of paper bleached by multistage CEH sequence. The conditions of different bleaching stages are summarized in Table 5.7. The lignin content of the unbleached pulp was 4.3%, and its natural chlorine requirement was 6.9%. The effect of increasing the amount of active chlorine in the hypochlorite stage is

TABLE 5.7 Different process conditions during multistage bleaching of cornstalk (Abou-state et al., 1973).

Stage	Temperature (°C)	Reaction time (min)	Pulp consistency (%)	Chemical and dose (%)
Chlorination (C)	20	60	3	Chlorine water, –
Extraction (E)	40	90	5	NaOH, 2
Hypo (H)	40	90	3	NaOCl, 30% and 50% of the amount of chlorine used in first stage
Acidification	20	20	3	H_2SO_3, 0.5%

TABLE 5.8 Effect of increasing the amount of active chlorine in the hypochlorite stage (Abou-state et al., 1973).

Active Cl_2 (% of the natural Cl_2 requirement)	30	50
Active Cl_2 (g/100 g pulp)	2.07	3.45
Brightness (%)	69.5	74.0
Opacity (%)	84.0	79.0
Yield of pulp (%)	47.5	47.0
Tensile strength (kg)	7.5	10.0
Bursting strength (lb/in.^{2})	37.8	60.5
No. of double folds	1200	2400
Tearing strength (g)	24	33

given in Table 5.8, which shows that this increase in the active chlorine raised the brightness and improved materially all the physical strength characteristics.

Behin et al. (2008) studied the prospects of cornstalk in dissolving pulp production by carrying out water and acid prehydrolysis at 160°C for 30 minutes followed by kraft pulping at 170°C for 90 minutes and subsequent bleaching by HEH and HEHP sequences (process conditions summarized in Table 5.9). Figs. 5.3 and 5.4 present the effect of two parameters (prehydrolysis and sulfidity) on alpha-cellulose content of pulps obtained by

TABLE 5.9 The process conditions followed during HEH and HEHP bleaching of water and acid prehydrolyzed kraft cornstalk pulps (Behin et al., 2008).

Stage	Temperature (°C)	Reaction time (min)	Pulp consistency (%)	Chemical and dose (%)
Hypo (H)	60	60	10	NaOCl, 2.5, pH 10
Extraction (E)	60	90	10	NaOH, 5
Hypo (H)	70	70	10	NaOCl, 5%, pH 10
Peroxide (P)	60	60	10	H_2O_2, 2.2%, pH 9–11 adjusted by NaOH

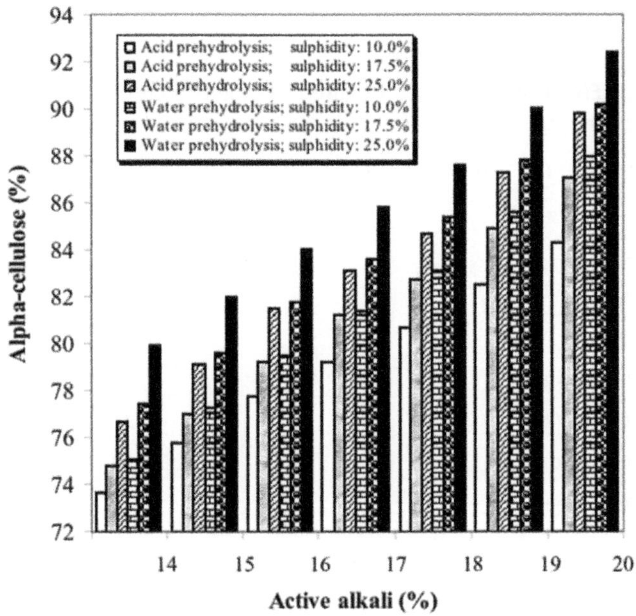

FIGURE 5.3 Alpha-cellulose content of cornstalk dissolving pulps obtained by HEH bleaching sequence (Behin et al., 2008).

HEH and HEHP bleaching sequence, respectively. The optimum conditions (water prehydrolysis, pulping with active alkali of 20%, sulfidity of 25% and HEH bleaching sequence) resulted in the alpha-cellulose content, degree of polymerization, and ash content of the produced dissolving pulp with

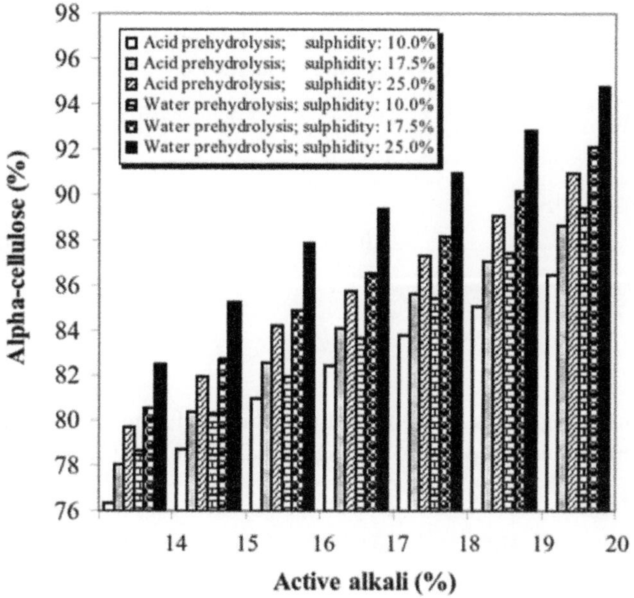

FIGURE 5.4 Alpha-cellulose content of cornstalk dissolving pulps obtained by HEHP bleaching sequence (Behin et al., 2008).

acceptable values of 94.7%, 269, and 1%, respectively, comparable with the pulps obtained by other sources.

5.4.1 ECF bleaching

Ates et al. (2015) pretreated cornstalk and other annual plants and agricultural wastes with the enzyme xylanase for the preliminary biological process (stalk to solution ratio of 1:7, enzyme dose of 2 IU/g, at pH 4.8, 48°C for 2 hours) and then bleached using the elementary chlorine-free bleaching sequence $D_1E_PD_2$ following the conditions summarized in Table 5.10.

The brightness values of handsheets bleached are given in Table 5.11. The soda–AQ rye straw pulp showed better bleachability followed by wheat straw, reed, and cornstalk.

Mishra et al. (2020) studied the bleaching behavior of soda–AQ cornstalk pulp. The unbleached pulp of kappa number 16.1 was easily bleached to the brightness and whiteness of 88.4% and 81.8 points, respectively, using elemental chlorine-free bleaching sequence $D_0E_{OP}D$ applying optimized kappa factor of 0.28 in the D_0 stage of bleaching and various conditions in different stages of bleaching as described in Table 5.12. The brightness and whiteness obtained with cornstalk soda–AQ pulp were significantly better than other agro-based pulps. The brightness reversion measured as a postcolor number of 0.78 was also comparatively lower than other agro residue–based raw material pulps.

TABLE 5.10 Bleaching plan applied to cornstalk pulp following preliminary biological bleaching (Ates et al., 2015).[a]

Particulars	D_1	E_P	D_2
ClO_2 (%)	3	–	1.5
$NaOH/MgSO_4/H_2O_2$	–	1.0/0.5/0.5	–
Time	120	180	120

[a]*Temperature 75°C and pulp concentration 7%.*

TABLE 5.11 Brightness properties of cornstalk and other annual plant soda–anthraquinone pulps following XDE_PD bleaching (Ates et al., 2015).

Raw material	Control	X stage	D_1 stage	E_P stage	D_2 stage
Rye straw	47.8	48.6	72.5	85.1	98.3
Wheat straw	34.5	36.7	62.3	76.1	92.0
Reed	38.3	37.8	53.5	75.8	88.6
Cornstalk	44.0	43.4	60.7	70.1	87.6
Rice straw	40.7	40.3	54.4	66.2	87.5
Barley straw	39.0	42.1	62.1	74.2	86.0
Hemp	44.4	38.6	56.8	78.4	80.1
Sunflower stalk	28.3	25.7	34.9	55.8	71.9
Tobacco stalk	31.2	30.9	44.4	56.8	68.4
Cotton stalk	19.3	17.0	35.4	42.7	61.0

Bleached pulp viscosity 21.1 cP was significantly higher to other straw pulps (Ates et al., 2015). Earlier the researchers were reported to achieve up to 80.0 point whiteness while using organosolv pulping followed by TCF bleaching sequences for cornstalk (Barbash et al., 2012). Detailed ECF bleaching result of cornstalk is given in Table 5.13. Development of pulp brightness during bleaching of corn straw pulp was found comparable with bagasse pulp and superior to wheat straw and rice straw pulps as shown in Fig. 5.5.

5.4.2 Total chlorine-free bleaching

Barbash et al. (2012) investigated the possibility of elemental chlorine-free bleaching of cornstalk pulps obtained from organosolvent process using hydrogen peroxide to a higher whiteness level ($>80\%$). To prevent the catalytic

TABLE 5.12 Process conditions during bleaching of cornstalk pulp using elemental chlorine-free bleaching sequence $D_0E_{OP}D$ (Mishra et al., 2020).

Particulars	D_0 stage	E_{OP} stage	D stage
Consistency (%)	10	10	10
Temperature (°C)	55	80	75
Time (min)	45	120	180

TABLE 5.13 Bleaching of cornstalk pulp using $D_0E_{OP}D$ bleaching sequence (Mishra et al., 2020).

Particulars	Results
Kappa number	16.1
D_O stage	
End pH	2.3
Residual ClO_2 (ppm)	12.1
E_{OP} stage	
End pH	11.3
Residual H_2O_2 (ppm)	3.4
Kappa no.	1.2
Brightness (% ISO)	82.0
D stage	
End pH	3.3
Residual ClO_2 (ppm)	20
Brightness (% ISO)	88.4
Whiteness (CIE)	81.8
Viscosity (cP)	21.1
Postcolor number	0.70
Shrinkage (%)	6.9

decomposition of hydrogen peroxide from metal ions present in pulp, chelating reagents were used during the hydrogen peroxide stage of bleaching. The cellulose obtained by alkali-sulfiteantragunone-ethanol method at a temperature of 170°C and cooking time of 60 minutes have the breaking length of 7300 m, burst index of 4.7 and tear index of 6.4, residual lignin of 2.6%, and whiteness

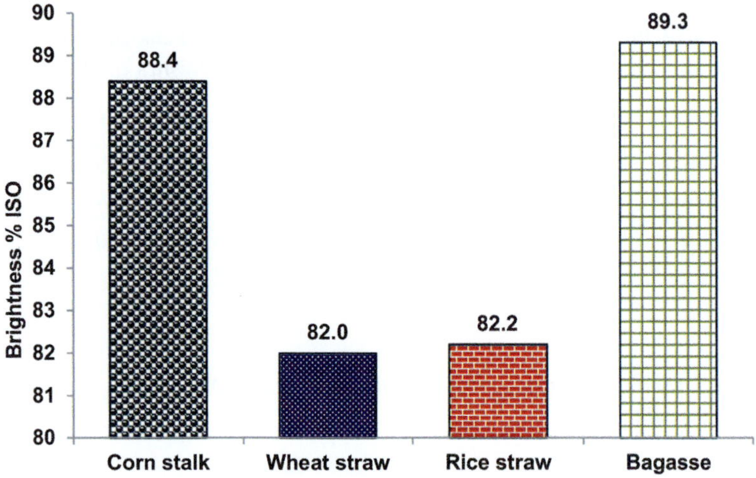

FIGURE 5.5 Brightness development of different agro residues pulps using $D_0E_{OP}D$ sequence (Tripathi et al., 2013; Kaur et al., 2018; Mishra et al., 2020).

TABLE 5.14 Characteristics of bleached organosolvent pulp from cornstalks using various reagents for chelating processing (bleaching scheme Q–P–A) (Barbash et al., 2012).

Particulars	UB pulp	Trilon B	Boric acid	Copperas	Bluestone
Yield of bleached pulp (%)		96.6	97.1	95.6	85.4
Residual lignin (%)	2.6	0.7	0.5	1.1	1.2
Breaking length (m)	7300	7100	7000	6200	6700
Burst index (kN/g)	4.7	4.1	4.3	3.7	3.5
Tear index [mN (m^2/g)]	6.4	6	6.1	5.3	5.2
Whiteness (%)	53	82.2	80.7	78.7	77.1

UB, Un-bleached.

of 53.0% was used for investigation. The effect of using trilon B, boric acid, copperas, and bluestone compound as a chelating reagent at a dose level of 0.2% BD cellulose on different properties is given in Table 5.14. Trilon B was found the most effective compound as a chelating agent followed by boric acid copperas and bluestone (Barbash et al., 2012).

Fagbemi et al. (2014) produced soda pulps of 11.9 and 30.0 kappa numbers using corn husk and kenaf bark, respectively, and subsequently bleached separately in a mixture of hydrogen peroxide and acetic acid (2:1) to study blending of bleached pulps in different ratios, 50%–50%, 75%–25%, and 25%–75% and 100% corn husk. Depending on targeted end use, papers of differing strength properties could be obtained from the combination of kenaf bark and corn husk. Such blending has the potential of producing paper of differing qualities and can find use in a wide variety of applications.

5.5 Papermaking properties

Papermaking is a complex process of producing paper from fibers of different origins, which involves the removal of water from an aqueous slurry consisting of fibers, fines, and other chemical additives on a machine wire by forming a wet mat. This mat is subsequently pressed and dried to make the paper, often with desirable physical strength properties and lowest possible cost.

Papermaking fibers are generally subjected to mechanical treatment known as beating or refining that is an important link in the pulp and papermaking process and results in structural changes of the fibers, including fiber shortening and internal/external fibrillation, so that the binding forces between the fibers are improved and paper strength properties are improved. As different types of fibers react differently to beating or refining, this process is carried out so as to take into account the type of fibers concerned. Mild chemical treatment during the pulping and mechanical processing of papermaking fibers will leave a substantial portion of the primary layer of cell wall intact and it will control the surface properties of fiber, whereas severe processing may disrupt the outer cell wall structure and the chemical properties will be predominantly those of the S2 layer.

Paper strength properties are largely a function of fiber length and interfiber bonding. Interfiber bonding increases with specific surface area by fibrillation and the presence of materials that facilitate surface interactions. To improve the final paper quality, various industrial papers may contain more than one fiber type along with other nonfibrous materials in the furnish. The various methods to improve the physical strength properties of paper include mechanical treatment, chemical additives, fractionation, blending, chemical treatments, enzymatic treatments, and paper machine modifications. Blending of different pulps is one of the most important physical treatment processes for improving the final paper quality.

5.5.1 Papermaking using cornstalk unbleached pulps and physical strength properties

The various studies of the papermaking quality of cornstalks were made earlier with the aim to find the possibilities of the utilization of cornstalks as a raw material for paper. To get the desired degree of pulping, the mechanical treatment of cornstalks was considered essential. The cortex or the outside shell was pulped using kraft and soda pulping processes, and the use of unbleached pulps for making brown wrapping paper was investigated; however, the papers did not compare favorably with wood fiber wrapping papers (Weber et al., 1935).

Abou-state et al. (1973) compared one- and two-stage pulping and the effect of depithing after the first stage was also studied. The yield and strength properties tend to improve when depithing is applied after the first stage of pulping. Although two-stage pulping results in a slightly better yield when depithing is applied after the first stage, the one-stage process is superior in that it results in much better strength properties (Abou-state et al., 1973).

Fahmy et al. (1992) studied cornstalks as a source of fine paper pulp. Pulps of reasonable yield and chemical composition were obtained from cornstalks at relatively reduced alkaline pulping conditions than those required for wood. Reduction of sulfidity from conventional 25%–10%, in order to lessen pollution resulted in almost the same improvement over soda pulps as using 25% sulfidity. Soda pulping using 15% NaOH based on raw material at a maximum temperature of 130°C for 30 minutes gave acceptable results. The strength properties of the obtained pulp compared favorably with pulps from other agricultural residues, namely, rice straw and bagasse. Applying oxygen under pressure from 5 to 15 kg instead of using sulfide in soda pulping enhanced delignification; however, the pulp yield and hemicellulose content decreased to a more remarkable extent and strength properties also deteriorated. This could be attributed to the open structure of cornstalks, which make them more susceptible to degradation by oxygen than wood (Fahmy et al., 1992).

Alkaline pulping and papermaking of cornstalks were studied in soda, soda–AQ, kraft, and kraft–AQ processes by Jahan et al. (2004). Breaking length was better in soda–AQ process than soda process, but tear strength was inferior. Strength properties in kraft processes were better than the soda and soda–AQ processes. Table 5.15 shows the comparison of cornstalk pulps produced in soda–AQ, high-sulfidity kraft and low sulfidity kraft–AQ processes and also compares these pulps with bamboo and mixed hardwood pulps in kraft process.

As shown in Table 5.16, the pulp produced from hydrolyzed cornstalk shows lower kappa number, higher brightness, and higher paper strength properties than the pulp from direct pulping of cornstalk (Ahmed and Zhu, 2006).

The papermaking properties are very good except tear strength. Because of corn pulp's high fine content, its water retention is very high (Jahan et al., 2009).

Latibari et al. (2011) investigated alkaline sulfite pulping of cornstalks by applying different chemical charges and pulping temperatures to generate supplementary pulp for the manufacture of corrugating board. Researchers reported the suitability of cornstalks for alkaline sulfite pulping. Cornstalk pulp was produced using 16% active alkali for 30 minutes cooking time at 145°C temperature using a 50:50 ratio of the sodium sulfite and sodium hydroxide. The pulp has 62.4 N (m/g) of tensile index, 3.80 kPa (m^2/g) of burst index, and 10.53 mN (m^2/g) of tear index at 50.2% total yield. The strength values of the alkaline sulfite pulps generated using cornstalks are shown in Table 5.17. The cornstalk pulp has superior properties than wheat

TABLE 5.15 Comparison of cornstalks pulps with bamboo and mixed hardwood—strength properties measured at about 40 °SR (Jahan et al., 2004).

	Soda–AQ	Kraft-high	L-Kraft–AQ	Bamboo	Mixed hardwood
Kappa number	19.0	22.7	18.8	24.6	45.0
Breaking length (m)	7307	7812	7809	5511	8600
Burst index [kPa (m^2/g)]	5.1	4.4	4.3	4.9	6.3
Tear index [mN (m^2/g)]	4.7	6.6	6.2	18.1	10.5

AQ, Anthraquinone.

TABLE 5.16 Paper properties after pulping of hydrolyzed and nonhydrolyzed cornstalk (Ahmed and Zhu, 2006).

Sample	Nonhydrolyzed	Hydrolyzed		
Prehydrolysis conditions	–	Sulfuric acid 0.5%, liquor to solid ratio 4:1, Temperature 130°C for 15 min and a ramp time of 30 min		
Pulping conditions	12% active alkalinity, 160°C, 0.5% AQ	12% active alkalinity, 150°C, AQ was not added	12% active alkalinity, 150°C and 0.5% AQ	
Pulping time (min)	60	30	15	30
Screened yield, % of initial cornstalk	44.0	38.7	41.6	41.7
Kappa no.	32.8	9.2	6.5	5.9
CSF (mL)	600	600	550	552
ISO brightness (%)	23.5	39.5	39.4	39.1
Scatt. coeff. (m^2/kg)	23.1	21.3	20.5	21.2
Tensile index [N (m/g)]	98	97	106	104
TEA (J/m^2)	97	90	108	103
Burst Index [kPa (m^2/g)]	4.9	5.0	5.7	5.8
Tearing index [mN (m^2/g)]	6.1	6.1	6.4	6.0

AQ, Anthraquinone; *CSF*, Canadian Standard Freeness.

TABLE 5.17 Pulp and paper properties of cornstalk soda and alkaline sulfite pulps (Latibari et al., 2011).

Active alkali (%)	Na_2SO_3/NaOH		Pulp freeness		Tear index [mN (m^2/g)]	Tensile index [N (m/g)]	Breaking length (km)	Burst index [kPa (m^2/g)]
	Na_2SO_3 (%)	NaOH (%)	Unrefined	Refined				
16	30	70	622	436	8.45	62.6	6.39	3.38
	40	60	606	436	9.88	63	6.43	3.81
	50	50	575	422	10.53	62.4	6.07	3.80
	60	40	575	422	9.75	53.2	5.43	2.60
	70	30	575	411	9.83	55.5	5.66	2.84
12	–	100	588	333	9.40	62.17	6.54	4.5
18	–	100	629	333	9.95	72.74	7.41	5.47
6.61[a]	–	100	–	180	3.5	–	5.61	4.4

[a] *Petit-Conil et al. (2001).*

straw CMP pulp (Petit-Conil et al., 2001), poplar wood CMP pulp (Boeva-Spiridonova et al., 2007), canola soda pulp (Enayati et al., 2009), and recycled fiber pulps (Petit-Conil et al., 2001).

Organosolvent pulping of cornstalk was reported to be more selective to react with lignin compared to cellulose and hemicellulose and resulted with higher yield with applying low energy (Jimenez et al., 2004; Barbash et al., 2012). The quality indices of unbleached organosolvent pulps from cornstalks are presented in Table 5.18.

Aremu et al. (2015b) investigated mechanical properties of the soda pulp of corn straw as shown in Table 5.19. The surface morphology of the papers produced was analyzed using scanning electron microscopy, which showed that there is a condensed and packed arrangement of fibers in corn straw. The results obtained from this study showed that corn straw is a suitable nonwood raw material for papermaking (Aremu et al., 2015a).

Ates et al. (2015) studied pulp and papermaking properties of different nonwood materials using soda–AQ pulping process. Researchers reported the pulp and papermaking properties of pulp obtained after refining at fixed Papirindustriens Forskningsinstitutt (PFI) revolution of 3000 revolutions (9000 revolutions for hemp). The pulp and paper properties of unbleached pulps are summarized in Table 5.20.

Daud et al. (2016) studied the chemical pulping and papermaking properties of corn (*Zea mays*) stalk from Malaysia. The soda pulping was carried out following the conditions: NaOH 15%, temperature 170°C, liquor-to-wood ratio 7, and cooking time 90 minutes. The paper handsheets made were tested for different mechanical properties as reported in Table 5.21.

Chesca et al. (2018) investigated the mechanical properties of cornstalks unbleached pulp obtained from different pulping processes. Researchers carried out six pulping experiments using different cooking agents, as shown in Table 5.22 along with pulping and pulp characteristics. The cornstalk pulp was easy to refine because of the presence of the higher amount of hemicellulose and the morphology of the pulp fiber. Authors reported that during papermaking, cornstalk may require very less refining energy compared to hardwood pulps, though most of the nonwood-based raw material pulps are processed without refining in commercial paper production.

Chesca et al. (2018) also investigated the possibility of replacing pulpwood with cornstalks, an abundant agricultural waste in Romania. It was found that cornstalks required lower temperature, less chemicals, and shorter cooking time because of the presence of less lignin than the hardwood or softwood. Natron pulping of cornstalks is reported to produce pulps with reasonable strength properties suitable for making corrugated board. Pulps having different kappa numbers following their paper properties are shown in Table 5.23. Different correlations of cornstalk pulp lignin content (kappa

TABLE 5.18 Quality indices of unbleached organosolvent pulps from cornstalks (Barbash et al., 2012).

Cooking temperature (°C)	Cooking time (min)	Yield of pulp (%)	Residual lignin (%)	Breaking length (m)	Burst index (kN/g)	Tear index [mN (m^2/g)]
110	60	78.5	11.6	3800	3.1	2.6
	90	75.0	8.8	4200	3.2	2.9
	120	74.8	8.2	4900	3.7	3.8
	150	78.5	7.6	5600	4.1	4.0
130	60	67.4	7.1	5000	3.5	3.8
	90	65.3	6.6	5100	3.7	4.0
	120	62.5	5.5	5700	3.9	4.3
	150	59.2	5.3	6500	4.3	4.9
150	60	66.8	4.0	6700	4.5	5.9
	90	64.2	3.5	7400	4.7	6.1
	120	61.8	2.8	7600	4.8	7.0
	150	57.3	2.7	7900	5.1	7.4
170	60	62.0	2.6	7300	4.7	6.4
	90	61.7	2.4	7900	5.1	7.6
	120	56.0	2.1	7800	4.5	5.7
	150	53.9	2.0	6500	4.3	5.2

TABLE 5.19 Result of paper characterization from corn straw (Aremu et al., 2015a).

Agro waste	Grammage	Thickness	Tensile index	Tear index	Modulus of elasticity	Elongation at break
Corn straw	74.01	0.42	5.03	12.29	0.13	1.08

TABLE 5.20 Properties of unbleached cornstalk pulp (Ates et al., 2015).

Particulars	Cornstalk	Rice straw	Cotton stalk	Hemp	Wheat straw
Kappa no.	10.4	11.2	17.6	14.8	12.6
Viscosity (cP)	32.01	26.75	6.07	65.87	24.62
PFI (rev)	3000	3000	3000	9000	3000
°SR	41.5	65.5	45	37	57.5
Grammage (g/m^2)	68.68	69.76	75.4	77.42	67.2
Density (kg/m^3)	735	611	591	645	896
Bulk (cm^3/g)	1.36	1.64	1.69	1.55	1.12
Tear index [mN (m^2/g)]	5.36	4.89	4.75	26.61	3.58
Burst index [kPa (m^2/g)]	3.16	3.42	2.11	2.59	4.84
Breaking length (km)	5.92	6.02	3.79	3.11	7.95
Tensile index [N (m/g)]	58.01	59	37.13	30.51	78

PFI, Papirindustriens Forskningsinstitutt.

TABLE 5.21 Properties of cornstalk chemical pulping fiber (Daud et al., 2016).

Particulars	Cornstalk	Date palm rachis	Palmyra fruit
Tensile index [N (m/g)]	9.1	1.1	13.8
Tear index [mN (m^2/g)]	1.2	4.4	1.1
Burst index [kPa (m^2/g)]	6.6	1.3	n.a.
Folding (Nm)	2.5	n.a.	n.a.

TABLE 5.22 Pulp and beating characteristics of different types of cornstalk pulps (Chesca et al., 2018).

Sample notation	Pulping method	Cooking agent	Total yield (%)	Kappa number	Viscosity (mL/g)	Schopper Riegler (°SR)	Beating duration (min)	Beating rate (°SR/min)
ZM Soda	Soda	NaOH	35.9	19.2	866	19	0–2.5	3.6
							2.5–5	2.8
							5–10	3.4
ZM Soda B	Soda–butanol	NaOH + butanol	41.0	22.4	799	22	0–2.5	2.0
							2.5–5	4.8
							5–10	2.8
ZM (KOH + NH_4OH)	KOH/NH_4OH	KOH + NH_4OH	47.1	99	750	26	0–2.5	3.2
							2.5–5	1.6
							5–10	2.4
PH170(30) ZM Soda	Prehydrolysis (170 °C/30 min) + soda pulping	Hot water/NaOH	33.2	46.9	946	20	0–2.5	3.6
							2.5–5	0.8
							5–10	2.4
PH100(60) ZM Soda	Prehydrolysis (100 °C/60 min) + soda pulping	Hot water/NaOH	34.9	22.5	1022	20	0–2.5	4.0
							2.5–5	3.2
							5–10	2.4
ZM Kraft	Kraft	Kraft NaOH + Na_2S	37	32.4	1100	24	0–2.5	3.6
							2.5–5	4.0
							5–10	2.2

TABLE 5.23 Natron pulping condition of cornstalks and pulp properties (Chesca et al., 2018).

Particulars	Values						
Alkali addition (%)	16	16	14	16	14	16	14
Temperature (°C)	160	140	140	140	160	120	140
Time (min)	30	60	90	30	60	90	60
Yield (%)	29.7	30.8	31.7	32.3	32.9	33.1	35
Kappa number	21.4	23.6	26.8	27.5	29.3	30.1	35.7
Tensile index [N (m/g)]	54.0	60.5	65.0	60.2	64.0	64.0	65.0
Burst index [kPa (m^2/g)]	2.3	2.6	3.0	2.8	2.8	2.8	2.9
Corrugating medium test index [N (m^2/g)]	1.7	1.8	1.8	1.9	2.0	2.0	2.1
Short-span compressive test index [N (m/g)]	25	27	32	29	33	30	33

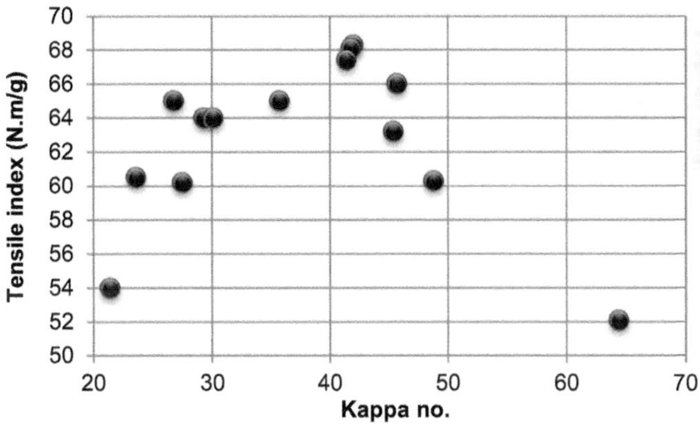

FIGURE 5.6 Influence of cornstalk pulp kappa number on tensile index (Chesca et al., 2018).

number) on strength properties are shown in Figs. 5.6–5.8. It can be concluded that cornstalk pulp can be utilized for making corrugated boards due to its relatively high-strength properties.

Mishra et al. (2020) studied the soda–AQ pulping of cornstalk in autoclave digester. Initial temperature of glycol bath was kept at 150°C. The temperature was raised to 168°C from 150°C in 20 minutes. Cooking at maximum temperature was done for 20 minutes using a fixed dose of 14.5% NaOH and 0.05% AQ on cornstalk to get the unbleached pulp of kappa number 16.1.

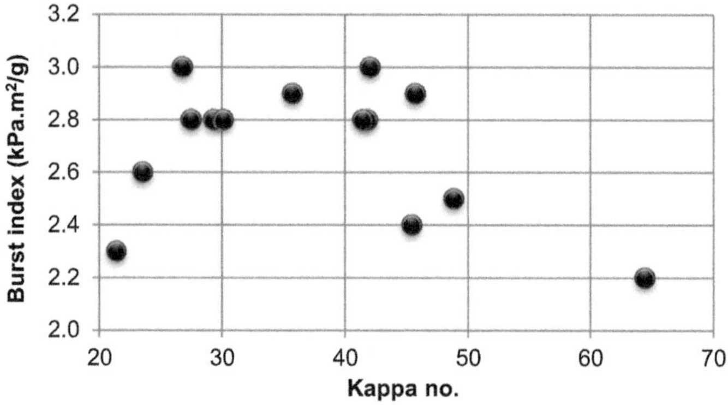

FIGURE 5.7 Influence of cornstalk pulp kappa number on burst index (Chesca et al., 2018).

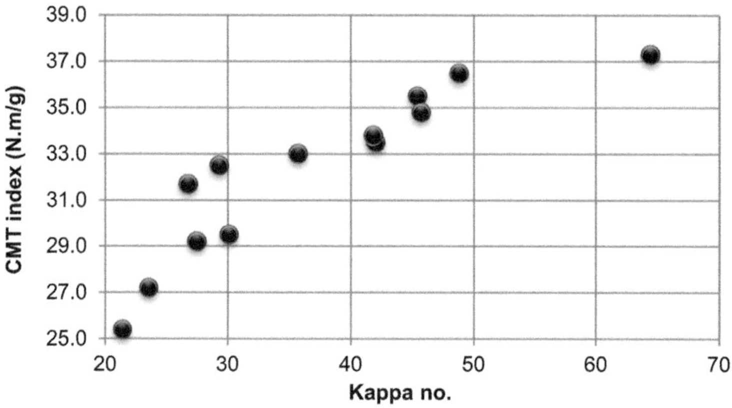

FIGURE 5.8 Influence of cornstalk pulp kappa number on Concora medium test (CMT) index (Chesca et al., 2018).

Fairly good pulp yield of 53.7% and pulp viscosity of 26.0 cP were obtained. The effect of beating on different pulp and paper characteristics are summarized in Table 5.24. Physical strength properties of unbleached cornstalk pulp were improved with the refining of pulp to 405 mL Canadian Standard Freeness (CSF). Further increase in the extent of refining improved physical strength properties except for tear index.

5.5.2 Papermaking using cornstalk bleached pulps and physical strength properties

Papers made from the bleached soda pulps of cornstalks showed fair strength and satisfactory color, formation, and cleanliness producing the satisfactory quality of medium grades of writing papers and bleached wrapping

TABLE 5.24 Physical strength properties of cornstalk unbleached pulps (Mishra et al., 2020).

Parameter	Unbleached pulp		
No. of revolutions	0	150	550
CSF (mL)	490	405	345
Bulk (cc/g)	1.32	1.30	1.32
Tensile index [N (m/g)]	49.1	58.7	60.8
Burst index (kN/g)	3.80	4.82	5.09
Tear index [mN (m^2/g)]	7.53	7.94	7.69
Double fold (no.)	86	319	349

CSF, Canadian Standard Freeness.

TABLE 5.25 Effects of water pretreatments—physical strength characteristics at 50 °SR (Abou-state et al., 1973).

Water pretreatment	–	Tap water	Boiling water for 0.5 h	Boiling water for 1 h
Ash (%)	1.00	0.73	0.64	0.43
Tensile strength (kg)	7.0	7.5	6.3	5.8
Bursting strength (lb/in^2)	32.2	37.8	29.6	22.45
No. of double folds	1000	1200	730	250
Tearing strength (g)	23	24	23	22

specialties such as butter papers. However, the higher costs due to low pulp yields limited the usage of cornstalks in comparison to wood as a raw material (Weber et al., 1935).

Abou-state et al. (1973) utilized the whole corn plant effectively for producing pulp and paper by employing comparatively simple methods of multistage bleaching after atmospheric pressure digestion. Pretreatment with tap water before pulping was found essential to remove the dust particles. However, the use of boiling water was not suggested. Table 5.25 shows that washing the raw material thoroughly with tap water before pulping lowered the ash content of bleached pulp to 0.73% and also improved the physical strength characteristics. Pretreatment with boiling water for 0.5 hour lowered the ash content of bleached pulp to 0.64% and also lowered tensile strength, bursting strength, tear strength, and folding resistance. When the treatment with boiling water was for 1 hour, the ash content of bleached pulp reduced to 0.43%, and further deterioration in strength properties took place.

After obtaining the pulp utilizing annual plants and agricultural waste following soda–AQ pulping method, it was subsequently pretreated with xylanase enzyme and then bleached using the elementary chlorine-free (ECF) process utilizing the DE_PD bleaching sequence (Ates et al., 2015). Each bleached pulp sample was beaten at 3000 revolutions PFI (9000 rev. for hemp), and changes in pulp freeness were determined. Paper sheets from bleached pulps are tested for the effects of bleaching and beating of on several paper properties. The pulp and paper properties of bleached pulps are summarized in Table 5.26.

Mishra et al. (2020) studied the bleaching behavior of soda–AQ cornstalk pulp. The unbleached pulp of kappa number 16.1 was easily bleached to the brightness of 88.4% using elemental chlorine-free bleaching sequence $D_0E_{OP}D$. The bleached pulp has an average fiber length of 0.93 mm and width of 19.0 μm and a viscosity of 21.1 cP. Physical strength properties of bleached cornstalk pulp were superior specifically tensile index compared to unbleached pulp perhaps due to the interference of residual lignin during fibrillation as well

TABLE 5.26 Properties of bleached pulps from different nonwood materials in constant beating stages (Ates et al., 2015).

Particulars	Cornstalk	Rice straw	Cottonstalk	Hemp	Wheat straw
Brightness (% ISO)	87.6	87.5	61.0	80.1	92.0
PFI (rev)	3000	3000	3000	9000	3000
°SR	30	54	27	11	45
Viscosity (cP)	29.7	29.9	5.9	32.5	25.0
Grammage (g/m^2)	71.1	62.9	92.3	58.1	69.4
Density (kg/m^3)	762	551	724	484	925
Bulk (g/cm^3)	1.31	1.81	1.38	2.07	1.08
Tear index [mN (m^2/g)]	5.42	5.77	4.04	22.81	4.28
Burst index [kPa (m^2/g)]	3.5	4.6	1.69	1.6	4.93
Breaking length (km)	6.19	6.93	3.22	2.13	8.16
Tensile index [N (m/g)]	60.7	68.0	31.6	20.8	80.0

PFI, Papirindustriens Forskningsinstitutt.

TABLE 5.27 Physical strength properties of cornstalk bleached pulps (Mishra et al., 2020).

Parameter	Bleached pulp		
No. of revolutions	0	150	500
CSF (mL)	480	405	345
Bulk (cc/g)	1.30	1.29	1.30
Tensile index [N (m/g)]	54.2	61.5	63.9
Burst index [kN/g]	4.16	4.85	5.37
Tear index [mN (m^2/g)]	8.73	8.28	7.29
Double fold (no.)	89	303	356

as poor fiber bonding in unbleached pulp refining and sheet making. The effects of the beating of bleached pulp on different pulp and paper characteristics are summarized in Table 5.27. The bleached cornstalk pulp had a tensile index of 54.2 N (m/g), burst index of 4.16, tear index of 8.73, and double fold number of 89. Upon refining of the pulp, the tear index of the bleached pulp was reduced even when refining the pulp to 405 mL CSF. However, other properties, such as tensile and burst indices, as well as double fold numbers, were improved.

5.5.3 Papermaking using cornstalk pulps blended with other pulps and physical strength properties

Several studies have been carried out with blending cornstalk pulp and other nonwood/wood pulps. Jahan et al. (2009) reported increase in tear strength when cornstalk pulp was blended with jute pulp. Similarly, Fagbemi et al. (2014) considered the effect of blending kenaf bark and corn husk pulps. Researchers reported with the increase in percentage of kenaf pulp, the physical strength properties were improved, and the highest values were obtained with a blend of 75% kenaf bark and 25% corn husk. The strength properties of handmade papers obtained from the pulp blending of long-fiber kenaf bark and short-fiber corn husk in different ratios are shown in Table 5.28. Blending of pulps from agricultural wastes such as corn husk and pulps from other nonwood materials such as kenaf contributes to the improvement of the paper strength properties and has the potential of producing paper of differing qualities and can find use in wide variety of applications (Fagbemi et al., 2014).

Balea et al. (2016) attempted to valorize cornstalk for the production of cellulose nanofibers (CNF) and used the same for improving the properties of recycled paper. To improve the tensile index of recycled paper by 20%

TABLE 5.28 Properties of paper samples from pulp mixture of kenaf bark and corn husk (Fagbemi et al., 2014).

Particular	Kenaf bark:corn husk				
	100:0	75:25	50:50	25:75	0:100
Grammage (g/m^2)	82.06	127.14	121.92	130	93.21
MOE (N/mm^2)	41.76	42.05	22.62	34.33	12.41
Elongation at break (mm)	2.46	3.35	5.38	2.75	3.49
Tensile index (Nm/g)	7.79	4.53	5.49	3.23	2.77
Tear index [mN (m^2/g)]	20.72	7.87	8.2	10	10.73

TABLE 5.29 Physical strength properties of different blends of cornstalk and wheat straw pulps (Mishra et al., 2020).

Particulars	Wheat straw:corn stalk			
	100:0	**70:30**	**65:35**	**60:40**
No. of revolutions	200	200	250	300
°SR	420	430	430	425
Bulk (cc/g)	1.31	1.29	1.30	1.28
Tensile index [N (m/g)]	53.2	56.5	57.5	60.6
Burst index (kN/g)	3.89	4.54	4.64	4.75
Tear index [mN (m^2/g)]	7.32	7.43	7.52	7.72
Double fold (no.)	85	101	126	134

requirement of corn CNF was only 0.5%, whereas 1.5% eucalyptus-CNF was required to get similar improvement in tensile index. Mishra et al. (2020) studied the effect of blending cornstalk pulp with wheat straw mill pulp on physical strength properties for evaluating potential of using cornstalk by agro-based paper industries. Bleached unrefined wheat straw mill pulp was blended with bleached unrefined cornstalk pulp in the ratio of 70:30, 65:35, and 60:40. The mill bleached wheat straw pulp and different blends were refined to obtain pulp having CSF of about 425 mL, and the effect of blending on physical strength properties of handsheets was analyzed. Detailed results of physical strength properties of different blends of cornstalk and wheat straw pulps are given in Table 5.29. The wheat straw mill bleached pulp has a tensile index of 52.3 N (m/g), burst index of 3.89, tear index of 7.32, and double fold of 85 number. Physical strength properties like tear, tensile, burst indices and double fold of wheat straw mill pulp were improved by 5.5%, 13.9%, 22.1%, and 57.6%, respectively, with the blending of 35% cornstalk pulp. The reason behind improvement in physical strength properties may be the higher average fiber length and coarseness of cornstalk pulp as compared to that of wheat straw pulp. The blending of cornstalk pulp with wheat straw pulp improved the physical strength properties of the resulting blend, which may help agro residue–based paper mills to improve their product quality.

The effective use of nonwood fiber resources, especially cornstalks, would play a major role in optimizing papermaking raw materials (Liu et al., 2018).

5.6 Conclusions

Cornstalk has potential to be used as raw material for pulp and papermaking. The morphological properties of cornstalk are comparable with hardwood pulp and bagasse pulp. Pulping behavior of cornstalk is similar to the agro residue raw materials like sugarcane bagasse and wheat straw. It can be cooked using Pandia-type continuous digester along with wheat straw or sugarcane bagasse to improve physical strength properties. Properties of cornstalk are varied to great extent from region to region therefore; careful evaluation is required prior to adopting it in the commercial scale production.

Acknowledgments

Authors thankfully acknowledge the support of scientific and nonscientific colleagues of ACIRD. Authors are highly thankful to family members as major portion of this chapter is written during lockdown period of pandemic Covid-19.

References

Abou-State, M.A., Zimaity, T., El-Masry, A.M., 1973. Pulp and paper from corn stalks. J. Appl. Chem. Biotechnol. 23, 511–516.

Ahmed, A., Zhu, J.Y., 2006. Cornstalk as a source of fibre and energy. In: Proc. 3rd Int. Symposium Emerg. Technol. Pulping Papermaking, Guangzhou, China, pp. 1–4.

Alagbe, E.E., Bassey, E.S., Daniel, O.E., Shongwe, M.B., Ojewumi, M.E., Igwe, C.C., 2019. Physical, chemical and mechanical properties of corn sheath as pulp and paper raw material. J. Phys. 1378, 032083.

Aremu, M.O., Aperolola, S.O., Dabonyan, O.O., 2015b. Suitability of Nigerian corn husk and plantain stalk for pulp and paper production. Eur. Sci. J. 11, 146–152.

Aremu, M.O., Rafiu, M.A., Adedeji, K.K., 2015a. Pulp and paper production from Nigerian pineapple leaves and corn straw as substitute to wood source. Int. Res. J. Eng. Technol. 2 (4), 1180–1188.

Arnold, L.K., 1930. Cornstalks as a raw material for paper production (Ph.D. thesis). Department of Chemical Engineering, Iowa State University, Ames, Iowa, US. Available from: https://lib.dr.iastate.edu/rtd/14205.

Assumpcao, R.M.V., 1992. Non-wood fibre utilization in pulping and papermaking – UNIDO's activities. In: TAPPI Non-wood Plant Fibre Pulping Progress Report No. 20:191–201.

Ates, S., Deniz, I., Kirci, H., Atik, C., Okan, O.T., 2015. Comparison of pulping and bleaching behaviors of some agricultural residues. Turkish J. Agric. For. 39, 144–153.

Bajpai, P., Tripathi, S.K., Sharma, N., Mishra, O.P., Bajpai, P.K., 2008. Production of rayon grade pulp from bamboo by modification in the bleaching sequence. Int. Pulp Bleaching Conf. 279–283. Quebec, Canada, 2–5 June 2008.

Balea, A., Merayo, N., Seara, M., Fuente, E., Blanco, A., Negro, C., 2016. "Effect of NFC from organosolv corn stalk pulp on retention and drainage during papermaking. Cellulose Chem. Technol. 50 (3–4), 377–383.

Barbash, V., Trembus, I., Nagorna, J., 2012. "Pulp obtaining from corn stalks. Chem. Chem. Technol. 6 (1), 83–87.

Behin, J., Mikaniki, F., Fadaei, Z., 2008. Dissolving pulp (alpha-cellulose) from corn stalk by kraft process. Iran. J. Chem. Eng. 5 (3), 14–28.

Behin, J., Zeyghami, M., 2009. Dissolving pulp from corn stalk residue and waste water of Merox unit. Chem. Eng. J. 152, 26–35.

Boeva-Spiridonova, R., Petkova, E., Georgieva, N., Yotova, L., Spiridonov, I., 2007. Utilization of a chemical-mechanical pulp with improved properties from poplar wood in the composition of packaging papers. BioResources 2 (1), 34–40.

Byrd, M., Jameel, H., Warby, S., Johnson, W., 2006. Chemical and pulping characteristics of corn stalk fractions. In: Zhan, H.Y., Chen, F.G., Fu, S.Y. (Eds.), New Technologies in Non-wood Fibre Pulping and Papermaking. South China University of Technology Press, Guangzhou, pp. 5–8.

Chesca, A.M., Nicu, R., Tofanica, B.M., Puitel, A.C., Vlase, R., Gavrilescu, D., 2018. Pulping of corn stalks – assessment for bio-based packaging materials. Cellulose Chem. Technol. 52 (7–8), 645–653.

Chesca, A.M., Tofanica, B.M., Puitel, A.C., Nicu, R., Gavrilescu, D., 2018b. Environmentally friendly cellulosic fibres from corn stalks. Environ. Eng. Manage. J. 17 (7), 1765–1771.

Daud, Z., Hatta, M.Z.M., Kassim, A.S.M., Awang, H., Aripin, A.M., 2013. Analysis the chemical composition and fibre morphology structure of corn stalk. Austr. J. Basic. Appl. Sci. 7 (9), 401–405.

Daud, Z., Hatta, M.Z.M., Kassim, A.S.M., Awang, H., Aripin, A.M., 2014. Exploring of agro waste (pineapple leaf, corn stalk, and napier grass) by chemical composition and morphological study. Bioresources 9 (1), 872–880.

Daud, Z., Hatta, M.Z.M., Latiff, A.A.A., Awang, H., 2016. Corn stalk fibre material by chemical pulping process for pulp and paper industry. Adv. Mater. Res. 1133, 608–611.

El-Sakhawy, M., Lonnberg, B., Fahmy, Y., Ibrahim, A.A., 1996. Organosolv pulping. 3-Ethanolo pulping of wheat straw. Cellulose Chem. Technol. 30, 161–174.

El-Sayed, E.S.A., El-Sakhawy, M., El-Sakhawy, M.A., 2020. Non-wood fibres as raw material for pulp and paper industry. Nordic Pulp Pap. Res. J. Available from: https://doi.org/10.1515/npprj-2019-0064.

Fagbemigun, T.K., Fagbemi, O.D., Otitoju, O., Mgbachiuzor, E., Igwe, C.C., 2014. Pulp and paper-making potential of corn husk. Int. J. AgriSci. 4 (4), 209–213.

Enayati, A.A., Hamzeh, Y., Mirshokraei, S.A., Molaei, M., 2009. Papermaking potential of canola stalks. BioResources 4 (1), 245–265.

Fagbemi, O.D., Fagbemigun, T.K., Otitoju, O., Mgbachiuzor, E., Igwe, C.C., 2014. Strength properties of paper from pulp blend of kenaf bark and corn husk: a preliminary study. Br. J. Appl. Sci. Technol. 4 (28), 4124–4129.

Fahmy, Y., Haikal, S., El-Samahy, M., 1992. Corn stalks as a source for fine paper pulp part I: Influence of sulfidity and oxygen in alkali pulping on pulp properties. Res. Ind. 37 (12), 218–222.

FAO, 2020. World Food and Agriculture - Statistical Yearbook 2020. Rome.

Ferdous, T., Quaiyyum, M.A., Salam, A., Jahan, M.S., 2020. Pulping of bagasse (*Saccharum officinarum*), kash (*Saccharum spontaneum*) and corn stalks (*Zea mays*. Curr. Res. Green. Sustain. Chem. 3, 100017. Available from: https://doi.org/10.1016/j.crgsc.2020.100017.

Flandez, J., Pelach, M.A., Vilaseca, F., Tijero, J., Monte, C., Pérez, I., et al., 2010. Lignocellulosic fibre from corn stalks as alternative for the production of brown grade papers. Proc. 21st TECNICELPA Conf. Exhibition/VI CIADICYP. Lisbon, Portugal.

Gould, K., 2007. Corn stover harvesting. Cattle Call. 12 (2), 1–2.

Guler, C., Sahin, H.I., Yeniay, S., 2016. The potential for using corn stalks as a raw material for production particleboard with industrial wood chips. Wood Res. 61 (2), 299–306.

Hart, Rudie, 2014. Anthraquinone a review of the rise and fall of a pulping catalyst. TAPPI J. 13 (10), 23–31.

Hunsigi, G., 1989. Agricultural fibres for paper pulp. Oulook Agric. 18 (3), 96–103.

Hurter, A.M. 1988. Utilization of annual plants and agricultural residues for the production of pulp and paper. In: Procedings TAPPI Pulping Conf. 1988. New Orleans, LA. Book 1, pp. 139–160.

Igathinathane, C., Womac,A.R., Sokhansanj, S., Pordesimo, L.O., 2004. Vertical mass and moisture distribution in standing corn stalks. In: ASAE/CSAE Annual International Meeting Sponsored by ASAE Fairmont Chateau Laurier, The Westin, Government Centre Ottawa, Ontario, Canada, pp. 1–18.

Jahan, M.S., Chowdhury, D., Russel, M.A.N., Mun, S.P., Quaiyyum, M.A., 2006. Alkaline sulfite - anthraquinone - methanol (ASAM) pulping of corn stalks. Cellulose Chem. Technol. 40 (7), 531–536.

Jahan, M.S., Gunter, B.G., Ataurrahman, A.F.M., 2009. Substituting wood with nonwood fibres in papermaking: a win-win solution for Bangladesh. In: Bangladesh Development Research Working Paper Series, BDRWPS No. 4.

Jahan, M.S., Rahman, M.M., 2012. Effect of pre-hydrolysis on the soda-anthraquinone pulping of corn stalks and *Saccharum spontaneum* (kash). Carbohydr. Polym. 88 (22), 583–588.

Jahan, M.S., Russel, M.A.N., Shamim, S.A.N., Mostafa, A.I., Abdul-Quaiyyum, M., 2004. Alkaline pulp of corn stalks. IPPTA J. 16 (1), 29–34.

Jimenez, L., Rodrignez, A., Resez, I., et al., 2004. Wood Fibre Sci. 36, 423.

Kadam, K.L., McMillan, J.D., 2003. Availability of corn stover as a sustainable feedstock for bioethanol production. Bioresour. Technol. 88, 17–25. Available from: https://doi.org/10.1016/S0960-8524(02)00269-9.

Kaur, D., Bhardwaj, N.K., Lohchab, R.K., 2018. A study on pulping of rice straw and impact of incorporation of chlorine dioxide during bleaching on pulp properties and effluents characteristics. J. Clean. Prod. 170, 174–182. Available from: https://doi.org/10.1016/j.jclepro.2017.09.111.

Lal, P.S., Tripathi, S.K., Sharma, A., Subrahmanyam, S.V., Kulkarni, A.G., 2004. ECF bleaching of Assam bamboo pulp. In: IPPTA Convention Issue, pp 91–95.

Lal, P.S., Tripathi, S.K., Sharma, A., Subrahmanyam, S.V., Kulkarni, A.G., 2005. Bleaching of bamboo pulp to high brightness. Int. Pulp Bleaching Conf. 281–283. Stockholm, Sweden, 14–16 June 2005.

Latibari, A.J., Hossein, M.A., Hosseinpour, R., 2011. Application of alkaline sulfite pulping on corn stalks. BioResources 6 (1), 48–58.

Leponiemi, A., 2008. Non-wood pulping possibilities – a challenge for the chemical pulping. Appita J. 61 (3), 234–243.

Liu, Z., Wang, H., Hui, L., 2018. Pulping and Papermaking of Non-Wood Fibres, Pulp and Paper Processing. Available from: https://doi.org/10.5772/intechopen.79017. Salim Newaz Kazi, IntechOpen.

Li, Z., Zhai, H., Zhang, Y., Li, Y., 2012. Cell morphology and chemical characteristics of corn stover fractions. Ind. Crop. Products 37 (1), 130–136.

Mayowa, A.A., 2018. Pulping of Non-Woody Biomass, Pulp and Paper Processing. Available from: https://doi.org/10.5772/intechopen.79749. Salim Newaz Kazi, IntechOpen.

McDougall, G.J., Morrison, I.M., Stewart, D., Weyers, J.D.B., Hilman, J.R., 1993. Plant fibres: botany, chemistry and processing. J. Sci. Food Agric. 62, 1–20.

Mesa, L., Albernas, Y., Morales, M., Corsano, G., González, E., 2016. Integration of Organosolv process for biomass pretreatment in a biorefinery. Chapter 11. In: Mussatto, S.I. (Ed.), Biomass Fractionation Technologies for a Lignocellulosic Feedstock Based Biorefinery. Elsevier, Amsterdam, pp. 229–254.

Mishra, O.P., Tripathi, S.K., Bhardwaj, N.K., 2020. Suitability of corn stalk pulp for improving physical strength properties of agro residue pulp. Cellulose Chem. Technol. 54 (1–2), 65–71.

Park, Y.C., Kim, J.S., Kim, T.H., 2018. Pretreatment of corn stover using organosolv with hydrogen peroxide for effective enzymatic saccharification. Energies 11 (5), 1301. Available from: https://doi.org/10.3390/en11051301.

Petit-Conil, M., Brochier, B., Labalette, F., Combette, P., 2001. Potential of wheat straw to produce chemimechanical pulps suited to corrugating papers manufacture, TAPPI 2001 Pulping Conference, 11 pp.

Pfaffli, I., Sisko, M., 1995. Fibre Atlas: Identification of Papermaking Fibres. Springer-Verlag, New York.

Reddy, N., Yang, Y., 2005. Structure and properties of high quality natural cellulose fibres from cornstalks. Polymer 46 (15), 5494–5500.

Sadawarte, N.S., 1995. Better technology needed to clean up nonwood fibre. Pulp Pap. Int. 37 (6), 84–95.

Saijonkari-Pahkala, K., 2001. Non-wood plants as raw material for pulp and paper. Academic Disertation, MIT Agrifood Research Finland.

Shakhes, J., Marandi, M.A.B., Zeinaly, F., Saraian, A., Saghafi, T., 2011. Tobacco residuals as promising lignocellulosic materials for pulp and paper industry. BioResources 6, 4481–4493.

Sharma, N., Tripathi, S.K., Bhardwaj, N.K., 2020. Utilization of sarkanda for making pulp and paper using elemental chlorine free and total chlorine free bleaching processes. Ind. Crop. Products 149, 112316. Available from: https://doi.org/10.1016/j.indcrop.2020.112316.

Singh, M., Thakur, S.S., Chandel, R., Sharma, A., 2015. Development and field evaluation of corn-stover collector attachment for maize combine harvester. Agric. Eng. Today 39 (3), 3–9.

Snelders, J., Dornez, E., Benjelloun-Mlayah, B., Huijgen, W.J.J., Courtin, C.M., 2014. Biorefining of wheat straw using an acetic and formic acid based organosolv fractionation process. Bioresour. Technol. 156, 275–282.

Sridach, W., 2010. The environmentally benign pulping process of non-wood fibres. Suranaree J. Sci. Technol. 17 (2), 105–123.

Sundquist, J., 2000. Organosolv pulping. In: Gullichsen, J., Fogellholm, C.J. (Eds.), Chemical Pulping. Papermaking Science and Technology Book 6B. Fapet Oy, Finland, pp. 411–427.

Theng, D., El Mansouri, N.E., Arbat, G., Ngo, B., Delgado-Aguilar, M., Pèlach, M., et al., 2017. Fibreboards made from corn stalk thermomechanical pulp and kraft lignin as a green adhesive. BioResources 12 (2), 2379–2393.

Tripathi, S.K., Alam, I., Bhardwaj, N.K., 2017. Effect of blending banana stem and hardwood pulps on sizing, ash retention, physical strength and optical properties of paper. Appita J. 70 (4), 378–384.

Tripathi, S.K., Bhardwaj, N.K., Chechi, S.K., Varadhan, R., 2019a. Suitability of banana stem pulp as replacement of softwood pulp for making superior grade unbleached paper from agro residue – lab and plant scale trial results. Appita J. 72 (3), 163–178.

Tripathi, S.K., Bhardwaj, N.K., Ghatak, H.R., 2019b. Effect of introducing ozone prior to elemental chlorine free bleaching of wheat straw pulp on pulp, paper and effluent properties. Cellulose Chem. Technol. 53 (1–2), 105–112.

Tripathi, S.K., Mishra, O.P., Bhardwaj, N.K., Varadhan, R., 2018. Pulp and papermaking properties of bamboo species *Melocanna baccifera*. Cellulose Chem. Technol. 52 (1–2), 81–88.

Tripathi, S.K., Mishra, O.P., Sharma, N., Bhardwaj, N.K., Varadhan, R., 2021. Impact of storage time on chemical constituents, pulp and papermaking properties of bamboo. IPPTA Pap. 360, 1–11.

Tripathi, S.K., Mishra, O.P., Sharma, N., Chakrabarti, S.K., Varadhan, R., 2010. Fractionation of bamboo pulp – improving paper strength. TAPPSA J. 6, 20–25.

Tripathi, S.K., Singh, S., Gangwar, A., Mishra, O.P., Chakrabarti, S.K., Bhardwaj, N.K., et al., 2013. Blending of banana stem with wheat straw and bagasse to enhance physical strength properties of paper. IPPTA J. 25 (2), 121–125.

Weber, C.G., Shaw, B.M., Martin, J.O., 1935. Papermaking Quality of Cornstalks. National Bureau of Standards Miscellaneous Publication M147. U.S. Department of Commerce, Washington.

Weber, H.A., 1929. Cellulose from corn stalks. Ind. Eng. Chem. 21 (3), 270–275. Available from: https://doi.org/10.1021/ie50231a024.

Weiss, N.D., Farmer, J.D., Schell, D.J., 2010. Impact of corn stover composition on hemicellulose conversion during dilute acid pretreatment and enzymatic cellulose digestibility of the pretreated solids. Bioresour. Technol. 101 (2), 674–678.

Won, J.M., Ahmed, A., 2004. Cornstalk as a raw material for papermaking. In: Proceedings of the 58th Appita Annual Conferences and Exhibition, Melbourne, Australia. Appita. vol. 1, pp. 5–11.

Wu, J., Zhang, X., Wan, J., Ma, F., Tang, Y., Zhang, X., 2011. Production of fibreboard using corn stalk pretreated with white-rot fungus trametes hirsute by hot pressing without adhesive. Bioresour. Technol. 102, 11258–11261.

Chapter 6

Pulping and papermaking of esparto grass

Abdelhak Maghchiche
Department of Pharmacy, Faculty of Medicine, University of Batna 2, Batna, Algeria

Research objectives

With the passage of time, papermaking has grown significantly, which now is considered one of our most vital industries. Cellulosic fiber, or pulp, is the primary ingredient in the production of paper, and large quantities are required each year. This fibrous pulp can be made from a wide range of raw materials, including specific wood species. Due to climate change drought and fires occurred in large areas of the world. Massive fires are also fueled by droughts that caused a continuous disappearance of forests in many regions, and their shrinkage in other parts of the world. Efforts are now focused on valuing herbs, lignocellulosic materials, and biomass to replace wood for the manufacture of various types of paper for various uses. The use of these nonwood materials may balance the ecosystem and maintain the exploitation of these materials to achieve sustainable development in the areas where these nonwood materials are found where they grow in abundance.

6.1 Introduction

Rags of cotton and linen were the first fibrous raw resources for the production of paper. As a result of the widespread consumption of paper, the pulp and paper industry worldwide witnessed rapid growth. With global warming, drought, fires, and environmental pollution, paper manufacture faced a shortage of raw material resources, mainly wood; the research was turned to other nonwood materials to compensate for the shortage of raw resources. As a result, the paper industry has witnessed an increasing trend of using nonwood plant fibers such as esparto grass and other plant raw materials to manufacture paper and, for this purpose, has become more competitive than wood as

Pulping and Papermaking of Nonwood Plant Fibers. DOI: https://doi.org/10.1016/B978-0-323-91625-7.00006-0

a raw resource. The pulp and paper industry's main raw materials can be classified into three types: wood, nonwood, and nonplant (mainly wastepaper), in which nonwood fiber content is a significant fiber source in locations with a scarcity of forest resources, actually, in the world's overall pulp intake, the proportions of wood pulp are 63%, waste paper pulp is 34%, and nonwood pulp is 3% (Liu et al., 2018). Local sources of cellulosic fibers include agricultural wastes and perennial plants, in some countries like North Africa countries, where there is limited forestry. The use of these nonwoody plant's fibers may preserve the forest resources, reducing the dependency of those countries on the imported pulp woods (Marrakchi et al., 2011). In recent years, annual grass has received greater attention for the production of pulp, paper, paperboard, and cellulose derivatives. Nonwood raw materials, such as bagasse, cereal straw, bamboo, reeds, esparto grass, jute, flax, and sisal, from emerging countries now provide more cellulose fibers (Bouiri and Amrani, 2010a,b).

Nonwoody plants provide many benefits, including short growth cycles, removal of irrigation needs and chemical needs, with low lignin content resulting in lower energy and chemical use during pulping (Hurter and Riccio, 1998). Agricultural waste has emerged as one of the most important new sources for the pulp and paper sector. Napier grass (*Pennisetum purpureum*) can be used, which is considered agricultural waste and has a high potential to become an alternative fiber in the papermaking industry (Zawawi, 2018). Esparto grass is a perennial Mediterranean herb that grows in tufts of 1 m to 1m20 high, making a large coverage area. It grows on its own, especially in arid and semiarid climates; it stops in the limits of the desert. Esparto grass is considered one of the most important annual grass plants for the production of paper fibers. In the past four decades, it has been used in such applications (Nadji et al., 2006). Esparto grass is an herbaceous plant with a high fibrosity. Moreover, the fiber properties (cellulose content, suppleness, and thinness) make this plant a useful raw material for papermaking, in terms of both quality and quantity. In Algeria and Tunisia, the esparto grass is the only source of paper-made raw material (Belkhir et al., 2013).

6.2 Nonwood plants

For more than 1900 years, the paper was made from grass and other nonwood materials. Wood is a relatively young papermaking fiber, dating back only 100 years. Commercial nonwood pulp production now accounts for 6% of total pulp production worldwide. The principal source for the raw material is the Gramineae family plants (Pahkala et al., 1997). Plant fibers are processed for use in papermaking from the pulp that consists mainly of cellulose; plant fiber pulp made from nonwood can be divided into two categories (Hurter, 2002; Mohieldin, 2014), according to the form of pulp that can be produced:

1. Popular nonwood fibers like grain straws, herbs, bagasse, reeds, and bamboo.
2. Cotton stables and linters, flax, hemp, and kenaf, bast fibers, sisal, bamboo, and abaca are examples of softwood fibers.

Nonwood pulp is made from the phloem fibers found in the stem of many kinds of grass, which are environment-friendly material available in huge amounts in nature (Liu et al., 2018).

Nonwood fibers are a potential source of pulping materials. However, these fibers, which are generated from chosen tissues of monocotyledonous and dicotyledonous plants, exhibit a wide range of chemical and physical properties. The chemical and physical properties of nonwood fibers are enormously variable, for pulping, lignin content, and content in cellulose, which are important factors.

6.2.1 Classification of nonwood raw materials

Nonwood fiber resources used in paper manufacture can be divided into four groups: grass fibers, bast fibers, seed husk fibers, and leaf fibers (Liu et al., 2018; Azeez, 2018), as described in the following subsections.

6.2.1.1 Grass fibers

The fiber of the grass plant is divided into various sections, each of which is made up of nodes and internodes. Such raw materials include esparto grass, bamboo, bagasse, rice straw, wheat straw, sorghum residue, corn straw, reed, and so on, where the cellulose content is typically about 50%, closing to wood. The average length of grass fibers is of 1–3 mm.

6.2.1.2 Bast fibers

Bast fibers that typically refer to plant materials contain a large distribution phloem. This material comes in two forms: One is bark, which is known as cotton stalk bark. Due to its huge fibers in the cortex, it has shown to be extremely useful in the pulp and paper industry. The other is hemp, which includes kenaf, jute, cotton, flax, and other varieties.

6.2.1.3 Seed hull fibers

Seed hull fibers are extracted from single-cell seeds or from fruit hairs, often known as the seed hair fibers. Most notably, cotton fiber and palmyra palm are suitable as alternative raw materials for papermaking cellulosic pulp.

6.2.1.4 *Leaf fiber materials*

Leaf fibers have ample cellulose, such as sisal, abaca, and sugarcane leaves; the leaves of certain plants are valuable for pulp and papermaking. Fiber has become increasingly significant in the pulp and paper sector in recent years.

6.2.2 Physical characteristics of nonwood plants

Nonwood plants have a wide range of physical and chemical qualities, and they all contain a variety of fibers and cell types in varying degrees.

Cereal straw monocots, sugarcane, bagasse, and maize stalks are similar to hardwoods, but they are significantly more heterogeneous, with a high proportion of very thin-walled cells, barrel-shaped parenchymatic cells, fine epidermal cells, and capillaries of various sizes. Hemp, flax, kenaf, and straw are dicots that contain two types of fibers: Longer layers of bast fibers surround short internal structure fibers. The pulp is more challenging because core fibers contain more lignin. The fiber dimensions of *Stipa tenacissima* and *Lygeum spartum* compared at different nonwood fiber lengths are important for comparing different kinds of nonwood fibers. The length and width of some common nonwood fibers (Han, 1998; Rowell et al., 2000) are given in Table 6.1.

6.2.3 Chemical characteristics of nonwood plants

The chemical properties of nonwood plants vary greatly depending on the source of fiber, fiber age, climate, soil conditions, and composition. Plant cell walls are mostly made up of lignin-mixed sugar-based polymers, with extractives, protein, starch, and inorganic compounds making up minor proportions. The chemical characteristics of certain nonwood fibers are depicted in Table 6.2.

6.3 Esparto grass

Esparto grass is a hardy, perennial grass that is endemic to the western Mediterranean, which grows in North Africa's arid regions. It is a member of the gramineous family and grows to a height of about 1 m; it consists of cylindrical shaped stems with a maximum height of around 1 m. Esparto grass is rhizomes (rhizomatous) that form large bouquets, grow formally circular, and slow as it takes the shape of a ring when it dies. The middle part of the tuft and the leaves are stiff, and when it dies, it stays in the bundle for several years. In addition, as a mechanism to combat water deficiency, leaves tend to fold and tilt to avoid sunlight and decrease water loss by evaporation. The leaf cell walls are rigid and exposed for a loss of chlorophyll during the summer to adapt to the dry condition's characteristic of semiarid regions, esparto grass possesses superficial roots that give them the ability to use

TABLE 6.1 Length and width of esparto grass compared to some nonwood fibers (Han, 1998; Rowell et al., 2000).

	Fiber length (mm)		Fiber length (mm)	
Common name (scientific name)	Avg.	Range	Avg.	Range
Esparto (*Stipa tenacissima*)	1.2	0.2–3.3	13	6–22
Albardine (*Lygeum spartum*)	1.1	0.2–3.1	12	6–21
Ramie (*Boehmeria nivea*)	120	60–250	50	11–80
Flax (*Linum usitatissimum*)	33	9–70	19	5–38
Hemp (*Cannabis sativa*)	25	5–55	25	10–51
Sunn (*Crotalaria juncea*)	8	4–12	30	25–50
Abaca (*Musa textilis*)	6	2–12	24	16–32
Kenaf (*Hibiscus cannabinus*)	5	2–6	21	14–33
Sisal (*Agave sisalana*)	3	1–8	20	8–41
Bamboo (*Dendrocalamus arundinacea*)	2.7	1.5–4.4	14	7–27
Raphia (*Raphia hookeri*)	2.4	–	30	17–46
Sabai (*Eulaliopsis binata*)	2.1	0.5–4.9	9	4–28
Common reed (*Phragmites communis*)	2.0	1.0–3.0	16	10–20
Jute (*Corchorus capsularis*)	2	2–5	20	10–25
Papyrus (*Cyperus papyrus*)	1.8	1.0–4.0	12	8–25

small amounts of available water to combat drought and desertification. Esparto grass reproduces by roots as well as by seeds; it spreads by the wind. Esparto grass blooms between April and June. They bear fruits between June and July, but esparto can produce flowers and fruits in other seasons for its ability to take advantage of good environmental conditions (rains in winter and summer). The distinction of fibers is tightly correlated to the vegetative cycle. At the base of the leaf (basal level), the fibers are short and broad below the leaf, but they grow longer and thinner above it (Belkhir et al., 2012). Alfa stem is made up of solid, rigid, and light cellulosic fibers that are generally used to make high-quality papers, as well as for decorating, cigarettes, and dielectric uses (Brahim et al., 2007). Approximately 3 million hectares of a covered surface by esparto grass in Algeria were estimated (Ghebalou, 2001). The esparto grass tends to prefer the calcareous soil, not very dense and permeable the quality of its cellulose, that is, its flexibility, smoothness, and mechanical resistance, imparts the properties

TABLE 6.2 Chemical composition of some nonwood fibers (% total) (Hurter, 1988).

Fiber source	Cellulose (alpha)	Lignin	Ash	Silica
Grasses				
Esparto	50–54	17–19	6–8	2–3
Arundo donax	–	21	4–6	1.1–1.3
Sabia	54–57	17–22	5–7	3–4
Switchgrass	–	34–36	1.5–2	–
Bast fibers				
Bast hemp	55–65	2–4	5–7	<1
Jute	39–42	21–26	0.5–1	<1
Bast kenaf	31–39	15–18	2–5	–
Seed hull fibers				
Cotton staple	85–90	3–3.3	1–1.5	<1
Cotton linters	80–85	3–3.5	1–1.2	<1
Leaf fibers				
Abaca	61	9	1	<1
Sisal	43–56	8–9	0.6–1	<1

Source: Adapted from Hurter, A.M., 1988. Utilization of annual plants and agricultural residues for the production of pulp, and paper. In: Pulping Conference: [Proceedings] (USA).

required for papermaking (Omar et al., 2007). Esparto grass is lignocellulosic fibers produced from two species of long, coarse perennial grass: (*S. tenacissima* and *L. spartum*) which they used for the production of crafts, such as ropes, baskets, cordage, shoes, and the production of high-quality paper, and as reinforcement in the production of biodegradable composites. Esparto grass steppes are considered open formations with vegetation cover; it ranges between 18% and 60% with esparto representing the species dominant vegetarian. Biomass accumulates in the steppes of the esparto grass to varying degrees. It ranges between 0.17 and 13 t of dry matter per hectare, depending on water availability and degree of degradation. *S. tenacissima* thrives mostly on Algeria's high plateaus, where it coexists with *L. spartum* in a vegetation alternation (Chifu and Meziani, 1978). The esparto grass is made up of two gray–green needlegrass species (*S. tenacissima* and *L. spartum*); both species of *L. spartum* and *S. tenacissima* are used to make fibers products and papermaking.

6.3.1 Stipa tenacissima L

S. tenacissima L. is a perennial herbaceous plant of the Poaceae family, growing in semiarid zones of Algeria and Mediterranean countries. It helps to prevent desertification, which is beneficial to the environment. Their fiber produces high-quality paper and is also utilized in the confections of many craft products.

6.3.1.1 Nomenclature (synonyms and related terms)

Scientific name: *Stipa tenacissima* L.
English name: Alfa, esparto, esparto grass, halfah grass, and needles grass
Arabic name: Halfah
French name: l'alfa veritable

6.3.1.2 Botanical classification

Plant kingdom: Plant
Phylum: Spermaphytes
Class: Monocotyledonous
Order: Poales
Family: Poaceae
Gender: Stipa
Genus/species: *Stipa tenacissima* L.
Common name: Alfa
Arabic name: Halfah

Esparto grass type *S. tenacissima* L. (Fig. 6.1) is a tall, evergreen perennial tussock grass with long, narrow leaves; it's one of the widespread species vegetation in the Mediterranean driest Basin and is plentiful if Annual rainfall varies from 100 to 500 mm, forming the typical tussock grasslands (Fig. 6.2) between 200 and 400 mm (Haase et al., 1999; White, 1983). Its fiber-rich leaves can grow to be 1 m long. When *S. Tenacissima* tufts are young, they are circular and uniform, but as they age and die, they become hollow in the center. The leaves are thin, ribbon-like, smooth, shining, solid, and covered at the base with a hairy sheath. *S. tenacissima* leaves mature in a period between 4 and 11 months after budding, depending on location and climatic conditions (Ghrab, 1981). The main characteristic of this plant is maintaining the ecosystem balance and economic interests, especially in pulp and papermaking.

6.3.1.3 Botanical description of Stipa tenacissima

S. tenacissima is a herbaceous, perennial plant, which appears as a roughly circular tuft whose diameter varies greatly and depends on multiple interactions with the climatic and soil conditions that it encounters. Alfa is made up

FIGURE 6.1 *Stipa tenacissima* L. (esparto grasslands).

FIGURE 6.2 *Stipa Tenacissima* L. tussocks.

of two parts: under- and aboveground. The first is formed of a rhizome and the second of a leaf composed of blades, sometimes up to a meter long (Bousmaha, 2012).

6.3.1.4 *Underground section of* Stipa tenacissima

The underground portion of the *S. tenacissima* tuft consists of very thick rhizomes distinguished by nodes and internodes, roots, and bushy rootlets, going down to different depths (up to 50 cm in the soil), depending on the quality of the rock and the density of the soil (Zeriahène, 1987).

6.3.1.5 *The aerial part of* Stipa tenacissima

The aerial part of *S. tenacissima* L. (Fig. 6.3) consists of twigs with sheaths surmounted by 30–120 cm blades, which, under the effect of drought, curl into gutters and take the appearance of a leaf (Bensid, 1990). According to Ghrab (1981), *S. tenacissima* is a xerophilic Poaceae; it varies from 20 to 150 cm in height and from 10 to 180 cm in diameter. The aerial part of the tuft includes the leaves, carrying nested sheaths one in the other, surmounted by blades that can reach 100 cm at maturity, and by inflorescence culms carrying the panicles at their top during the flowering period.

6.3.1.6 *Chemical composition of* Stipa tenacissima

Many authors studied the chemical composition of *S. tenacissima*; the results are summarized by El-Abbassi et al. (2020), in Table 6.3; the results indicate that the major component of *S. tenacissima* is cellulose, followed by hemicellulose and lignin as appearing in Table 6.3. Chemical analysis results vary widely with site, season, and leaf level. Several studies have evaluated the impact of various actions undergone by this fiber on the properties of the pulp and paper.

FIGURE 6.3 *Stipa tenacissima* L. plant (Bousmaha, 2012).

TABLE 6.3 Chemical composition of *Stipa tenacissima* fiber by El-Abbassi et al. (2020).

Chemical composition of Alfa/fibers in % of dried matter						
Cellulose	Lignin	Ashes	Silica	Hemicellulose/Pectin	Wax	Others
38.8	20.5			33.5		
39.5	19.5			27.6		
43.8	18.7	4.6	1.7	28.4		2.6
47.6	17.7	5.1		22.1		7.3
46.1	19.9			30.2		
46	20	7.2		24	1.5	1.1
43.9	17.9			27.6		
45.4	14.9			38.5	2	
45	23	2		25	5	
45	24	2		24	5	

Thermal analysis was conducted on sodium hydroxide-extracted Alfa fibers and sodium hypochlorite bleached. A substantial weight loss takes place at 70°C due to the small amount of sampling moisture. The weight-loss rate then increased with the temperature rise to 350°C before degradation (Maghchiche et al., 2013).

6.3.1.7 *Mineral composition of* Stipa tenacissima

The dried plant was milled to less than 1 mm in diameter for the determination of raw fiber. The sample was first boiled in dilute acid (H_2SO_4) and then in dilute alkali (KOH). The residue, which is not soluble in the acid–alkali process, was measured gravimetrically, and the results were given as a percentage of dry matter in dry mass. The mineral composition of *S. tenacissima* is described in Table 6.4.

S. tenacissima fibers have a remarkably high quality with SiO_2 and CaO. The silica is present in the formulation of grass esparto and annual plants act as plant skin to defend themselves against external factors is one of the reasons why alkaline treatments using NaOH are used to delignify this grass (Maghchiche et al., 2013).

TABLE 6.4 Mineral composition of *Stipa tenacissima* ashes (Maghchiche et al., 2013).

Element	%
SiO_2	32.5
CaO	7.25
MgO	2.40
K_2O	1.32
Na_2O	0.40
P_2O_5	0.60
Fe_2O_3	2.60

TABLE 6.5 Morphological proprieties of *Stipa tenacissima* L.

Morphological proprieties of *S. tenacissima* L. (Alfa)	
Length (mm)	Diameter (μm)
0.2–3	6–22
0.4	8.65
1.2	13
1.5	12
1.5–1.9	9–15
2–5	5–10

6.3.1.8 *Morphological proprieties of* Stipa tenacissima

El-Abbassi et al. (2020) synthesized the literature data on the morphological features of elementary *S. tenacissima* fibers tested by various researchers; the results (Table 6.5) show that the fiber length is between 0.2 and 5 mm and the diameter is between 5 and 22 m.

6.3.2 *Lygeum spartum* L

Lygeum is a genus of the grass family; the species is *L. spartum* (Fig. 6.4), commonly known as esparto grass, the *L. spartum* spread in North Africa and the western Mediterranean basin's arid areas, especially in the high plateaus of Algeria; its leaves are made up of cellulose-rich fibers. *L. spartum*

FIGURE 6.4 *Lygeum spartum* grass.

L. is a perennial grass (*Poaceae*) that colonizes limestone or gypsum crust soils (Le Houérou, 1995). *Lygeum* is a genus with only one species: a creeping rhizome, a cylindrical, and a squalid leaf define the *L. spartum*. The stems are terminated by a wide spathe with one-to-four flowers, each consisting of three stamens and one pistil (Quézel and Santa, 1962).

6.3.2.1 Nomenclature (synonyms and related terms)

Scientific name: *Lygeum spartum*
English name: Albardine, Lygeum, esparto grass, false esparto grass
Arabic name: ḥalfah mahbūla, sennāġ, senhac, senagh
French name: fol-alpha, sparate, spart, faux sparte

6.3.2.2 Botanical classification

Botanical name: *Lygeum spartum*
Kingdom: Plantae
Order: Poales
Family: Poaceae
Subfamily: Pooideae
Species: *L. spartum*
Genus: *Lygeum*
Life cycle: perennial height: up to 100 cm

L. spartum is a naturally occurring species in Algerian steppes (Belouadah et al., 2015), which is one of the most common rhizomatous perennial grasses (Nedjimi, 2014). *L. spartum* forms a natural barrier in the

Algerian highlands against the advancement of sand and desertification, due to its highly developed root system.

L. spartum has a tolerance to environmental stresses plant, salt stress, and its use as a fodder grass for livestock steppe regions (Nedjimi, 2009). New cellulosic fibers were extracted from *L. spartum* plant fibers; these fibers are rich in cellulose content with a crystallinity index of 46.19%; the fiber stem anatomy showed a strong presence of fiber cells. *L. spartum* fibers are thermally stable until 220°C with an activation energy of 68.77 kJ/mol (Belouadah et al., 2015).

6.3.2.3 *Difference between* Stipa tenacissima *and* Lygeum spartum

Two fairly similar plants are usually confused under the name esparto grass: *S. tenacissima*, the sheet folds into a half-sheath, and the tuft somewhat reminiscent of that of barley. The *L. spartum*, with dimensions, is more reduced. The first inhabits mountainous areas and highlands. The trained eye can distinguish the green–yellow complexion of *S. tenacissima* from the green–blue complexion of the *L. spartum* (Louis and Despois, 1986). *L. spartum* inflorescence consists of two-to-three conical spikes at the base to form a tube covered with long silky white hair shown in Fig. 6.5 (Zeriahene et al., 1998), while the *S. tenacissima* florescence is a contracted panicle of 25–35 cm long, which is indicated in Fig. 6.6 (Schöning et al., 2004).

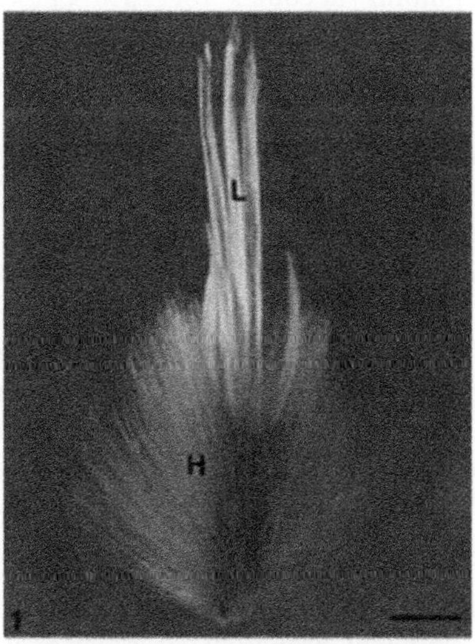

FIGURE 6.5 Florescence of *Lygeum spartum* (Zeriahene et al., 1998).

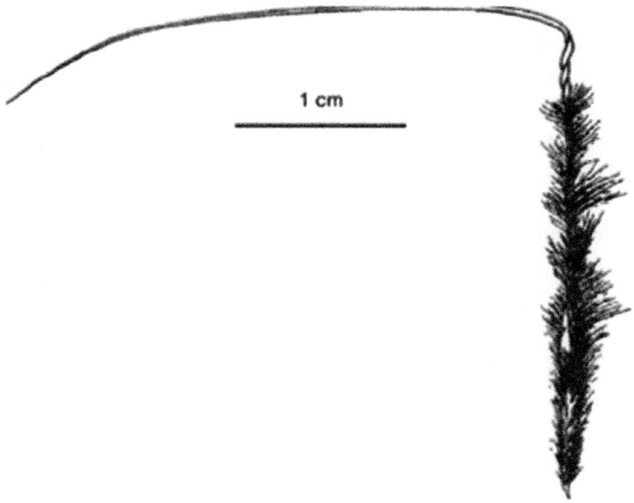

FIGURE 6.6 Florescence of *Stipa tenacissima* (Schöning et al., 2004).

6.3.2.4 Extraction methods of esparto grass fibers

Esparto grass is often used in the form of pulp or in its pure state. Depending on the final application of these extracted fibers. The researchers try to optimize these parameters in the most appropriate way depending on the final application of these extracted fibers, so we will present known separation methods: mechanical (including thermomechanical), chemical–mechanical, or all-chemical, and, more recently, enzymatic extraction. Currently, chemical pulping accounts for 75% of world pulp production (Passas, 2012). A well-detailed study on the structure, chemical composition, and morphological characteristics of the esparto grass fibers has been described by many researchers. The acid pretreatment of the esparto Alfa fibers makes it possible to remove the lignin much more easily than in the case of the untreated fibers (Bouiri and Amrani, 2010a,b). In enzymatic extraction, enzymes are added directly to the bioreactor for fibers extraction. In terms of time-savings and environmental friendliness, this technology has been proved to be a promising replacement for standard-setting procedures (Sisti et al., 2018). Chemical extraction method based on cellulose separation from noncellulosic components using a soda process to dissolve noncellulosic compounds, such as lignin, pectin, and hemicellulose, as well as the numerous ingredients constituting the stems plant reserve and outer wall.

6.3.2.5 Papermaking of esparto grass

Uses of the various esparto grasses in papermaking are due to the wide variety of physical and chemical properties of esparto grass fibers. The most used methods in pulping esparto species include alkaline processes such as kraft pulping or sulfate process and soda (NaOH) methods, and sulfite methods (Table 6.6) also show esparto grass pulping methods, the soda pulping, the only method used yet in countries producing nonwood pulps. Formic acid and hydrogen peroxide are used in the bleaching process (Saijonkari-Pahkala, 2001).

6.3.2.6 Economic values of esparto grass

The main use of esparto grass is in the manufacture of paper of the best quality. Leaves of esparto grass have a large number of fibrous elements rich in cellulose, which after bleaching will be excellent for the paper industry. *S. tenacissima* L. mixed with *L. spartum* is used for artisanal purposes (making basketwork, mats, rugs, shoes, etc.).

6.4 Papermaking

Humanity has known papermaking; since ancient times, the Chinese have been the first to make and use paper. The manufacture of paper does not require complicated or expensive raw materials. As the materials that are used are wood, lignocellulosic plants, grass, agricultural waste, remnants that result from wood manufacturing, and recycled paper.

Paper is of great importance in human life; it's one of the basic materials in human life, in writing, in the field of printing and packaging, and it is also included in many industries such as building materials. The raw material for the paper industry is cut and transported to special paper factories, where they are manufactured mechanically and chemically, until they become white laminate papers or manufactured to other papers. In the mechanical method, it enters the raw material in a rotating machine that removes the bark and then cuts the raw material into very thin sheets less than 2 cm and shines these chips, disintegrates the wood from each other, and then kneads until it becomes chips from white paper.

TABLE 6.6 esparto grass pulping methods (Saijonkari-Pahkala, 2001).

Process	Chemicals used	Usual uses of pulp isolation
Soda	NaOH	Ideal for every form of paper use
Kraft/sulfate	$NaOH + Na_2S$	Paper bags, making boxes
Sulfit	$NaHSO_3$ and/or Na_2SO_3	Paper for special purposes
Organo-solvent	Ethanol	Possible application in all form of paper

Chemically manufacturing paper method is divided into two basic stages. The first stage is the manufacture of fibrous blocks, which is to cut trees' roots until they are about 1–2 m in length, and remove bark from them. Then, it is washed well, cut into small pieces, dipped in a chemical solution and then crushed to be turned into a paper that is ready for writing and printing.

6.4.1 Esparto grass plant collection

Esparto grass is a natural plant that serves as raw materials for the production of esparto pulp. After harvesting esparto, which is dried, pressed, and arranged in bales then, is sent to the factory for pulp processing.

6.4.2 Esparto grass as raw materials for papermaking

Esparto grass has distinctive properties due to its natural composition on the one hand, in addition to being short and hard fibers. It is also a rare raw material for producing high-quality paper. Thanks to the short fibers and thin films of this plan, it provides a significant number of fibers per gram compared to the short fibers of other origins. Fiber dimensions, particularly the length, are largely related to pulp quality (Janin, 1981).

Esparto grass fibers have thick walls and are short (less than 3 mm) in length, with an average length of 1.5 mm. The diameter of the fiber varies from an average of about 0.012 mm, resulting in a form factor of about 125 (Bouiri and Amrani, 2011).

The characteristics of bleached and unbleached esparto stems and pulp were characterized as being intermediate stem fibers between those woody plants and nonwoody. Often resembling eucalyptus in appearance and by using a refining procedure to the unbleached and bleached esparto grass pulp, during the refining process, the fibers are just slightly shortened. Finally, unrefined and refined pulp leaves are traditionally handcrafted. Paper made from esparto has excellent low-density values and good mechanical qualities, which are greatly improved by refining treatment, particularly for unbleached pulp. This study shows that these nonwood species have a lot of potential for papermaking (Marrakchi et al., 2011).

6.4.3 Grass for papermaking

Alternative resources are being explored as a result of the rising demand for paper and a scarcity of raw materials. Due to its abundance, fibers derived from nonwood can be a good alternative. For the pulp and paper industry, agricultural waste or annual nonwoody plants are a valuable supply of fiber. Many nonwood raw materials, such as cotton stalks, rice, and others, such as seagrass, which is a good source of cellulose, are available, due to the scarcity of its content of lignin, holocellulose and cellulose amounts are similar to those found in softwood and hardwood. There is a possibility of using the seagrass as an alternative in the manufacture of paper and can be used as a substitute for raw materials for nonwood paper pulp in the manufacture of food packaging (Menta, 2015). Cogon grass is a good alternate source of fibers among these nonwood fibers, especially for the pulp and paper industries, though it has good chemical and surface morphological properties (Mohd Kassim et al., 2015).

6.4.4 Esparto grass for papermaking

Esparto grass is used in paper production in a mild alkaline solution through a heating and grinding process and undergoes an antibacterial and antifungal treatment. Paper made from esparto grass is suitable for of all types printing because of its stability, natural color, and neutral pH. Esparto's high-quality paper is manufactured from its cellulose-rich fibers. Due to the short length of esparto fibers, cellulose fiber of esparto grass is combined with wood pulp in a ratio of 5%–10% to produce a dense and well-packed paper with excellent inking qualities and very good folding properties. Esparto grass paper holds its bulk and absorbs block letters well due to its short fiber length. Esparto grass has been used since the middle of the last century in papermaking; currently, only Algeria and Tunisia produce esparto pulp. Esparto grass renders outstandingly fine paper writing and printing. The paper made from esparto grass has strong resilience when in contact with the plate print, with strong opacity and relative lint-free feeling smooth (El-Abbassi et al., 2020).

6.4.5 Production of pulp from esparto grass

The kraft method was used to separate two cellulose pulps with different chemical compositions from Algerian esparto grass: One was made from raw esparto grass, and the other was made from esparto grass that had been prepared with dilute acid. It has also been shown that the kraft method and the acid pretreated method of esparto grass are desirable methods for processing high-quality cellulose pulp. Esparto grass has morphological and technical features ideal for the manufacture of bleached kraft pulp. The pulp's physico-mechanical qualities were not considerably harmed by the bleaching process, making this species an ideal supply of raw material for the pulp and paper industry (Bouiri and Amrani, 2010a,b).

A mixture of formic acid, acetic acid, and water to pulp esparto grass was investigated at atmospheric pressure; the Kappa number and degree of polymerization were calculated for the unbleached pulps obtained. Peroxy acids in an acidic organic medium and hydrogen peroxide in an alkaline medium were used to delignify the unbleached pulp (Ammar et al., 2009).

6.5 Conclusion

Esparto grass is a wild plant that grows naturally without human intervention. It's an evergreen plant that thrives in arid, stony, and calcareous environments, which requires very little water to develop, without the use of insecticides or pesticides.

The process of producing pulp from nonwood fibers is increasing increasingly due to increased demand for paper products and insufficient wood. Esparto grass is a strong raw fiber used for industrial paper pulp production. The properties of esparto grass fibers like the amount of cellulose, the durability, the smoothness, and the mechanical resistance make it a suitable raw material resource as a value-added product for papermaking and producing high-quality pulp and paper. Esparto grass can be exploited to supply raw materials for cellulose and papermaking by the multiple advantages available, renewable, cheap cost and represent a business opportunity for agriculture in arid and semiarid areas by fighting deforestation, improve soil quality, and increase biodiversity.

Esparto grass has similar characteristics to those obtained from paper pulp, is available, is cost-effective in comparison with wood products, and has nearly identical chemical and morphological properties to wood fibers. As a result, the future of nonwood plant fibers as a pulping and papermaking raw materials is bright. The use of esparto grass as an alternative source of paper production would fulfill environmental needs, such as deforestation and global warming.

References

Ammar, H., Mallek, H., Abdelkefi, F., Benjelloun-Mlayeh, B., El Gharbi, R., 2009. Separating, characterization and application of Alfa grass (*Stipa tenacissima*) chemical components 1. Pulping of Alfa grass with formic acid/acetic acid mixture. J. Soc. Chim. Tunis. 11, 69–75.

Azeez, M.A., 2018. Pulping of non-woody biomass. Pulp Pap. Process. 55–86. Available from: https://doi.org/10.5772/intechopen.79749.

Belkhir, S., Koubaa, A., Khadhri, A., Ksontini, M., Smiti, S., 2012. Variations in the morphological characteristics of *Stipa tenacissima* fiber: the case of Tunisia. Ind. Crops Prod. 37 (1), 200–206.

Belkhir, S., Koubaa, A., Khadhri, A., Ksontini, M., Nadji, H., Smiti, S., et al., 2013. Seasonal effect on the chemical composition of the leaves of *Stipa tenacissima* L. and implications for pulp properties. Ind. Crops Prod. 44, 56–61.

Belouadah, Z., Ati, A., Rokbi, M., 2015. Characterization of new natural cellulosic fiber from *Lygeum spartum* L. Carbohydr. Polym. 134, 429–437.

Brahim, S.B., Cheikh, R.B., 2007. Influence of fibre orientation and volume fraction on the tensile properties of unidirectional Alfa-polyester composite. Compos. Sci. Technol. 67 (1), 140–147.

Bensid, T., 1990. Structures spatiales et interférences entre individus dans deux populations d'alfa (Stipatenacissima L.) et d'armoise blanche (Artemisia herba alba Asso.) vivant dans les Hautes Plaines Steppiques de Tlemcen (Algérie). Mém. Mag. Bio., univ. Tlemcen.

Bouiri, B., Amrani, M., 2010a. Production of dissolving grade pulp from alfa. BioResources 5 (1), 291–302.

Bouiri, B., Amrani, M., 2010b. Elemental chlorine-free bleaching halfa pulp. J. Ind Eng. Chem. 16 (4), 587–592.

Bouiri, B., Amrani, M., 2011. Elemental chlorine-free delignification of kraft pulp produced from halfa (*Stipa tenacissima*). Fibres Text. East. Eur. 87, 142–147.

Bousmaha, T., 2012. Contribution à l'étude de l'évolution de la nappe alfatière dans la mise en défens De Nofikha. (Naâma) (Doctoral dissertation).

Chifu, T., Meziani, S.A., 1978. Contribution à l'étude de la flore du Nord de Tébessa.

El-Abbassi, F.E., Assarar, M., Ayad, R., Bourmaud, A., Baley, C., 2020. A review on alfa fibre (*Stipa tenacissima* L.): from the plant architecture to the reinforcement of polymer composites. Composites Part A 128, 105677.

Ghebalou, A., 2001. Colloque sur l'industrie papetière. CIPA, Algeria.

Ghrab, S., 1981. Étude de la variabilité éco-phénologique de l'alfa en Tunisie centrale. application en Vue de la sauvegarde et de l'amélioration des nappes alfatières.

Haase, P., Pugnaire, F.I., Clark, S.C., Incoll, L.D., 1999. Environmental control of canopy dynamics and photosynthetic rate in the evergreen tussock grass *Stipa tenacissima*. Plant Ecol. 145 (2), 327–339.

Han, J.S., 1998. Properties of nonwood fibers. In: Proceedings of the Korean Society of Wood Science and Technology Annual Meeting. The Korean Society of Science and Technology, Seoul, pp. 3–12.

Hurter, A.M., 1988. Utilization of annual plants and agricultural residues for the production of pulp, and paper. In: Pulping Conference: [Proceedings] (USA).

Hurter, R.W., 2002. Strategic Market Management: Pulp and Paper. Hurter Consult Incorporated, Ontario, Canada, p. 48.

Hurter, R.W., Riccio, F.A., 1998. Why CEO's don't want to hear about non-woods-or should they? In: Proceedings, 1998 North American Nonwood Fiber Symposium (Atlanta), pp. 1–10.

Janin, G., 1981. Technique de micro cuissons papetières: perfectionnement et précision des rendements En pâte. Ann. Sci. Forest. 38 (1), 107–126.

Le Houérou, H.N., 1995. Considérations biogéographiques sur les steppes arides du nord de l'Afrique. Science et changements planétaires/Sécheresse 6 (2), 167–182.

Liu, Z., Wang, H., Hui, L., 2018. Pulping and papermaking of non-wood fibers. Pulp and Paper Processing. InTech.

Louis, A., Despois, J., 1986. Alfa. Encyclopédie berbère (3), 440–442.

Maghchiche, A., Haouam, A., Immirzi, B., 2013. Extraction and characterization of Algerian Alfa, Grass short fibers (*Stipa tenacissima*). Chem. Chem. Technol. 7 (3), 339–344.

Marrakchi, Z., Khiari, R., Oueslati, H., Mauret, E., Mhenni, F., 2011. Pulping and papermaking properties of Tunisian Alfa stems (*Stipa tenacissima*)-effects of refining process. Ind. Crops Prod. 34 (3), 1572–1582.

Menta, B.M., 2015. 5th Euro-Global Summit and Expo on Food & Beverages. Alicante, Spain. J. Food Process Technol. 2015. Available from: https://doi.org/10.4172/2157-7110.S1.0186:5.

Mohd Kassim, A.S., Mohd Aripin, A., Ishak, N., Zainulabidin, M.H.B., 2015. Cogon grass as an alternative fibre for pulp and paper based industry: on chemical and surface morphological properties, Applied Mechanics and Materials, vol. 773. Trans Tech Publications Ltd, pp. 1242–1245.

Mohieldin, S.D., 2014. Pretreatment approaches in non-wood plants for pulp and paper production: a review. For. Prod. Ind. 3, 84–88.
Nadji, A., Brochier-Salon, M.C., Bruzzèse, C., Belgacem, M.N., 2006. Chemical composition and pulp properties of Alfa tenacissima. Cellul. Chem. Technol. 40 (1–2), 45–52.
Nedjimi, B., 2009. Salt tolerance strategies of *Lygeum spartum* L.: a new fodder crop for Algerian, saline steppes. Flora—Morphol. Distrib., Funct. Ecol. Plant 204 (10), 747–754.
Nedjimi, B., 2014. Comportement physiologique du sparte (*Lygeum spartum* L.) en milieu salé. Revue Nature et Technologie 6 (2), 30–33.
Omar, A., Khadra, M.B., 2007. Esparto grass (*Stipa tenacissima* L.), raw material of papermaking. First part. Химия растительного сырья (4), 25–30.
Pahkala, K., Paavilainen, L., Mela, T., 1997. Grass species as raw material for pulp and paper. In: Proceedings of the XVIII International Grassland Congress, June 8–17, Winnipeg and Saskatoon, Canada, pp. 55–60.
Passas, R., 2012. Natural fibres for paper and packaging. In: Kozlowski, R.M. (Ed.), Handbook of Natural Fibres: Processing and Applications. Woodhead Publishing Series in Textiles, Elsevier Science, pp. 367–400.
Quézel, P., Santa, S., 1962. Nouvelle flore de l'Algérie et des régions désertiques méridionales.
Rowell, R.M., Han, J.S., Rowell, J.S., 2000. Characterization and factors effecting fiber properties. In: Frollini, E., Leão, A., Mattoso L.H.C. (Eds.), Natural Polymers and Agrofibers Based Composites. USP, São Carlos, pp. 115–134. Available from: https://doi.org/10.5772/intechopen.79017.
Saijonkari-Pahkala, K., 2001. Non-Wood Plants as Raw Material for Pulp and Paper.
Schöning, C., Espadaler, X., Hensen, I., Roces, F., 2004. Seed predation of the tussock-grass *Stipa tenacissima* L. by ants (*Messor* spp.) in south-eastern Spain: the adaptive value of trypanocarpy. J. Arid. Environ. 56 (1), 43–61.
Sisti, L., Totaro, G., Vannini, M., Celli, A., 2018. Retting process as a pretreatment of natural fibers for the development of polymer composites. Lignocellulosic Composite Materials. Springer, Cham, pp. 97–135.
White, F., 1983. The Vegetation of Africa (Vol. 20).
Zawawi, B.D., 2018. Application of xylanase enzyme on Napier grass for chemical pulping process for hand sheet made papermaking. In: 2nd International Conference on Advances in Biotechnology July 23–25, at Kuala Lumpur, Malaysia. Available from: https://doi.org/10.29011/2574-7258-C1-005.
Zeriahène, N., 1987. *Étude du système racinaire de l'alfa en relation avec l'adaptation au xérophytisme* (Doctoral dissertation, Thèse de magister). Institut des sciences de la nature, Oran.
Zeriahene, N., Prat, R., Goldberg, R., Catesson, A.M., Harche-Kaid, M., 1998. Cell walls of seed hairs from *Lygeum spartum*: ultrastructure, composition and mechanical properties. Ann. Bot. 81 (1), 61–66.

Chapter 7

Pulping and papermaking of jute

M. Mostafizur Rahman[1], M. Sarwar Jahan[1] and Yonghao Ni[2,3]
[1]Pulp and Paper Research Division, Bangladesh Council of Scientific and Industrial Research (BCSIR), Dhanmondi, Dhaka, Bangladesh, [2]Department of Chemical Engineering, University of New Brunswick, Fredericton, NB, Canada, [3]Department of Chemical and Biomedical Engineering, University of Maine, Orono, ME, United States

7.1 Introduction

7.1.1 What is pulping?

The defibration of wood and nonwood into individual fibers by mechanical and or chemical treatment or in combination with both treatments is called pulping process. Chemical pulping involves the dissolution of lignin from the cell wall of lignocelluloses and liberates cellulosic fibers. The conventional chemical pulping processes are kraft, sulfite, neutral sulfite semichemical (NSSC), and soda processes, amongst the kraft process is the predominant pulping process, which requires sodium hydroxide and sodium sulfide as cooking chemicals. Pulp are two types: (1) paper grade pulp and (2) dissolving grade pulp, which is highly purified cellulose. Traditionally, dissolving grade pulp is produced by the prehydrolysis kraft and acid sulfite processes, and is used as the starting raw material for the production of cellulose derivatives.

Pulp processing steps are (1) raw material preparation and handling; (2) cooking, pulp washing, and screening; (3) chemical recovery; (4) pulp bleaching; and (5) papermaking steps, which includes, stock preparation, papermaking, and converting.

The main raw materials for pulp production are hardwood, softwood, nonwood, and recycle fibers.

7.1.2 Jute fiber production

Jute fiber is a bast fiber obtained from dicotyledonous plants of the genus *Corchorus* and family Tilaceae and grown mainly in Bangladesh, India, and

Pulping and Papermaking of Nonwood Plant Fibers. DOI: https://doi.org/10.1016/B978-0-323-91625-7.00005-9

other Asian countries. Fig. 7.1 shows a jute plantation in Bangladesh. Conventionally, the natural microbial process separates the bast fiber from the plants, which is known as retting process. In this process, jute bundles are submerged in clear slow flowing water in canals, tanks, ponds, or ditches with the minimum ratio of plant material to water 1:20. The retting process determines the quality of fiber and ultimately determines the price of fiber. Paridah et al. (2011) showed that 14–28 days were required to degrade the pectin materials, hemicellulose, and lignin from jute plant, and this period can be considered retting time. After separating bast fiber from the jute plant, the woody portion of the plant remains as residue, which is called jute stick. The different types of packaging materials are produced from the jute fiber as it is the cheapest textile fibers. Jute fiber was known as the "Golden fiber of Bangladesh" because of its color and prime importance in the economic

FIGURE 7.1 Jute plant.

development of Bangladesh. The economy of Bangladesh was dependent on the export of jute and related products, and it contributed to the country's workforce. Traditionally, jute is used in sacking, gunny bags, hessian, and so on. In recent years, jute product has faced hard competition with synthetic products. As a result, the demand for jute fiber has declined both in the local and overseas markets. Therefore, expanded usage is needed to recover the lost glory of jute. Researchers study different diversified products from jute fiber. Various grades of pulp from jute are the aim of diversification of jute products of this chapter.

In this chapter, pulping of jute fiber, jute sticks, and whole jute plant (without retting) in different processes for the production of paper grade and dissolving grade pulps have been discussed based on published literature. Different cellulose products from the jute pulp are also discussed.

7.1.3 Chemical and morphological characteristics

The suitability of a raw material for pulping depends on the chemical and morphological properties. Jute plant consists of bast fiber and woody core (jute stick). Table 7.1 shows the chemical characteristics of jute fiber and jute sticks from different literatures and also shows comparison with kenaf fiber and kenaf sticks since both kenaf (*Hibiscus cannabinus*) and jute are cultivated for their bast fiber. As shown in Table 7.1, jute fiber and jute sticks are very similar to kenaf fiber and kenaf sticks, respectively. As a bast fiber, jute fiber contains a higher amount of α-cellulose and lower amount of lignin. On the other hand, the jute sticks are chemically similar to hardwood (Sixta, 2006). The morphology and chemical composition of jute fiber and jute stick depend on harvesting time, geographical locations and retting process. Alam et al. (2013) showed that the chemical composition varied in different parts of the jute plant (GJP). The α-cellulose content increased, whereas hemicellulose, lignin, ash, and extractives decreased from the top to the bottom part of the jute plant. Jahan et al. (2008) also showed that the α-cellulose content in the middle part of both bark and core of jute plant was higher. The bark had higher α-cellulose and lower lignin than core of the plant. Little variation of chemical compositions especially the α-cellulose, hemicelluloses, and lignin was observed among the different varieties of jute. In the same variety, the middle portion contained the largest amount of α-cellulose and the top portion contained the lowest. Klason lignin content varied from 24.2% to 26.8% (Rahman et al., 2016).

The average fiber length was above 2 mm for jute fiber and less than 1 mm for jute stick (Table 7.2). The fiber width of jute sticks is larger than jute fiber, jute fiber consequently produced higher slender ratio. As shown in Table 7.2, jute fiber and sticks are very similar to kenaf fiber and sticks, respectively. Although the fiber length of different varieties of jute sticks

TABLE 7.1 Chemical characteristics of jute.

	Klason lignin (%)	Holocellulose (%)	α-Cellulose (%)	Pentosan (%)	Ash (%)	References
Jute fiber	12.7	87.6	63.1	13.5	–	Jahan et al. (2007)
Jute fiber	12–13	–	60–63	21–24	0.7–1.2	Roy and Lutfar (2012)
Jute stick	25.1	74.8	39.3	15.8	0.75	Matin et al. (2015)
Jute stick	24.1	–	45.7	22.7	2.7	Jahan et al. (2004)
Kenaf	14.7	82.6	56.4	13.5	2.2	Ashori et al. (2006)
Kenaf stick	22.1	75.8	46.1	20.7	1.6	
Kenaf	12.5	73.2	55.5	10.6	0.6	Udohitinah and Oluwadare (2011)

TABLE 7.2 Morphological characteristics of jute.

	Fiber length	Fiber width	Wall thickness	Lumen	References
Jute fiber	2.54	20			Jahan et al. (2007)
Jute fiber	2.5	18			Roy and Lutfar (2012)
Jute stick	0.66	33.3	1.7	23.4	Jahan et al. (2004)
Kenaf bast	2.48	24.24	4.85	15.44	Ashori et al. (2006)
Kenaf stick	0.72	31.88	4.08	25.77	
Kenaf bast	2.9	28.16	11.04	6.08	Udohitinah and Oluwadare (2011)

was similar, the width and slenderness ratios were different (Rahman et al., 2016). The chemical and morphological properties of jute fiber and sticks imply that both raw materials are suitable for pulping as softwood and hardwood substitutes, respectively.

7.2 Pulping

7.2.1 Jute fiber pulping

Retting has been used for a long time for the extraction of bast fibers from jute plants. Pulping of retted jute fiber was evaluated in different processes. Shafi et al. (1993) used whole length jute fiber (*Corchorus capsularis*) for neutral sulfite–anthraquinone (NS–AQ) pulping. At an alkali charge of 14% only, a fully defiberized pulp was obtained with no screening rejects; thus pulp yield was better. As compared to chopped jute fiber, pulp yield was higher and almost similar in quality; thus chopping cost was reduced in whole length jute pulping. Jute mills discard the bottom part of jute and also generate jute wastes which are called jute cutting (JC) and caddis, respectively. The amount of JC and caddis accounts for 15% and 3% of jute used in mills. Jahan et al. (2007) studied soda–AQ pulping of jute fiber, JCs, and jute caddis. As shown in Table 7.3, high pulp yield of 69.1% was obtained from jute fiber with a kappa number of 11. Jute fiber, cuttings, and caddis pulp show a high tear index similar to that of softwood pulp. In order to further evaluate the properties of JCs and caddis pulping, soda–AQ, alkaline sulfite–AQ (AS–AQ), and NS–AQ processes were evaluated by Jahan et al. (2013). The NS–AQ process was superior to the others as it showed enhanced physical properties and a favorable pulp yield of 66.1% for JC and 59.9% for caddis at a kappa number of about 11. At oSR 44, the tensile

TABLE 7.3 Pulping of jute.

Raw material	Process	Pulp yield (%)	Kappa number	Viscosity (DP)	Tensile [N (m/g)]	Tear [mN (m^2 g)]	Burst [kP (m^2/g)]	References
Jute fiber	Soda–AQ	69.1	11.0		42.7	28.3	5.1	Jahan et al. (2007)
Jute cutting	Soda–AQ	60.2	13.9		40.9	20.1	4.9	
Jute caddis	Soda–AQ	56.9	13.2		42.1	19.9	5.5	
Whole jute plant	Soda–EDA	50.3	21.3		7138	113	72.2	Jahan and Farouqui (2000)
	Soda–MEA	51.0	23.6		7469	102	78.6	
Jute fiber (whole length)	NSAQ	64.4	20.1		107.5	17.2	9.1	Shafi et al. (1993)
Jute fiber (chopped)	NSAQ	64.6	19.8		105.2	19.0	9.2	
Jute fiber	Bio-soda	68	7.5[a]		58.5	20.1	4.1	Sahin (2007)
Jute fiber	Bio-kraft	74	4.8[a]		61.1	19.4	4.7	
WJP	Kraft	45	22.7		60.8	15.5	6.7	Jahan et al. (2008)
Jute stick		43–45	14–24		–	–	–	Rahman et al. (2016)
Jute fiber	Soda	68	50		5100	–	–	Sahin (2003)
Jute fiber	Ethanol	79	66		3000	–	–	
Jute fiber	Ethanol-soda	66	29		6800			
Hardwood (Maple)	Kraft	51.4	19.7	52	80	7.9		Francis et al. (2001)
Softwood	Kraft	46.3	19.5		70	16.4		Rautiainen and Alen (2007)

AQ, Anthraquinone; *EDA*, ethylenediamine; *MEA*, monoethanol amine; *NSAQ*, neutral sulfite anthraquinone; *WJP*, whole jute plant.
[a]*Residual lignin.*

index of NS–AQ pulp was about 100 N m/g for JC and 70 N m/g for caddis, while their tear index was almost similar to each other and other softwood pulp (Labosky and Ifju, 2007) (Table 7.3). The bleachability NS–AQ pulp also was outstanding, and the brightness of JC and caddis pulps reached to 89% and 85%, respectively, using a total chlorine dioxide of about 15 kg/ton D0(EP)D1 bleaching sequences. The strength properties of NS–AQ pulp were still greater after bleaching.

The AS AQ methanol (ASAM) process is a modified conventional AS process. The target of the process was to get better strength like kraft pulp and easier bleachability like sulfite pulp. The delignification in ASAM process reached to very low kappa numbers (Patt et al., 1987). Jute fiber was cooked by the ASAM process and produced pulp of very low kappa number (7.6), high pulp yield (61.5%), and unbleached brightness (63%) (Jahan et al., 2005). ASAM pulp from jute fiber had very high-strength properties. The ASAM pulp also showed good bleachability in DED bleaching sequences with high papermaking properties as compared to kraft pulp from jute fiber.

Jute was subjected to the mechanical pulping process with or without pretreatment. Fungal and alkali-treated jute bast fibers were refined in an atmospheric refiner for mechanical pulping (Sabharwal et al., 1995). To produce mechanical pulp, a laboratory single-disc refiner was also used for hammer milled and cut jute bast fiber in under atmospheric conditions. The strength properties of the refined pulp of cut jute were higher than the hammer-milled jute fiber or thermomechanical pulps. The pretreatment influences a great deal of the mechanical pulp properties. The fungal treated jute required lower energy consumption in refining; however, papermaking properties were increased significantly. The fungal treatment of jute fiber required lower refining energy, while the alkali treatment prior to refining produced pulp with lower yield and higher strength. The energy consumption of fungal treated jute bast fiber required 450 kWh/ton of pulp, which was 1150 kWh/ton of pulp lower energy for wood. Thus jute fiber saves energy consumption if wood is replaced.

Bleaching of jute fiber pulp is easier in both elemental chlorine free (ECF) and total chlorine free (TCF) bleaching processes. With the consumption of only 8.5 kg ClO_2/ton of pulp, the final brightness of soda–AQ jute fiber pulp reached >89% (Jahan et al., 2010a,b).

The study proved that jute fiber produced the pulp of high yield and low kappa number and required less ClO_2 to get high brightness. The papermaking properties of jute pulp are close to softwood pulp. Alkaline or NS–AQ processes showed the best performance for chemical pulp from jute fiber.

7.2.2 Jute stick pulping

Jute stick is a by-product of the jute fiber extraction process. Compared to jute fiber, jute sticks require more drastic cooking conditions due to higher lignin content and their lower syringyl-to-guaiacyl ratio. The molar ratio of syringaldehyde to guaiacyl was found to be 2.0 for jute fiber and 1.4 for stick lignins (Islam and Sarkanen, 1993). Different varieties of jute sticks were evaluated and found nearly the same pulp yield of 44% at kappa number 15 in the soda–AQ process (Rahman et al., 2016). The chemical constituent of jute fiber and stick (Table 7.1) provides an assumption that jute sticks should yield lower pulp than jute fiber. Kraft pulping of jute sticks and JC was performed and obtained higher yield and lower permanganate numbers from JC than jute sticks (Ghosh et al., 1983b). Chemi-mechanical and soda pulping of jute sticks were carried out by Sanyal et al. (1981) and obtained pulp yield of 76%–81% and 44%–80%, respectively. The breaking length of soda and chemi-mechanical pulps was 3000–5000 and 3333–5222 m, and fold number was 14–220 and 8–400, respectively. Jute stick pulping was carried out with a mixture of JC or rice straw by kraft and chemi-mechanical processes and obtained pulp yield of 48%–70% with breaking length of 2330–6550 m and folding endurance of 20–480 (Ghosh et al., 1983a). Mechanical pulping of soda-impregnated jute stick–jute caddies mixture gave pulp yields of 68%–77% yield with BL of 3000–6260 m and folding endurance of 6–75.

Islam et al. (1993) produced pulp from jute sticks using the soda–AQ process at 16%, 18%, and 22% levels of alkali charge with 0.05% and 0.1% AQ on the raw material. The pulp obtained at 16% alkali charge with 0.05% AQ was enough to produce a good pulp yield and strength. AQ addition in soda liquor increased pulp yield from 38% to 48%, cooking time decreased from 3 to 1 hours, and alkali charge dropped from 22% to 16%. The properties of both the bleached and unbleached soda–AQ pulp were better than those of soda pulp. Pulp yield from jute sticks was lower or close to hardwood (Haque et al., 2019; Jahan et al., 2010a,b). JCs and jute sticks were digested separately in a pilot plant autoclave using the soda process followed by CEH bleaching sequences (Islam et al., 1989). Paper was produced in a pilot plant fourdrinier machine by blending jute stick and JC bleached pulp at the ratio of 50:50, 60:40, and 80:20. Bleached and blended pulps were mixed with 1% rosin, 6% china clay, and 1% TiO_2, refined, producing good-quality writing and printing paper.

It can be concluded from the above discussion that jute sticks require higher chemical charge than jute fiber to delignify in a bleachable grade pulp. Pulp yield is comparable with other nonwood and hardwood. Chemi-mechanical pulping of jute stick mixed with other raw materials produced high pulp yield with good strength properties. Writing and printing grade paper can be produced from the jute stick pulp by blending with JC/jute fiber pulp.

7.2.3 Whole jute plant

The whole jute plant contains both long and short fibers, which originated from bast and cores fraction, respectively. Different processes were investigated for whole jute plant pulping. Soda–AQ pulping processes of the whole jute plant yielded 5%–10% higher and better quality than pulp obtained using the soda process (Islam et al., 1988). Unbleached soda–AQ pulp showed improved papermaking properties than the soda pulp. Soda–AQ pulp from whole jute stalks was recommended for the production of printing and writing paper. Studies show that a bleachable grade kraft pulp of kappa number ~22 can be produced from the whole jute plant by using 17% Na_2O (Kishor et al., 1994). The pulp can be bleached to 76% ISO brightness by CEH bleaching sequences. Bleached pulp strength properties were much superior to those of eucalyptus and bagasse pulps.

The bleachable grade pulp with a yield of 48% can be obtained from whole jute/kenaf plant in both soda–AQ and kraft processes. In soda–AQ process, the bio pretreatment of whole jute had 15% lower kappa number at the same alkali charge (Mohiuddin et al., 2005). Bleaching of whole jute plant pulp reached an ISO brightness of 80%–85% by CEH and DED bleaching sequences. The soda/soda–AQ pulp of whole jute plant was oxygen delignified followed by bleaching with different ECF sequences, namely, ODED, ODEP, and ODEPD and compared CEHH and DEPD bleaching (without oxygen delignification) (Ghosh et al., 2006). The brightness of ODED, ODEP, and ODEPD bleached pulps was above 86% ISO. However, the brightness of CEHH and DEPD sequences bleached pulps reached to 80% with the sacrifice of strength properties and a higher effluent load.

ASAM pulping of the whole jute plant was optimized by varying alkali ratio (Na_2SO_3/NaOH), alkali charge, methanol ratio in the pulping liquors, and cooking time and temperature (Jahan et al., 2005). The pulp yield and kappa number decreased with lower alkali ratio and a high alkali charge and temperature. Addition of methanol (MeOH) in the pulping, however, increased the selectivity of delignification. Optimum conditions of ASAM pulping included alkali ratio at 80%, alkali charge at 22%, 25% MeOH and cooking time to be 1 hour at 170°C, which resulted in higher yield, greater strength properties, and unbleached brightness compared to kraft pulp. Further oxygen bleaching increased brightness to 56% but reduced kappa number by 35%.

The suitability of 16 tossa jute (*Corchorus olitorius*) cultivars and 14 white jute (*C. capsularis*) cultivars as sources of raw materials for pulp and paper production was evaluated in Barrackpore, West Bengal, India. In 1995, 1996, and 1997, the crops were harvested at 100, 120, or 140 days after planting (DAP) (Kumar et al., 2003). Kumar et al. (2003) obtained the highest pulp yields by cold soda pulping, followed by cold and hot soda pulping, for tossa jute cultivars JRO 66 and JRO 3690 (83%–87%), and white jute cultivars JRC 321 and JRC 698 (82%–92%), and the physical properties of the paper sheets were satisfactory. At 120 DAP, JRO 66 produced a dry biomass yield of 16 t/ha, of which 87% was converted into paper pulp by cold soda pulping. Thus JRO 66 is a prospective source of raw material for papermaking.

Alam (2013) showed that a good pulp yield and improved strength properties (except tear) were obtained at the bottom part of jute plant. The jute plant pulp required less refining energy as compared to muli bamboo and JCs.

Alkaline peroxide mechanical pulping (APMP) of whole jute (Xu, 2001) showed higher tensile than aspen APMP pulp and market-bleached chem-thermo mechanical pulp.

7.2.4 Reinforcement of jute pulp

The long fiber length of jute fiber pulp results in paper with high tear strength which is equivalent to softwood pulp. Therefore, many studies were carried out for the reinforcement of jute fiber pulp with short fiber pulp. *Trema orientalis* is a very fast-growing species with short fiber length suitable for pulping. Reinforcement of jute pulp increased papermaking properties of *T. orientalis* pulp (Jahan and Rawshan, 2009). It was also reported that jute pulp blending increased papermaking properties of cotton stalks pulp (Jahan and Farouqui, 2001). The tear index of mulberry plant pulp significantly increased by reinforcing long jute fiber pulp (Rahman and Jahan, 2014).

7.2.5 Nonconventional pulping process

Many researchers investigated different nonconventional pulping processes for jute fiber, jute stick, and whole jute plant. A high-temperature acetic acid (AA) (HAc) pulping process was investigated by Sahin and Young (2008) and obtained a pulp yield of 77% with a kappa number of 50 for 30 minutes of cooking at 150°C, by further increasing the cooking temperature to 200°C, the kappa number decreased to 42 without a sacrifice of pulp yield, but the tensile strength increased by increasing cooking time and temperature. The addition of ethanol to the soda liquor effectively improved both pulpability and the physical properties of jute pulp (Sahin, 2003). In ethanol–alkali process, the selectively fast delignification occurred. The pulp yield was higher in ethanol–alkali process than in the soda process at a given kappa number, and the papermaking properties were evidently improved. The best strength properties were obtained at 175°C and 2.5 hours reaction time.

AA, formic acid (FA), and peroxyformic acid pulping of jute fiber were evaluated at atmospheric conditions (Jahan et al., 2007). The resulting pulps were bleached by alkaline peroxide bleaching. The FA could delignify jute fiber alone, while AA could not delignify without an HCl catalyst. Peroxyformic acid was able to more selectively delignify jute fiber and showed good bleachability in two-stage alkaline peroxide bleaching than AA and FA. The organic acid pulping process dissolved hemicellulose; therefore, the process is suitable for producing dissolving pulp. In addition, dissolved lignin and hemicellulose in the spent acid liquor can be easily separated. The separated lignin had lower molecular weight and higher phenolic hydroxyl groups, which is suitable for phenolic resin (Jahan et al., 2007; Sutradhar et al., 2020). Therefore, the organic acid process is suitable for integrated biorefinery. Pulping of the jute stick was also carried out by HCl catalyzed AA process at atmospheric conditions (Karim et al., 2010). Pulp yield from jute sticks was 58% with permanganate number 21. Pulp was bleached to 76% brightness HEH bleaching sequences. The lignin from the pulping liquor was used for the preparation of lignin phenol formaldehyde resins in jute stick particle boards (Roy et al., 1989). It was observed that lignin could be substituted with 50% phenol in phenol formaldehyde resin of 50%. Bhaduri et al. (1995) isolated hemicellulose from the degumming liquor of ramie fiber and used it in beating operation, significantly reducing the beating energy consumption and consequently improving the strength properties of paper sheets made from the pulp of jute sticks.

Different additives are used in soda/kraft liquor to improve delignification and pulp yield (de Almeida and Gomide, 2013; Hamzeh et al., 2009; Jahan and Farouqui, 2000; Jahan et al., 2000; Kubes et al., 1980). This is a slight modification of the conventional process. Jahan (2001) investigated seven additives to the soda pulping process; ethylenediamine (EDA), monoethanolamine, hexamethylenetetramine, nitrobenzene, anthrone, AQ, and dihydrodihydroxyanthracene (DDA) for making paper from jute fiber and whole jute plants (Jahan, 2001). These pulps were tested for kappa number, yield, pentosans, and α-cellulose. Pulp from the soda–DDA process showed the lowest kappa number and the highest yield for both jute fiber and whole jute plants. The sheet density and papermaking properties of handsheets were tested. Jute pulp from the soda–EDA process exhibited the best tear index; however, the soda–DDA process produced pulp with the highest papermaking properties. It was observed that the delignification rate was enhanced by the addition of 2% of nitrobenzene in soda liquor after that the delignification rate was retarded with the increase of nitrobenzene (NB) percentage in the soda–NB process; on the other hand, hexamethylenetetramine addition in soda liquor increased delignification rate (Jahan et al., 2000). The improved physical properties of both pulps were observed than the soda pulp.

Turgut (2012) studied ethanol (EtOH)-water-based pulping of jute to produce better quality chemical and papermaking properties. To get sufficient delignification in ethanol (EtOH)-water pulping, cooking needs to be continued for longer cooking time at high temperature. However, the strength properties were affected with the extended delignification at increasing temperatures. Ethanol-based pulping requires a higher temperature and consequently needs a high-pressure digester, thus increasing the capital investment.

Sutradhar et al. (2018) investigated potassium hydroxide (KOH) pulping of jute sticks. The pulp yield from jute stick in this process was 48.8% with kappa number 32.5. The brightness of the produced pulp was 74.9% in D_0(EP) D_1 in bleaching sequences. Excellent papermaking properties were observed from jute stick KOH pulp. As the bleachability was not good, it could be used in brown pulp for cardboard. Recent studies showed that KOH pulping fits well in biorefinery, where all biomass components are used in biobased product development (Jahan et al., 2016b). KOH spent liquor can be used for soil amendment (Jahan et al., 2016a).

The above discussion supports the conclusion that organic acid and KOH pulping processes are suitable for small-scale pulp mills based on jute, which also fits well into integrated an biorefinery.

7.3 Dissolving pulp

Recently, cellulose derivatives and regenerated cellulose is the prime interest of scientists due to increasing demand for green and sustainable materials. The cellulose derivatives and regenerated cellulose is produced from dissolving pulp; therefore, dissolving pulp production has increased globally. The dissolving pulp production process also well fits into the integrated biorefinery concept. The dissolved biomass in prehydrolysis step can be used in biochemical and biomaterials, which makes the process cost-effective. Dissolving pulp is characterized with high-purity cellulose (90%–99%), low content of hemicellulose (<4%), and traces of lignin and resin (Sixta, 2006). Many studies were carried out for dissolving pulp production both from jute fiber and jute sticks. Table 7.4 shows dissolving pulp properties from jute fiber and jute stick in different processes. In an investigation, Matin et al. (2015) produced dissolving pulp from jute sticks using the prehydrolysis kraft process. In this investigation, 70% of hemicelluloses were dissolved with marginal loss of α-cellulose at 170°C for 1 hour of prehydrolysis. The prehydrolysed jute sticks were pulped in kraft process, where pulp yield was 36.2% with kappa number 18.5 (16% active alkali for 2 hours of cooking at 170°C). After $D_0(EP)D_1(EP)D_1$ bleaching, the purity of produced pulp was 92% α-cellulose and 89% brightness. Banerjee et al. (1981) reported that the prehydrolysis of jute sticks at 160°–180° for 2 hours removed 60%–80% of hemicellulose and 3%–5% of lignin. Prehydrolysis temperature and time affect subsequent pulp yield in kraft cooking. Bleaching of jute stick dissolving pulp gave product in 24%–35% yield with α-cellulose content 90%–94%, S_{18} 4.5%–10.5%, and viscosity 7–12 cP.

EDA in soda liquor accelerates delignification and also dissolves hemicellulose from the lignocelluloses (Jahan et al., 2000). Jahan et al. (2015) investigated dissolving pulp from jute stick by soda/kraft–EDA processes. EDA addition in soda and kraft liquor decreased pentosans content in the produced pulp. The produced pulp in kraft–EDA process had lower kappa number and pentosans content than that of produced pulp in the soda–EDA process. Prehydrolysis of jute stick in soda/kraft–EDA process further increased purity and decreased kappa number of pulp. The bleaching response of soda–EDA pulp was inferior to the kraft–EDA in ECF bleaching sequences. The final brightness of soda–EDA and kraft–EDA pulp was

TABLE 7.4 Dissolving pulp from jute.

Raw material	Process	Pulp yield (%)	Kappa number	Viscosity (DP)	α-Cellulose (%)	Pentosan/alkali solubility, 18 (%)	References
Jute fiber	Pre-kraft	59.9	10.9	6.3	91.8	8.0	Jahan et al. (2007)
Jute cutting	Pre-kraft	49.9	11.2	4.5	91.0	6.9	
Jute caddis	Pre-kraft	49.9	12.5	4.6	91.3	7.3	
Jute stick	Pre-soda–EDA	30.0	24.8	6.3	90.3	7.0	Jahan et al. (2015)
	Pre-kraft–EDA	29.9	17.2	3.8	94.2	5.7	
Jute fiber	FA/PFA	56.6	–	9.5	96.6	1.8	Jahan et al. (2008)
Jute cutting	FA/PFA	48.7	–	12.8	98.8	1.3	
Jute stick	Pre-kraft	24–35		7–12	90–94	4.5–10.5	Banerjee et al. (1981)
Eucalyptus grandis	Acid sulfite	47.7	8.9	31.1	90.7	6.8	Christov et al. (1998)
Birch	Pre-soda–AQ	34.1		507	94.2	5.8	Testova et al. (2014)

AQ, Anthraquinone; *EDA*, ethylenediamine; *FA*, formic acid; *PFA*, peroxyformic acid.

85.1% and 89.7%, respectively. Dissolving pulp with 93% purity can be produced without prehydrolysis by adding EDA in kraft liquor. The high α-cellulose and low lignin contents in jute fiber make it a good source of dissolving pulp production. The EDA in soda liquor was further investigated for producing dissolving pulp from jute fiber. Pulp produced from jute fiber was 90%–97% α-cellulose (Jahan and Rawshan, 2009). Similar to jute stick, the bleachability of soda–EDA pulp from jute fiber was also inferior to prehydrolysis soda and kraft pulp in ECF bleaching sequences. Dissolving pulp can also be produced from jute fiber based on a prehydrolysis kraft process (Jahan, 2009). The α-cellulose content in the produced pulp could be increased by increasing prehydrolysis temperature or the addition of H_2SO_4 in prehydrolysis liquor, but the pulp viscosity decreased. It is known that organic acid dissolves hemicellulose along with lignin (Jahan et al., 2006, 2007). Therefore, FA pulping of jute fiber, JCs, and jute caddis was done to prepare dissolving pulp (Jahan et al., 2008). The bleached pulp yields in this process were very high (49%–59%) with a high pulp viscosity, α-cellulose content and a good brightness (81%–87%) in totally chlorine-free bleaching. The prehydrolysed kraft pulp from jute was fractionated into short, medium, and long fibers to study the purity and reactivity of dissolving pulp (Nayeem et al., 2017). The shorter fiber fraction showed higher Fock reactivity and lower content of α-cellulose in pulp. The α-cellulose content in the pulp was increased by removing residual hemicellulose by alkaline extraction. The α-cellulose content even increased up to 96% for long fiber fraction pulp.

Jute mill generates JCs and jute caddis (CD) as waste, which was mentioned in the earlier section. Sarkar et al. (2018) showed that FA treatment of these jute mill wastes produced dissolving pulp, at the same time efficiently fractionated dissolved lignin and hemicellulose, which can integrate biorefinery. Alkali extraction of the produced dissolving pulp reduced residual lignin and hemicellulose. The beachability and Fock reactivity was very good in the $D_0(EP)D_1$ bleaching sequence.

Guha et al. (1986) showed that the HNO_3 method is a good method for the preparation of high-grade dissolving pulp. HNO_3 pulping of jute sticks and caddies yielded high α-cellulose pulp, which can be used for making rayon and other cellulose derivatives. The caddies pulp, upon xanthation, yielded clean viscose from which cellulose film was successfully made. Minor defects present in the pulp could be rectified by improved pulping and bleaching techniques.

7.4 Jute supply and store for pulp mill

Jute is completely new and nonconventional raw material for pulping, and it is a seasonal crop. Therefore, the studies on the growth of jute plant, harvesting, collection, transportation, and storage are extensively needed for the development of pulp mills based on jute. For setting up a pulp mill, an effective storage facility and a sustainable supply of jute to the mill site are needed. Akhtaruzzaman (1998) discussed extensively the storage and supply of jute for pulping. Jute plant is harvested in the monsoon season. Therefore, transportation of jute plant to the mill is a big challenge. Another drawback, jute plants and sticks are affected by fungal decomposition; thus fiber is damaged. As the density of jute plant and stick are very low, thus storing of these raw materials are difficult. Wong (1991) reported that by compressing kenaf stems to high-density bales of 500 kg/m^3, the volume of storage for 1-year supply of the stem is about 217,000 m^3 for a 50,000 t/year of chemical pulp mill. Mohiuddin and Rashid (2001) recommended that jute plant bundles of 12-in. diameter can be kept in vertical position with constant air flow for 10 days to remove water and moisture down to 15%–18% and also suggested to spray fungicide if necessary. Similar to kenaf storage in the USA, China, and El Salvador, the bundles of jute plant can be stored in horizontal multiple stories on frames or even crisscrossed directions.

In the case of retted jute fiber, the situation becomes different. The farmers bring jute fiber in bell for selling centers, which is a common practice in jute fabric mills; therefore, the collection and transportation mechanisms are already well accepted. The density of the retted bast fiber is much higher than stems (1500 vs. 185 kg/m^3) (Islam and Khan, 1984; Joseph, 1980), which facilitates easy transportation and storage of retted jute fiber. Joseph (1980) also showed that the dried fiber is protected from microorganisms and insects. Without any problem, jute fiber storage is being well practiced in jute fabric mills for a long time. Considering the supply and storage, the bast fiber is favorable as a pulping raw material.

7.5 Nanocellulose

Chemically, cellulose is a polydispersed linear biopolymer of (1,4)-D-glucose in which a single chain is as wide as 0.05 nm that forms microfibril through regular and irregular hydrogen bondings. Microfibrils are enclosed together with hemicelluloses and lignins by chemical bonding (Rong et al., 2001; Vignon et al., 2004). So, cellulose in the natural fibers has both crystalline and amorphous regions. Nanocrystalline cellulose (NCC), cellulose nanocrystals, nanofiber, and nanowhisker are generally a rod-shaped crystalline cellulose with a diameter in the region of 1–100 nm (Ruiz et al., 2000; Xia et al., 2003). Jute contains a very high amount of α-cellulose, which supports its use in NCC preparation with high yield. Jahan et al. (2011) produced NCC from jute cellulose with high yield having a width of 40 nm. The dissolved lignin and hemicellulose in the cellulose extraction process could be used for biomaterials, biofuel, and biochemical. Acid hydrolysis of jute cellulose liberated MCC/NCC and dissolved amorphous cellulose as sugars, which could be converted to bioethanol. Cao et al. (2012) prepared cellulose nanowhiskers from jute fibers with a diameter of 3-10 nm by TEMPO/NaBr/NaClO system oxidation. The yield and crystallinity was very high ($>80\%$ and 69.72%, respectively).

Steam explosion of jute fibers produced nanocellulose having an average diameter of 50 nm (Thomas et al., 2015). The cross-linked nanocomposite films were prepared using the isolated nanocellulose with natural rubber latex. The tensile strength and Young's modulus of the prepared nanocomposite were increased with the nanocellulose addition.

Baheti et al. (2012) produced nanocellulose from jute fibers using a ball milling process, and they produced nanocellulose with particle size refinement below 500 nm with narrow size distribution. It was applied in bioplastics production for automotive parts, packaging, and agriculture applications. Ball milling using deionized water was produced. But the produced NCC was contaminated from the milling media.

The nanocellulose was also prepared by acid hydrolysis from the cellulose of jute waste fibers (Jabbar, 2017); the prepared nanocellulose suspensions of different concentrations were coated over woven jute reinforcement to prepare composites. The nanocellulose coating increased tensile modulus, flexural strength, flexural modulus, fatigue life, and fracture toughness and decrease in tensile strength of the nanocomposite.

Cao et al. (2012) prepared a very high yield ($>80\%$) nanocellulose from jute fiber with a diameter of 15–40 nm by TEMPO (2,2,6,6-tetramethylpiperidine-1-oxyl radical) oxidation followed by homogenization. In the direct immersion-drying method, nanocellulose from jute fiber (diameter of about 3–10 nm) forms a spider-web-like nano-porous network on the electro-spun nanofibrous membrane (100–300 nm) (Cao et al., 2013). This nano-porous network has potential use in the nano-filtration process.

Ragheb et al. (1991) prepared carboxymethyl cellulose (CMC) from jute waste fibers and found that the degree of substitution (DS) of CMC increased with the increase in the degree of purity α-cellulose. Ria et al. (2022) prepared methylcellulose from the jute pulp in different stages of pulps. The highest methoxyl content was obtained in the bleached pulp after 3000 revolutions of refining and 5 hours reaction time (Ria et al., 2022).

NCC or cellulose derivatives represent other important options for high-value-added products from jute.

7.6 Conclusion

Jute fiber has high density, high cellulose, and low lignin, which make the fiber suitable as high-quality raw material for pulping. On the other hand, jute sticks, a by-product of jute extraction process, have characteristics of low density, moderate lignin, and cellulose contents like other nonwood, also suitable for pulping. NS–AQ and AS–AQ processes are the best choices for jute pulping with respect to high pulp yield, improved delignification, and papermaking properties. Jute fiber pulp showed excellent bleachability.

Pulp produced from jute sticks and whole jute plant also showed good pulp yield and papermaking properties like hardwood. Organic acid efficiently dissolved lignin and hemicellulose from jute fiber and solid residue (cellulose) could be bleached by alkaline peroxide bleaching. Dissolving pulp with high purity and reactivity is also produced from jute fiber, which could be used in NCC or cellulose derivative preparation.

For whole jute plant or jute stick pulping, special measures need to be taken for transportation and storage. As the collection and transportation mechanisms of jute fiber are well established in jute fabric mills, therefore, entire situation is much better for retted jute fiber pulping.

References

Akhtaruzzaman, A.F.M., 1998. Strategies for jute pulping. In: Tappi Pulping Conference, TAPPI press, pp. 661–668.

Alam, M.R., 2013. Pulping of different parts of whole green jute (*C. capsularis*) plant by Kraft process. Bangladesh J. Sci. Ind. Res. 48 (2), 105–108. Available from: https://doi.org/10.3329/bjsir.v48i2.15740.

Ashori, A., Harun, J., Raverty, W.D., Yusoff, M.N.M., 2006. Chemical and morphological characteristics of Malaysian cultivated kenaf (*Hibiscus cannabinus*) fiber. Polym. Technol. Eng 45 (1), 131–134.

Baheti, V., Abbasi, R., Militky, J., 2012. Ball milling of jute fibre wastes to prepare nanocellulose. World J. Eng. 9 (1), 45–50.

Banerjee, S.K., A., Day, Deb, Sarkar, N., 1981. Prehydrolyzed kraft Cook. jute stick. (Eff. prehydrolysis condition). IPPTA 18 (2), 67–73.

Bhaduri, S.K., Ghosh, I.N., Sarkar, N.D., 1995. Ramie hemicellulose as beater additive in paper making from jute-stick kraft pulp. Ind. Crop. Products 4 (2), 79–84. Available from: https://doi.org/10.1016/0926-6690(95)00018-8.

Cao, X., Ding, B., Yu, J., Al-Deyab, S.S., 2012. Cellulose nanowhiskers extracted from TEMPO-oxidized jute fibers. Carbohydr. Polym. 90 (2), 1075–1080. Available from: https://doi.org/10.1016/j.carbpol.2012.06.046.

Cao, X., Wang, X., Ding, B., 2013. Novel spider-web-like nanoporous networks based on jute cellulose nanowhiskers. Carbohydr. Polym. 92 (2), 2041–2047. Available from: https://doi.org/10.1016/j.carbpol.2012.11.085.

Christov, L.P., Akhtar, M., Prior, B.A., 1998. The potential of biosulfite pulping in dissolving pulp production. Enzyme Microb. Technol. 23 (1–2), 70–74.

de Almeida, D.P., Gomide, J.L., 2013. Anthraquinone and surfactant effect on soda pulping. O Pap. 74, 53–56.

Francis, R.C., Hausch, D.L., Granzow, S.G., Makkonen, H.P., Kamdem, D.P., 2001. Fiber yield for fully bleached kraft pulps from black locust (*Robinia pseudoacacia*) and silver maple (*Acer saccharinum*). Holz als Roh-und Werkst. 59 (1–2), 49–52.

Ghosh, U.K., Anand, V., Pokhara, P., 2006. Pulping and Environment Friendly Bleaching of Whole Jute. 한국펄프종이학회기타간행물 359.

Ghosh, I.N., Roy, A.K., Sanyal, A.K., 1983a. Mixed pulping of jute stick and other agricultural residues. IPPTA 20 (1), 65–67.

Ghosh, I.N., Roy, A.K., Sanyal, A.K., 1983b. Mixed pulping of jute stick and jute root cuttings. Indian. Pulp Pap. 37 (5), 17–20.

Guha, A.K., Mathew, M.D., Mukherjee, P.P., 1986. Nitric acid as pulping agent for jute wastes for high alpha pulp. Indian Pulp Pap. 41 (2), 5–8.

Hamzeh, Y., Abyaz, A., Mahsa, O., Niaraki, S.M., Abdulkhani, A., 2009. Application of surfactants as pulping additives in soda pulping of bagasse. BioResources 4 (4), 1267–1275.

Haque, M.M., Aziz, M.I., Hossain, M.S., Quaiyyum, M.A., Alam, M.Z., Jahan, M.S., 2019. Pulping of hybrid acacia planted in a social forestry program in Bangladesh. Cellulose Chem. Technol. 53 (7–8), 739–745. Available from: https://doi.org/10.1016/S0266-3538(01)00046-X.

Islam, M.A., Khan, M.A., 1984. Pilot plant production of writing paper from the kraft pulp of whole jute stalk (jute plant without retting). Bangladesh J. Sci. Ind. Res. 19 (1–4), 119–128.

Islam, M.A., Khan, M.A., Ali, E.M., Khorasani, M., 1989. Writing and printing paper from the blended pulp of jute stick and jute cutting by the soda process. Bangladesh J. Sci. Ind. Res. 24 (1–4), 1–10.

Islam, A., Sarkanen, K.V., 1993. The isolation and characterization of the lignins of jute (*Corchorus capsularis*). Holzforschung 47 (2), 123–132. Available from: https://doi.org/10.1515/hfsg.1993.47.2.123.

Islam, M.A., Khan, M.A., Ali, E.M., Khorasani, S.S.M.A., 1988. Pulping of whole jute stalk (jute plant without retting) by the soda-anthraquinone process. Bangladesh J. Sci. Ind. Res. 23 (1–4), 44–52.

Islam, M.A., Khan, M.A., Abdullah, S., Islam, M., Khalidul, 1993. Studies on the jute stick pulp by the soda-anthraquinone process. Bangladesh J. Sci. Ind. Res. 28 (1), 43–53.

Jabbar, A., 2017. Extraction of nanocellulose from waste jute fibers and characterization of mechanical and dynamic mechanical behavior of nanocellulose-coated jute/green epoxy composites. Sustainable Jute-Based Composite Materials. Springer, Cham, pp. 71–85. Available from: https://doi.org/10.1007/978-3-319-65457-7_5.

Jahan, M.S., 2001. Evaluation of additives in soda pulping of jute. Tappi J. 84 (8), 1–11.

Jahan, M.S., 2009. Studies on the effect of prehydrolysis and amine in cooking liquor on producing dissolving pulp from jute (*Corchorus capsularis*). Wood Sci. Technol. 43 (3–4), 213. Available from: https://doi.org/10.1007/s00226-008-0213-6.

Jahan, M.S., Akter, T., Nayeem, J., Samaddar, P.R., Moniruzzaman, M., 2016a. Potassium hydroxide pulping of Saccharum spontaneum (KASH). J-FOR 6 (1), 46–53.

Jahan, M.S., Haris, F., Rahman, M.M., Samaddar, P.R., Sutradhar, S., 2016b. Potassium hydroxide pulping of rice straw in biorefinery initiatives. Biores. Tech. 219, 445–450.

Jahan, M.S., Al-Maruf, A., Quaiyyum, M.A., 2007. Comparative studies of pulping of jute fiber, jute cutting and jute caddis. Bangladesh J. Sci. Ind. Res. 42 (4), 425–434. Available from: https://doi.org/10.3329/bjsir.v42i4.750.

Jahan, M.S., Chowdhury, D.N., Islam, M.K., 2005. Alkaline sulphite anthraquinone methanol (ASAM) pulping of jute. IPPTA J. 17 (3), 37–43.

Jahan, M.S., Chowdhury, N., Ni, Y., 2010a. Effect of different locations on the morphological, chemical, pulping and papermaking properties of Tremaorientalis (Nalita). Bioresour. Technol. 101 (6), 1892–1898. Available from: https://doi.org/10.1016/j.biortech.2009.10.024.

Jahan, M.S., Ni, Y., He, Z., 2010b. Chlorine dioxide bleaching of soda-anthraquinone jute pulp to a very high brightness. BioResources 5 (2), 870–880.

Jahan, M.S., Faisuland, F.I., Khalidul, M.I., 2000. Pulping of whole jute plant (Corchorus capsularis) by soda-additive methods. IPPTA J. 12 (1), 29–38.

Jahan, M.S., Farouqui, F.I., 2000. Pulping of whole jute plant (Corchorus capsularis) by soda-amine liquor. Holzforschung 54 (6), 625–630. Available from: https://doi.org/10.1515/HF.2000.105.

Jahan, M.S., Farouqui, F.I., 2001. Studies on the blending of cotton stalk plant pulp with jute pulp for paper making. J. Bangladesh Chem. Soc. 14 (1), 1–7.

Jahan, M.S., Gosh, S., Rahman, M.M., Ni, Y., 2013. Non-wood pulping: utilization of Bangladesh jute cutting and caddis. Tappi J. 12 (9), 41–46.

Jahan, M.S., Mun, S.P., Rashid, M., 2004. Fiber dimensions and chemical properties of various nonwood materials and their suitability for paper production. Journal of Korea Technical Association of the Pulp and Paper. Industry 36 (5), 29–35.

Jahan, M.S., Rahman, H., Samaddar, P.R., Rahman, M., 2015. Ethylenediamine in alkaline cooking of jute stick for producing dissolving pulp. Bangladesh J. Sci. Ind. Res. 50 (1), 7–14. Available from: https://doi.org/10.3329/bjsir.v50i1.23804.

Jahan, M.S., Rawshan, S., 2009. Reinforcing potential of Jute pulp with Trema orientalis (Nalita) pulp. BioResources 4 (3), 921–931.

Jahan, M.S., Rawsan, S., Chowdhury, D.N., Al-Maruf, A., 2008. Alternative pulping process for producing dissolving pulp from jute. BioResources 3 (4), 1359–1370.

Jahan, M.S., Saeed, A., He, Z., Ni, Y., 2011. Jute as raw material for the preparation of microcrystalline cellulose. Cellulose 18 (2), 451–459. Available from: https://doi.org/10.1007/s10570-010-9481-z.

Joseph, M.L., 1980. Essent. Text, second ed. Holt, Rinehart and Wilston, New York, p. 44.

Karim, M.S., Seal, H.P., Rouf, M.A., Rahman, M.I., Talukder, M.H.R., Karmaker, P.G., 2010. Acetic acid pulp from jute stick, rice-straw and bagasse. J. Agrofor. Environ. 3 (2), 171–174.

Kishor, H., Rawat, N., Endlay, N., Roy, T.K., 1994. Whole jute—a promising raw material for Indian pulp and paper industry. IPPTA 6, 61.

Krishnan, S.R.A., Sarkar, N.L.D., 1984. Plastic grade cellulose-acetate from jute stick. Res. Ind. 29 (2), 83–85.

Kubes, G.J., Fleming, B.I., MacLeod, J.M., Bolker, H.I., 1980. Alkaline pulping with additives. A review. Wood Sci. Technol. 14 (3), 207–228. Available from: https://doi.org/10.1007/BF00350570.

Kumar, D., Hazra, S.K., Day, A., 2003. Suitability of specific variety of jute (*Corchorus* species) whole plant as raw material for pulp and paper (No. RESEARCH). Indian J. Agric. Sci. 73 (1), 29–31.

Labosky, P., Ifju, G., 2007. A study of loblolly pine growth increments—Part IV. Papermaking properties. Wood Fiber Sci. 13 (2), 120–137.

Matin, M., Rahaman, M.M., Nayeem, J., Sarkar, M., Jahan, M.S., 2015. Dissolving pulp from jute stick. Carbohydr. Polym. 115, 44–48. Available from: https://doi.org/10.1016/j.carbpol.2014.08.090.

Mohiuddin, G., Rashid, M., 2001. Biotechnological application of enzymes for making pulp and paper from green jute/Kenaf (the whole plant): Storing Protocol of Jute Plants. Report prepared for: Int. Jute Study Group. (IJSG) – Bangladesh; Common Fund for Commodities (CFC) – Netherlands; UNIDO – Vienna International Centre, p. 24.

Mohiuddin, G., Rashid, M., Rahman, M., Hasib, A., Razzaque, A., 2005. Biopulping of whole jute plant in soda-anthraquinone (AQ) and kraft processes. Tappi J. 4 (3), 23–27.

Nayeem, J., Sarkar, M., Quadery, A.H., Jahan, M.S., 2017. High purity dissolving pulp from jute. Nordic Pulp Pap. Res. J. 32 (4), 620–626. Available from: https://doi.org/10.3183/npprj-2017-32-04_p620-629_jahan.

Paridah, M.T., Basher, A.B., SaifulAzry, S., Ahmed, Z., 2011. Retting process of some bast plant fibres and its effect on fibre quality: a review. BioResources 6 (4), 5260–5281.

Patt, R., Kordsachia, O., Knoblauch, J., 1987. The ASAM process alkaline sulfite, anthraquinone, methanol pulping. Int. Syrup. Wood Pulping Chem. Paris. I 355, 360.

Ragheb, A., Abd El-Thalouth, I., Elsayed, H., Hebeish, A., 1991. Preparation and characterization of carboxymethyl cellulose from jute wastes. Indian. J. Fib Tex. Res. 16, 263–269.
Rahman, M.M., Jahan, M.S., 2014. Evaluation of mulberry plant as a pulping raw material. Biomass Convers. Biorefinery 4 (1), 53–58. Available from: https://doi.org/10.1007/s13399-013-0095-1.
Rahman, M.M., Siddiqua, S., Akter, F., Jahan, M.S., Quaiyyum, M.A., 2016. Variation of morphological and chemical properties of three varieties of jute stick. Bangladesh J. Sci. Ind. Res. 51 (4), 307–312. Available from: https://doi.org/10.3329/bjsir.v51i4.30451.
Rautiainen, R., Alen, R., 2007. Papermaking properties of the ECF-bleached kraft pulps from first-thinning Scots pine (*Pinus sylvestris* L.). Holzforschung 61 (1), 8–13.
Ria, S.A., Ferdous, T., Yasin Arafat, K.M., Jahan, M.S., 2022. Pulp refining in improving degree of substitution of methylcellulose preparation from jute pulp. Biomass Conv. Biorefinery 12, 2431–2439. Available from: https://doi.org/10.1007/s13399-020-00741-x.
Rong, M.Z., Zhang, M.Q., Lui, Y., Yang, G.C., Zeng, H.M., 2001. The effect of fiber treatment on the mechanical properties of unidirectional sisal-reinforced epoxy composites. Compos. Sci. Technol. 61, 1437–1447.
Roy, S., Lutfar, L.B., 2012. Bast fibres: jute. ". In *Handbook of Natural Fibres*. Woodhead Publishing, pp. 39–59.
Roy, A.K., Sardar, D., Sen, S.K., 1989. Jute stick lignin-based adhesives for particle boards. Biol. Wastes 27 (1), 63–66. Available from: https://doi.org/10.1016/0269-7483(89)90030-X.
Ruiz, M.M., Cavaillé, J.Y., Dufresne, A., Gérard, J.F., Graillat, C., 2000. Processing and characterization of new thermoset nanocomposites based on cellulose whiskers. Composite Interfaces 7 (2), 117–131. Available from: https://doi.org/10.1163/156855400300184271.
Sabharwal, H.S., Akhtar, M., Blanchette, R.A., Young, R.A., 1995. Refiner mechanical and biomechanical pulping of jute. Holzforschung 49 (6), 537–544.
Sahin, H.T., 2003. Base-catalyzed organosolv pulping of jute. J. Chem. Technol. Biotechnol.: Int. Res. Process, Environ. Clean. Technol. 78 (12), 1267–1273. Available from: https://doi.org/10.1002/jctb.931.
Sahin, H.T., 2007. Caustic soda and bio-soda pulping of jute. J. Appl. Biol. Sci. 1 (1), 63–67.
Sahin, H.T., Young, R.A., 2008. Auto-catalyzed acetic acid pulping of jute. Ind. Crop. Products 28 (1), 24–28. Available from: https://doi.org/10.1016/j.indcrop.2007.12.008.
Sanyal, A.K., Roy, A.K., Ghosh, I.N., 1981. Low cost jute stick pulp for mini paper plants. Indian Pulp Pap. 36 (3), 3–7.
Sarkar, M., Nayeem, J., Popy, R.S., Quadery, A.H., Jahan, M.S., 2018. Dissolving pulp from jute wastes. Bioresour. Technol. Rep. 4, 96–100. Available from: https://doi.org/10.1016/j.biteb.2018.09.008.
Shafi, M., Akhtaruzzaman, A.F.M., Mian, A.J., 1993. Pulping of whole length jute by neutral sulphite anthraquinone (NS-AQ) process. Holzforschung. 47, 83–87. Available from: https://doi.org/10.1515/hfsg.1993.47.1.83.
Sixta, H., 2006. Handbook of Pulp, Vol. 1. P. 1352.
Sutradhar, S., Arafat, K.M.Y., Nayeem, J., Jahan, M.S., 2020. Organic acid lignin from rice straw in phenol-formaldehyde resin preparation for plywood. Cellul. Chem. Technol. 54 (5–6), 463–471.
Sutradhar, S., Sarkar, M., Nayeem, J., Jahan, M.S., Tian, C., 2018. Potassium hydroxide pulping of four non-woods. Bangladesh J. Sci. Ind. Res. 53 (1), 1–6. Available from: https://doi.org/10.3329/bjsir.v53i1.35903.
Testova, L., Borrega, M., Tolonen, L.K., Penttilä, P.A., Serimaa, R., Larsson, P.T., et al., 2014. Dissolving-grade birch pulps produced under various prehydrolysis intensities: quality, structure and applications. Cellulose 21 (3), 2007–2021.

Thomas, M.G., Abraham, E., Jyotishkumar, P., Maria, H.J., Pothen, L.A., Thomas, S., 2015. Nanocelluloses from jute fibers and their nanocomposites with natural rubber: preparation and characterization. Int. J. Biol. Macromol. 81, 768–777. Available from: https://doi.org/10.1016/j.ijbiomac.2015.08.053.

Turgut, S.H., 2012. Properties of jute pulps prepared from ethanol-water pulping process. Res. J. Chem. Environ. 16 (1), 40–43.

Udohitinah, J.S., Oluwadare, A.O., 2011. Pulping properties of kraft pulp of Nigerian-grown kenaf (*Hibiscus cannabinus* L.). BioResources 6 (1), 751–761.

Vignon, M.R., Heux, L., Malainine, M.E., Mahrouz, M., 2004. Arabinan-cellulose composite in Opuntiaficus-indica prickly pear spines. Carbohyd Res. 339 (1), 123–131. Available from: https://doi.org/10.1016/j.carres.2003.09.023.

Wong, A., 1991. Nonwood Plant Fibre Progress Report. TAPPI Press. 20, 203–210.

Xia, Y., Yang, P., Sun, Y., Wu, Y., Mayers, B., Gates, B., 2003. One-dimensional nanostructures: synthesis, characterization, and applications. Adv. Mater. 15, 353–389. Available from: https://doi.org/10.1002/adma.200390087.

Xu, E.C., 2001. APMP pulps from nonwood fiber part 2: jute. TAPPI 84 (11), 1–11.

Chapter 8

Pulping and papermaking of kenaf

Latifah Jasmani[1] and Z.M.A Ainun[2]
[1]*Biomass Technology Programme, Forest Products Division, Forest Research Institute Malaysia (FRIM), Kepong Selangor, Selangor, Malaysia,* [2]*Program of Pulp and Paper & Pollution Control, Institute of Tropical Forestry and Forest Products (INTROP), Universiti Putra Malaysia, Serdang, Selangor, Malaysia*

8.1 Introduction

8.1.1 Status of kenaf in Malaysia

Kenaf (*Hibiscus cannabinus* L.), native to sub-Saharan Africa, is from Malvaceae family (Fig. 8.1). It is a fast growing tropical plant that can be harvested between 4 and 5 months. It was first introduced to India followed by Africa, Latin America, and Malaysia. Early on, it was interesting to note that kenaf has mostly been grown widely for cordage fiber in Asia and Central America. Kenaf is an annual herbaceous plant that produces fibers for many uses like rope, automotive dashboard, and pulp and paper. In addition, its leaves can also be used for animal feed. It is an annual plant that grows a straight stem to a height of up to 6 m on a 5- to 7-month growth cycle in suitable temperate and tropical climatic conditions. As a summer crop, the kenaf seed is planted during the spring, and the crop is harvested in early autumn. In tropical areas, kenaf can produce two crops per year. Yields from 6 to 8 metric tons per acre per year (air dry) were demonstrated in the southern United States. Due to this, kenaf has become a popular plant for the production of fibres. Basically, a kenaf stem consists of two types of materials: long, stiff bast fibers and short and sponge-like fibers known as core materials. Both bast and core fibers are significantly different in their anatomical structures as well as in their basic properties. Since kenaf consists of two different materials, the separation of kenaf core and bast is technically possible but not economically attractive.

Kenaf was introduced in Malaysia in 2000 in which the Malaysian Agricultural Research and Development Institute (MARDI) was the leading

Pulping and Papermaking of Nonwood Plant Fibers. DOI: https://doi.org/10.1016/B978-0-323-91625-7.00010-2

FIGURE 8.1 Kenaf flower.

agency for research and development activities. It was only in 2010 the government through National Kenaf and Tobacco Board (NKTB) introduced kenaf as a new crop to public to partially replace tobacco. The National Tobacco Board (NTB) was first established in 1973 to aid and monitor the tobacco production industry. However, as result of trade competition impact on tobacco industry in response to ASEAN Free Trade Area (AFTA) in 2010, government has identified a new source of growth for income generation by the introduction of kenaf as a new crop. The development was also in-line with the governmental policy of reducing tobacco use in compliance with the World Health Organization (WHO) Tobacco Control Framework (FCTC). The NTB was then replaced by NKTB in 2010 to provide better allocation for kenaf and tobacco industry.

8.2 Kenaf for pulp and paper applications

Kenaf has a long history of usage for pulp production in which its popularity spreads across the globe. Liu (2002) reported that the paper made from kenaf that was used in India has higher quality than the paper that was used for making England bank note. Kenaf was brought to United States in the 1940s as an alternative to jute (Fisher, 1994). Twenty years later, the US Department of Agriculture (USDA) recognized kenaf as the most promising crop for papermaking (Fisher, 1994; Nieschlag et al., 1960). Research that was carried out to produce pulp from kenaf showed that the cost was cut to half compared to the pulp produced from wood. It was also observed that kenaf can produce almost five times dry materials for the purpose of pulping per unit in a year. In the 1970s the USDA no longer pursued research on kenaf as a result of energy crisis. Nevertheless, private sectors still conducted the related research (Liu, 2002). There were two known companies in the world that had established pulp and paper mills using kenaf as their raw materials, namely, Phoenix Pulp Paper that was established in Thailand (Leehka and Thapar, 1983; Mital and Maheshwari, 1994) with an annual capacity of 70,000 t and Vision Paper in the United States. Vision Paper (Rymsza, 2001) managed to produce paper made from 100% kenaf using chlorine-free bleaching process that appeared to have a very environment-friendly attractive feature.

Kenaf has been explored in various developed and developing countries as potential raw materials for pulp and paper manufacture as well as wood composites. The countries, such as India, China, and Thailand, were those that produced 90% of global kenaf production. Out of 0.33 million kenaf produced in 2005−2006, the highest producer was India followed by China and Thailand accounting for 42%, 25%, and 11%, respectively (FAO, 2006).

Kenaf was studied for its papermaking potential for both of its bast and core (Maddern and French, 1989). Soda−anthraquinone (AQ) pulping followed by bleaching was carried out for both fibers. The morphology of kenaf bast fiber was described as thin, long, and stiff with an average fiber length of 2.6 mm, whereas core fibers are short with an average fiber length of only 0.6 mm with thin cell walls. The fibers showed distinct behavior in pulping process in which core was found to be difficult to pulp that leads to low pulp quality. Paper produced from bast pulp gave a bulky sheet but with good tear strength, light scattering, and bonding stiffness. On the other hand, core pulp gave denser paper, smooth and good optical properties. The well-bonded sheet produced from core is a result of its thin wall so that they collapse easily to form good fiber network. The pulp also requires low beating and can be used to enhance smoothness of other end products.

Kenaf fiber was also studied for use in linerboard (Schroeter, 1994). Kenaf bast and core fibers underwent soda−AQ pulping and pulp, which were converted to paper for linerboard evaluation. Smoothness was improved by the addition of kenaf core pulp. The same observation was also noted on both fibers in terms of its refining behavior in which core fiber needs no beating, whereas kenaf fiber needs little beating. Core pulp can be used as a top liner for board as it enhances the smoothness and could reduce energy cost.

8.2.1 Pulping and papermaking

Pulping and papermaking is the required step to form a paper. Kenaf has been identified as the most promising crop for pulp production in the early 1960s by the USDA. Australia has also followed the suit by initiating research on kenaf from agronomy and pulping perspectives (Wood, 1981). Other countries, such as Malaysia, Thailand, Iran, and China, have also shown a strong interest on kenaf for pulp and paper applications. The increase in interest is mostly to complement the continuous supply of timbers (Ang et al., 2010; Dutt et al., 2010; Mossello et al., 2010a,b,c). Malaysia government allocated a substantial amount for research and development in kenaf (Edeerozey et al., 2007). Various developments are carried out by numerous authors (Resalati, 2009; Shakhes et al., 2011; Zeinaly et al., 2013) to study kenaf from the perspectives of chemical and mechanical pulping on whole kenaf as well as its respective bast and core fibers.

8.2.1.1 Pulping processes

Pulping is a process of liberating the fiber from its structure by subjecting wood or cellulosic fibers to cooking process using chemicals to convert wood to pulp for papermaking purpose. There are different types of pulping processes used industrially to produce cellulosic fibers. The pulping that is most commonly used includes kraft, soda, and sulfite. Pulping is also known as a delignification process in which lignin as an undesirable component in the pulp is removed from the wood structure during the chemical pulping process. Chemical pulping usually gives a yield of about 50%. The pulp produced from this type of pulping is also known as chemical pulps that can find application in packaging industry. The bleaching process is applied after pulping for its use in printing and writing applications. The paper produced from chemical pulping is stronger than mechanical pulps. On the other hand, during mechanical pulping, mechanical force is applied on the wood chip that results in the release of fibers. As much as this process is straight forward and relatively simple to perform, it does come with some drawbacks. Due to the mechanical treatment, it leads to the production of short and imperfect fibers with lots of fines. To improve the quality of mechanical fibers, process optimization is usually employed by using some chemicals for pretreatment and increasing the process temperatures. The yield of mechanical pulp is between 85% and 95%. This high yield pulp has still contained a substantial quantity of lignin due to the absence or little amount of chemical amount used for lignin removal. Mechanical pulp is commonly used for newsprint. The yellowing effect on newsprint after few months takes place due to the presence of lignin in the pulp that reacts with light. Paper manufacturers usually apply little coating to selected paper products to avoid yellowing. The use of fiber in papermaking depends on the desired end product and its specifications as many types of paper can be produced. Selection and mixing of pulp from many choices of various raw materials are conducted by paper manufacturers to meet specifications in the most economical way. Kenaf can always be the choice of pulping feedstock if it meets the specification of product and is commercially attractive.

8.2.1.1.1 Kenaf chemical pulping

As mentioned previously, kenaf has properties comparable to many hardwoods. Thus, kenaf can become an alternative for wood for conversion to various paper products. Kenaf had undergone pilot papermaking trial as early as in 1970 in which it was added at 40% in a mixture furnish of hardwood and softwood pulp (Clark, 1970). Kenaf pulp produced via kraft pulping was reported to give yield in the range of 45%–48% (Clark and Wolff, 1965). Yu et al. (2011) studied kenaf whole stem using continuous kraft pulping and found that it gave good yield of pulp with good properties comparable to hardwood pulp. It was also found to be easily bleached. Mossello et al. (2010a,b,c) prepared kenaf whole stem pulp from soda–AQ and observed that its properties was in between of those pulps prepared from bast and core fibers. Paper made from whole stem was also found by the author to show better properties than that of core and bast. In addition, the production cost is considered to be more economic if the whole stem is used rather than separating the core and bast (Mossello et al., 2010a,b,c; Ibrahim et al., 2011). The properties of paper from whole stem kenaf prepared via soda and kraft pulping are shown in Table 8.1. Core pulps prepared via soda and kraft pulping generally exhibited low paper strength, which can only be improved upon beating. On the other hand, bast pulp had good tear and tensile strength. Han et al. (1999) reported that paper made of core fibre was thinner and smoother than paper made of bast fibre (Table 8.2). Gartside (1981) observed that kenaf bast pulp produced from neutral sulfite semichemical (NSSC) had good bonding strength as well as high tear strength comparable with pulp prepared chemically. This characteristic can lead kenaf bast to be used for linerboard. Core pulp, on the hand, displayed weak tear strength but has good bonding strength. This type of pulp could only have potential for sole use in a corrugating board to be used without blending with other stronger pulps. Law et al. (2003) suggested separate processing for kenaf as they observed a distinct performance of the core and bast fibers when chemithermomechanical pulping (CTMP) processes were employed as well as their beating behavior using PFI mill. Bleaching was also carried out on kenaf soda-AQ kenaf with a biobleaching technique using white rot fungi

TABLE 8.1 Paper properties of kenaf whole stem using soda and kraft pulping processes (Touzinsky et al., 1980).

	Soda	Kraft
Yield (%)	62	55
Alpha-cellulose (%)	68	71
Pentosans (%)	19	20
Kappa number	45	27
Yield bleached (%)	53	48
Burst factor (g/cm^2/gsm)	54	50
Tear factor (g/gsm)	102	93
Breaking length (m)	9600	10,300

TABLE 8.2 Paper properties from bast and core of kenaf via kraft pulping (Han et al., 1999).

	Bast	Core
Density (kg/m^3)	571	906
Freeness, CSF (mL)	631	279
Caliper (mm)	0.121	0.075
Strain (elongation) (%)	2.29	2.38
Tensile strength (kN/g)	5.36	7.22
ISO brightness (%)	25.3	20.5
Printing opacity (%)	97.0	94.4
Burst strength (kPa)	333.8	381.8
Burst index (kPam^2/g)	4.92	5.71
Tear resistance (mN)	1446.3	263
Tear index [mN (m^2/g)]	20.9	3.9
Smoothness (mL/min)	329.3	72.6
Fiber length (mm)	2.8	0.81

(*Phanerochaete chrysosporium*). Nezamoleslami et al. (1998) conducted biobleaching with a combination of extraction and peroxide stage to produce pulps with high yield, higher opacity, and brightness as well as strength than that of traditional bleaching using chlorine.

8.2.1.1.2 Kenaf mechanical pulping

Mechanical pulping gives high yield pulp but requires high-energy consumption. Because of this situation, mechanical pulp mills are commonly integrated in a paper mill. The availability of mechanical pulp in the market is thus in scarcity due to this pulp and paper mill integration. The excess steam produced during refining process is then used for paper mill for cost saving. It was reported that mechanical pulp only contributed to 8% of global pulp production in 2011 worth of USD 39 million tonnes (FAO, 2011). Another factor is also due to the reason that a high production scale of mechanical pulp process is not as profit-attractive as that of chemical pulp. The refiners used in the process are also limited in size and pulp mill thus can scale down the production size in comparison to chemical process (Table 8.2).

Kenaf bast pulp prepared from refiner mechanical pulping (RMP) process gives acceptable pulp quality that is suitable for newsprint applications.

Pretreatment with sulfite slightly improved the pulp strength. Mohta et al. (2004) reported that whole kenaf stem pulp from RMP resulted in an increase of strength properties of newsprint. Kenaf pulps processed via thermomechanical pulping (TMP) lead to weak pulp properties. Nevertheless, adding peroxide to the TMP pulp as a bleaching agent enhanced the strength properties to some extent (Touzinsky et al., 1980) and even increased the brightness for newsprint requirement (Cunningham et al., 1979).

Alkaline sulfite CTMP process conducted on bast and core of kenaf showed that bast has stronger pulp than core. Puri and Higgins (1981) conducted the same process on kenaf whole stem and observed that the resulting pulp showed intermediate quality between core and bast especially on strength and drainage properties. Alkaline peroxide mechanical process (APMP) was also conducted on kenaf. APMP is a mechanical pulp. It was reported that the use of such chemicals, with an addition of sodium hydroxide and hydrogen peroxide, can save the energy consumption to half in comparison to TMP pulping (Myers and Bagby, 1995; Lawford and Tombler, 1982). The strength of the paper produced from this type of pulping has properties between printing and writing papers and newsprint. The brightness, however, is more suitable for newsprint as it is quite low. Xu (2001) observed that whole stem kenaf form APMP process showed better tensile strength at the same density than that of aspen pulp.

Pulping with fungal pretreatment or also known as biomechanical pulping resulted in the reduction of energy requirement up to 38% and the improvement of paper strength in comparison to RMP pulp (Ahmed et al., 1999). However, this technique also resulted in brightness reduction and thus led to an increase of bleaching agent requirement during bleaching (Table 8.3).

TABLE 8.3 Paper properties using thermomechanical pulping (TMP) and chemithermomechanical pulping (CTMP) processes using kenaf whole stem (Touzinsky et al., 1980).

	TMP	CTMP
Brightness (%)	65	70
Burst index (MN/kg)	1.4	1.7
Tear index (N m^2/kg)	8.1	8.0
Breaking length (km)	3.5	4.3
Apparent density (kg/m^3)	318	388
Long fiber (%)	36	33
Fines (%)	49	53
Opacity (%)	95	90

8.2.2 Kenaf paper commercialization

A commercialization trial of kenaf was first started by the Americans. Kenaf was attempted for use as a raw material for pulp and paper during a commercial scale trial in 1978 using CTMP processes at Ohio. During the pulping processes, the challenges emerged when the refiner parts needed replacement after long production time. This was due to the presence of metal, sands, and stones from the uncleaned raw material (Hodgson et al., 1981). This is one of the most important factors to take into consideration when kenaf is selected as a raw material as it tends to bring along those unwanted materials in the early process of harvesting and handling. The resulting kenaf pulp was converted into paper successfully at paper mill. The paper was made from a mixture of kenaf and other kraft pulp (8%–25%). The physical strength and optical properties of the resulting paper was compared with the control sample. It was found that the kenaf paper showed higher strength and brightness, similar smoothness but lower opacity and porosity than that of control sample. Due to lower opacity, print-through was a challenge for this kenaf paper. However, the printing for newsprint was conducted with much success. Overall, the paper machine and printing trials for kenaf were successfully conducted regardless some issues during pulping. In this case, kenaf needs to be sufficiently clean and free from contaminants prior to pulping process. Adequate washing needs to be carried out to lessen the metal disturbance during bleaching (Lawford and Tombler, 1982).

Kenaf is bulky therefore there was a tendency to reduce the harvested kenaf into chip-size before it was transported. Although it could reduce the transport cost, this practice creates more issues of unwanted contaminants removal before pulping. Furthermore, ash content in kenaf was reported to be higher, 2%–6% (D'Agostino et al., 1996; Neto et al., 1996), than that of wood. A second and third commercial scale trial took place using kenaf for newsprint production. The second trial was run in a continuous processing system started from handling, pulping until papermaking. The trial showed that it was feasible to produce newsprint from kenaf (Kugler, 1981; Taylor, 1993). The third trial was used kenaf whole stem (82%) with some addition of balsam and spruce kraft pulp (18%). CTMP pulping was employed using Sprout Bauer followed by papermaking and newsprint production. The paper showed good newsprint quality in terms of strength, brightness, and opacity if not better than the normal newsprint at that time (Kugler, 1981). Through this trial, economic and technical feasibility was evaluated. The issue was still mostly on the bulkiness of kenaf that pose a big challenge to transport the raw material to the mill. Furthermore, kenaf needs big space for long-time storage, which is taken into consideration as well by the pulp and paper mill (Young, 1987). This new product faces big hurdles in pulp and paper industry as the existing player has already supplied good pulp quality with good price and readily available in the market. New product needs to be more competitive than what already is in the market demand. There were also some adventures began in 1990 in which core and bast fibers were used and sold for various products that include high-quality printing and writing paper as well as adsorbents (Kugler, 1996). Vision Paper started producing printing and writing paper from kenaf bast fibers in 1992. The company produced paper using kenaf completely. Besides paper from 100% kenaf, they also produce paper made from kenaf and recycled pulp mixture. The price, however, was more expensive than the readily available printing and writing paper due to small capacity and batch process run by the mill. The use of kenaf pulp has increased the price by three times than the price using wood pulp. The difference was due to small-scale pulping capacity by the company (Rymsza, 1998a,b). Even though Vision Paper produced tree-free paper coupled with environment-friendly bleaching that could be environmentally attractive, but not all people were willing to pay more. Some people are still concerned about the price if there is no significant value added on the product. The price could be reduced if improvement on pulping technology and yield is looked into.

Furthermore, to lower the price, the volume must be higher. It was reported that kenaf was only used for 0.05% of total global production that accounts for about USD 15,000 t (Sato, 2001). The new embarkment on kenaf had also an effect on the price difference as the conventional market in pulp and paper industry is not ready to embrace new products. Thailand embarked on the use of kenaf in 1981 in which Phoenix paper mill used whole stem kenaf for the production of chemical pulp for pulp and paper. The kraft pulping was employed with 200 t/day capacity. The problem arose when they could not get a continuous supply of quality kenaf at a reasonable price. The problem has in turn made the price go up tremendously. The company had to find other alternative materials later in which they used planted eucalyptus and bamboos. These materials were converted to chemical pulps. Even though kenaf was still being used, but it only accounts for 5% of total mill production (Monti, 2013). It was also reported that using whole stem kenaf can reduce the cost rather than separated them into bast and cores (Paul, 1991).

8.2.2.1 Malaysian experience: kenaf for pulp and paper

Malaysian government initiated a project on kenaf with the objective to use it as raw materials for pulp and paper application in 2006. The project involved a comprehensive study encompassing chipping, pulping, bleaching, and papermaking at both laboratory (Latifah et al., 2010) and pilot scale trials. In this project, whole stem kenaf was used for all the activities involved. From these activities, market and financial analysis looking into the potential commercialization of kenaf for pulp and paper manufacture was also conducted. The properties of kenaf paper produced at pilot scale were found to be comparable with, if not better than those commercial printing and writing grades. The attractiveness of kenaf from its technical perspectives, however, did not live up to its name when it comes to its financial analysis assessment. It was found that in this project the price of kenaf as feedstock was too high for it to be financially feasible if it is to be used 100% in pulp and paper manufacture. A combination with other lower cost raw materials, such as wood or abundantly available oil-palm empty fruit bunches, was recommended to provide a breakeven pulp and paper business venture. Other alternative is to use kenaf as a reinforcement material for recycled pulps. The findings from this project also recommended that any future endeavor on kenaf related to pulp and paper should also address agronomics and production economics of kenaf cultivation (Mahmudin et al., 2012).

8.2.3 Specialty paper and other value-added products

Specialty papers are commonly produced for niche applications and usually in small amounts. Specialty paper is available in the market as filter paper, security paper, and cigarette paper. Cotton, abaca, and flex are some of the examples of materials used for the production of specialty paper. Sholton (1981) reported that kenaf can have potentials for use in specialty cigarette and filter papers. However, kenaf has not been known for any commercial use in specialty paper to date.

8.2.3.1 Cellulose dissolving pulp and nanowhiskers

Dissolving pulp prepared from kenaf bast was reported by Collier et al. (2000) using *N*-methylmorpholine-*N*-oxide (NMMO) and evaluated against other pulps. They observed that the kenaf viscosity was lower than other pulps, which is mostly due to the lignin and hemicellulose.

Ionic liquid was also used with a combination of sonication for enzymatic saccharification (Ninomiya et al., 2012). Pure cellulose of kenaf was also isolated via ultrasound irradiation giving 25% yield (Pappas et al., 2002). The crystallinity of the cellulose was only slightly lower than that using conventional way.

Kenaf bast was also converted to cellulose nanocrystals via sulfuric acid hydrolysis with an acid concentration of 65 wt.% at 45°C for 40 minutes. The kenaf nanocrystals were found to have a crystallinity of 80% and having 12 nm width and 150 nm length in dimension (Kargarzadeh et al., 2012). Jonoobi et al. (2009) prepared cellulose nanofiber from kenaf. The nanofiber had 81.4% crystallinity as opposed to raw kenaf of having only 48.2%. The diameter of the kenaf nanofiber was found to be in the range of 10–90 nm. Shi et al. (2011) also isolated cellulose nanocrystals from kenaf bast having a diameter between 7 and 84 nm and a length between 100 and 1400 nm.

8.3 Conclusions

Kenaf can be converted to pulp and paper products successfully via various pulping methods from individual bast, core, or whole stem fibers. Nevertheless, papers made from the corresponding fibers exhibit distinct properties that could lead to the creation of different market demands for kenaf. If kenaf to be used for pulp and paper industry, economical evaluation should be given utmost consideration as it is bulky and requires size reduction prior to transportation to mill. In addition, the price and quality of kenaf and its related end product must be competitive enough to compete with the existing and readily available of pulp and paper products from wood.

References

Ahmed, A., Akhtar, M., Myers, G.C., Scott, G.M., 1999. Biomechanical pulping of kenaf. In: Proceedings of the 1999 TAPPI Pulping Conference, Oct. 31–Nov. 4, 1999, Orlando, FL. Volume 1. Atlanta, GA: TAPPI Press, pp. 285–290.

Ang, L.S., Leh, C.P., Lee, C.C., 2010. Effects of alkaline pre-impregnation and pulping on Malaysia cultivated kenaf (*Hibiscus cannabinus*). BioResources 5 (3), 1446–1462.

Clark, T.F., Wolff, I.A., 1965. A search for new fiber crops, VIII. Sulfate pulping of kenaf (*Hibiscus cannabinus*). TAPPI J. 48 (6), 381–384.

Clark, T.F., 1970. First bond paper made using kenaf pulp in the furnish. Pap. Trade J. 158 (46), 28.

Collier, B.J., Dever, M., Petrovan, S., Collier, J.R., Li, Z., Wei, X., 2000. Rheology of lyocell solutions from different cellulose sources. J. Polym. Environ. 8 (3), 151–154.

Cunningham, R.L., Touzinsky, G.F., Bagby, M.O., 1979. Brightening of kenaf thermomechanical pulp. TAPPI J. 62 (4), 69–70.

D'Agostino, D., Richard, J.A., Taylor, J.D., 1996. Continuous steam explosion pulping: a viable alternative for pulping of non-woody fibers. In: TAPPI Pulping Conference, 377–386.

Dutt, D., Upadhyaya, J.S., Tyagi, C.H., 2010. Studies on *Hibiscus cannabinus*, *Hibiscus sabdariffa*, and *Cannabinus sativa* pulp to be a substitute for softwood pulp-part 1: AS-AQ delignification process. BioResources 5 (4), 2123–2136.

Edeerozey, A.M.M., Hazizan, M.A., Azhar, A.B., Zainal Ariffin, M.I., 2007. Chemical modification of kenaf fibers. Mater. Lett. 61 (10), 2023–2025.

FAO, 2006. Jute, Kenaf, Sisal, Aabaca, Coir and Allied Fibers Statistics. Available from: http://www.fao.org/es/esc/en/20953/21005/highlight_51023en.html.

FAO, 2011. FAO Paper and Paperboard Capacity Survey 2010–2015. Available from: http://www.fao.org/docrep/014/i2285t/i2285t00.pdf.

Fisher, G.L., 1994. Availability of kenaf fibers for the US paper industry. In: TAPPI Pulping Conference: 91. TAPPI Press.

Gartside, G., 1981. Kraft, Soda and NSSC Pulping of Kenaf. In: Kenaf as a Potential Source of Pulp in Australia: Proceedings of the Kenaf Conference, pp. 144.

Han, J.S., Miyashita, E.S., Spielvogel, S.J., 1999. Properties of kenaf from various cultivars, growth and pulping conditions. In: Kenaf Properties, Processing and Products. Mississippi State, MS: Mississippi State University, Ag & Bio Engineering, pp. 267–283.

Hodgson, P.W., Lawford, W.H., Perrault, J., Thompson, C.A., 1981. Commercial paper machine trial of CTMP kenaf in newsprint. TAPPI J. 64 (9), 161–164.

Ibrahim, M., Wan Daud, W.R., Law, K.N., 2011. Comparative properties of soda pulps from stalk, bast and core of Malaysian grown kenaf. BioResources 6 (4), 5074–5085.

Jonoobi, M., Harun, J., Mishra, M., Oksman, K., 2009. Chemical composition, crystallinity and thermal degradation of bleached and unbleached Kenaf bast (*Hibiscus cannabinus*) pulp and nanofiber. BioResources 4 (2), 626–639.

Kargarzadeh, H., Ahmad, I., Abdullah, I., Dufresne, A., Zainudin, S.Y., Sheltami, R.M., 2012. Effects of hydrolysis conditions on the morphology, crystallinity, and thermal stability of cellulose nanocrystals extracted from Kenaf bast fibers. Cellulose 19 (3), 855–866.

Kugler, D.E., 1996. Kenaf commercialization 1986–1995. In: Janick, J. (Ed.), Progress in New Crops. ASHS Press, Alexandria, pp. 129–132.

Kugler, D.E., 1981. Kenaf Newsprint: Realizing Commercialization of a New Crop After Four Decades of Research and Development. A Report on the Kenaf Demonstration Project, United States. Cooperative State Research Service. Special Projects and Program Systems, p. 13.

Latifah, J., Mahmudin, S., Rushdan, I., Mohd Nor, M.Y., 2010. Kenaf Paper: From laboratory to pilot plant production. In: Wan Asma, I., Mahanim, S.M.A, Rafidah, J., Zulkafli, H., Jailani, S. (Eds.), Proceedings of the Seminar on Biomass for Biofuels and Value Added Products. 27–28 October 2009, Kuala Lumpur, pp. 98–104.

Law, K.N., Robert, L., Nde, D.M.S., 2003. Characterization of kenaf by chemimechanical pulping. TAPPI J. 2 (6), 14–18.

Lawford, W.H., Tombler, G., 1982. Kenaf pulp for newsprint. Pulp Paper Can 83 (12), 99–103.

Leehka, V.P., Thapar, S.K., 1983. Experience in kenaf pulping in Thailand, non-wood plant fiber pulping. Progress Report No. 14. TAPPI Press, Atlanta, GA.

Liu, A., 2002. World Production and Utilization of Jute, Kenaf, and Allied Fibers. Available from: http://www.chinaconsulting.com/kenafworkshop.htm (accessed 8 March 2009).

Maddern, K.N., French, J., 1989. Papermaking properties of some Australian non-wood fibers. APPITA J. 42 (6), 433–437.

Mahmudin, S., Ainun, Z.M.A., Latifah, J., Mohd Shahwahid, O., Paridah, M.T., Jalaluddin, H., et al., 2012. Kenaf: A Potential Fibre for Pulp and Paper Manufacture. MTIB.

Mital, S.K., Maheshwari, S., 1994. A few technical aspects of pulp production from kenaf. Proc. Pulping Conference, TAPPI, San Diego, 105.

Mohta, D.C., Roy, D.W., Whiting, P., 2004. Refiner mechanical pulp from kenaf for newsprint manufacture. TAPPI J. 3 (4), 9–14.

Monti, A., 2013. Kenaf: A Multi-Purpose Crop for Several Industrial Applications. Springer-Verlag Berlin Heidelberg.

Mossello, A.A., Harun, J., Md Tahir, P., Resalati, H., Ibrahim, R., Shamsi, S.R.F., et al., 2010a. A review of literatures related of using kenaf for pulp production (beating, fractionation, and recycled fiber). Mod. Appl. Sci. 4 (9), 21.

Mossello, A.A., Harun, J., Ibrahim, R., Resalati, H., Shamsi, S.R.F., Md Tahir, P., et al., 2010b. Evaluation of linerboard properties from Malaysian cultivated kenaf soda-anthraquinone pulps versus commercial pulps. BioResources 5 (3), 1595–1604.

Mossello, A.A., Harun, J., Resalati, H., Ibrahim, R., Shamsi, S.R.F., Md Tahir, P., 2010c. New approach to use of kenaf for paper and paperboard production. BioResources 5 (4), 2112–2122.

Myers, G.C., Bagby, M.O., 1995. Feasibility of using kenaf chemithermomechanical pulp in printing and writing paper. TAPPI J. 78, 5.

Neto, C.P., Seca, A., Fradinho, D., Coimbra, M.A., Domingues, F., Evtuguin, D., et al., 1996. Chemical composition and structural features of the macromolecular components of *Hibiscus cannabinus* grown in Portugal. Ind. Crop. Products 5 (3), 189–196.

Nezamoleslami, A., Suzuki, K., Nishida, T., Ueno, T., 1998. Biobleaching of kenaf bast fiber, soda-AQ pulp using white-rot fungus. TAPPI J. 81 (6), 179–183.

Nieschlag, H.J., Nelson, G.H., Wolff, I.A., Perdue, R.E., 1960. A search for new fiber crops. TAPPI J. 43, 193–201.

Ninomiya, K., Kamide, K., Takahashi, K., Shimizu, N., 2012. Enhanced enzymatic saccharification of kenaf powder after ultrasonic pretreatment in ionic liquids at room temperature. Bioresour. Technol. 103 (1), 259–265.

Pappas, C., Tarantilis, P.A., Daliani, I., Mavromoustakos, T., Polissiou, M., 2002. Comparison of classical and ultrasound-assisted isolation procedures of cellulose from Kenaf (*Hibiscus cannabinus* L.) and Eucalyptus (*Eucalyptus rodustrus* Sm.). Ultrason. Sonochem. 9 (1), 19–23.

Paul, P.K., 1991. Phoenix and Its Views about Kenaf. Phoenix Pulp & Paper Co., Ltd., Bangkok.

Puri, V.P., Higgins, H.G., 1981. Chemimechanical and chemithermomechanical pulping of kenaf. In: Kenaf as a Potential Source of Pulp in Australia, Brisbane.

Resalati, H., 2009. Kenaf CMP pulp; an alternative potential to supplement wood shortages for making newsprint. In: Proceedings TAPPI Engineering, Pulping and Environmental Conference, vol. 3, pp. 1462–1488.

Rymsza, T.A., 2001. Commercial paper making with kenaf. In: Paper Presented at the PIRA Nonwoods Conference, Amsterdam, NL.

Rymsza, T.A., 1998a. Creating High Value markets for kenaf paper: A commercial experience. Available from: http://www.visionpaper.com/PDF_speeches_papers/982aks.pdf.

Rymsza, T.A., 1998b. Utilizing kenaf for high value paper applications in the USA: A commercial experience. Available from: http://www.visionpaper.com/PDF_speeches_papers/hppsjapan.pdf.

Sato, N., 2001. Kenaf popular as nonwood alternative for paper. Kyodo News Service Japan Economic Newswire, Japan.

Schroeter, M.C., 1994. Use of kenaf for linerboard quality enhancement. In: Proceeding, Paper Presented at the Pulping Conference, San Diego.

Shakhes, J., Zeinaly, F., Marandi, M.A.B., Saghafi, T., 2011. The effects of processing variables on the soda and soda-AQ pulping of kenaf bast fiber. BioResources 6 (4), 4626–4639.

Shi, J., Shi, S.Q., Barnes, H.M., Pittman, J.C.U., 2011. A chemical process for preparing cellulosic fibers hierarchically from kenaf bast fibers. BioResources 6 (1), 879–890.

Sholton, E.J., 1981. Specialty Paper from Kenaf. Kenaf as a Potential Source of Pulp in Australia, Brisbane.

Taylor, C.S., 1993. Kenaf: An Emerging New Crop Industry. Wiley, New crops, New York, pp. 402–407.

Touzinsky, G.F., Cunningham, R.L., Bagby, M.O., 1980. Papermaking properties of kenaf thermomechanical pulp. TAPPI J. 63 (1), 53.

Wood, I.M., 1981. Kenaf for Paper Pulp: Research and Development around the World. Kenaf as a Potential Source of Pulp in Australia, Brisbane.

Xu, E.C., 2001. APMP pulps from non-wood fibres Part 1: Kenaf and straw. APPITA J. 54 (5), 444–448.

Young, J., 1987. Kenaf newsprint is a proven commodity. TAPPI J. 11, 81–83.

Yu, D.M., Chen, K.F., Zhao, C.F., Xu, J., 2011. The study of Kraft continuous cooking of kenaf. Adv. Mater. Res. 236, 1212–1215.

Zeinaly, F., Jalal, S., Zeinali, N., 2013. Multi stage peroxide and activated peroxide bleaching of kenaf bast pulp. Carbohydr. Polym. 92 (2), 976–981.

Chapter 9

Pulping and papermaking from roselle (*Hibiscus sabdariffa* L.)

Binita Baishya Kalita and Sunita Boruah
Department of Textiles and Apparel Designing, College of Community Science, Assam Agricultural University, Jorhat, Assam, India

9.1 Introduction

Roselle (*Hibiscus sabdariffa* L.) is an annual shrub cultivated in subtropical as well as tropical regions, which traces its lineage to the Malvaceae family. It is brought to India, other parts of Asia, Central America, and the United States of America from West Africa in the midst of the slave trade. Roselle is cultivated in well-drained soils with a pH of 5.5−6.0 at a temperature of 25°C−35°C or even higher. It can grow annually and attain a height of about 3−3.5 m with a diameter of 1.0−2.0 cm within 130−150 days after sowing.

In the past few decades, a considerable rise in the demand for the use of roselle in the industrial application has been observed. Roselle is currently being explored in developed and developing countries as a viable raw material for pulp, paper, and wood composite manufacturing. The bark and core fiber of the roselle stem have been classified as long and short fiber, which is a source of pulp for the paper industry. Fiber length is an important factor to consider in pulping. Roselle fiber has been identified as a composite material since it constitutes mainly of cellulose, hemicelluloses, lignin, and other components. Cellulose is the organic white fibrous substance present in most plants that is insoluble in water and neutral organic solvents but has a strong affinity toward water making it extremely suitable for pulp and paper manufacturing. The lignin and cellulose content of the raw plant materials determine the strength of the paper. The mechanical strength and the tensile strength of the pulp are directly proportional to the cellulose content of the plant. On the contrary, the presence of lignin can reduce the strength of the paper as it is an undesirable polymer, and its removal during the pulping process demands the utilization of a high amount of energy and chemicals.

Roselle has lower lignin content than other nonwoody raw materials, which reduces the pulping time and chemical change compared to those of other nonwoody raw materials. The removal of hemicellulose results in

Pulping and Papermaking of Nonwood Plant Fibers. DOI: https://doi.org/10.1016/B978-0-323-91625-7.00012-6

lowered strength properties since hemicelluloses aid in the internal cohesion of the cell wall and their removal will in turn weaken the fiber by reducing interfiber adhesion. The chemical composition and physical properties of fiber often influence its performance of the end product. Cellulose in the form of pulp is separated from plant materials chemically or mechanically in a process known as pulping. The pulp is the major raw material for papermaking and the production of other paper products.

The three major processing steps in pulp and paper production include pulping, bleaching, and paper production. The characteristic of the feedstock and the desired qualities of the end product determines the type of pulping and the amount of bleaching used. The pulping techniques are categorized as mechanical, thermal, semichemical, or fully chemical methods. Some common hybrid methods that include different chemical and biopulping methods have been used in pulping both wood and nonwood plant materials including thermomechanical pulping (TMP) and chemical TMP. There are two primary processes involved in the papermaking: dry-end and wet-end operations. The wet-end operations consist of wet paper sheets formed by cleaned and bleached pulp. Those wet sheets are then dried, and various surface treatments are applied to the paper, hence constituting the dry-end operations. Roselle bark is a viable substitute for paper manufacturers in pursuit of superlative quality, an economic source of nonwood fiber that has the potential to become a widely accepted resource in the pulp and paper industry (Fig. 9.1).

9.2 Roselle (*H. sabdariffa*) for pulping and papermaking

The pulp is the raw material that is obtained from the woody or nonwoody plants through the pulping process. It is a semifinished lingo cellulosic product that has been degraded physically and/or chemically such that the

(A) (B)

FIGURE 9.1 (A) Roselle plant and (B) roselle stem.

separated fibers can be spread in water and converted into a web. Plant fibers can be classified into two major botanical classifications—core and bast fiber. The lignin content of roselle is higher in the core than in the bast fiber and has proven to be one of the most important factors affecting the pulping of nonwood fibers. In the present times, information or literature is scarce on the potentials of roselle for pulp and handsheets.

9.2.1 Physical characteristics of the raw materials

The plant roselle (*H. sabdariffa* L.) is a tropical annual shrub, herbaceous and hermaphrodite, belonging to the Malvaceae family. The plant grows to a height of 1.6–3.2 m or more and a basal diameter range of 1.16–1.23 cm. The stem is usually red or green, practically unbranched or with branches near the base, and is glabrous or slightly hairy with minute tubercles. The leaves are serrated, lower leaves are ovate and undivided, and upper ones are 3–5 lobed. The plant bears large yellow flowers with a dark crimson eye, and the plant produces a large number of fruits that are 1.25–2 cm long, green when immature, and 5-valved, with each valve containing 3–4 seeds. The physical dimensions of the stem, bark, and central core of the plant are tabulated as below (Table 9.1).

9.2.2 Raw material preparation

After harvesting the roselle plant, leaves and unwanted portions are removed manually. Cleaning is necessary to ensure that the pulp is free of foreign matter and dirt. Roselle stems are cut into required length and prepared for their bast and core or without debarking. Besides that, roselle chips are also prepared, which are air-dried to about 15% moisture content and then powdered in a Wiley mill. The powder is then made to pass through 40 mesh and retain on 80 mesh. The chemical constituent is analyzed with the TAPPI standard method and tabulated in Table 9.2.

9.3 Pulp manufacturing

9.3.1 Pulping process

Pulp manufacturing from roselle starts with raw material preparation. The pulping techniques have been broadly classified into mechanical, thermal, biopulping, and semichemical or fully chemical methods. Few of the chemical pulping methods consist of Kraft, sulfite, soda, and organosolv pulping. The process of chemical pulping requires the disintegration of the lignin and hemicelluloses into minute water-soluble molecules which can be washed away without depolymerizing the cellulose fibers. The properties of the isolated pulp rely a lot on the processing techniques of the chemical methods used.

TABLE 9.1 Physical characteristics of roselle and few other selected plant species.

Sl. No.	Particulars	*Hibiscus sabdariffa*	*Buddleja nivea*	*Alpinia allughas*	*Cordia dichotoma*	*Hibiscus cannabinus*
1.	Height of the plant, cm	310.00	250.00	280.00	316.00	324.76
2.	Diameter of the stem, cm					
	Bottom	2.90	2.50	3.10	4.00	3.00
	Middle	2.35	1.75	2.14	3.20	2.40
	Top	1.54	1.30	1.75	2.50	1.82
3.	Moisture content, %					
	Whole plant	81.06	78.00	86.70	78.00	76.00
	Bark/sheath	88.10	79.70	89.31	88.00	78.60
4.	No. of leaves in stem	19.00	17.00	15.00	12.00	18.00
5.	Weight of the whole plant, g	530.00	380.00	470.00	560.00	570.00
	Weight of the bark/sheath, g	290.00	120.00	309.00	250.00	260.00
	Weight of core, g	240.00	140.00	105.00	320.00	300.00
6.	Bark density of the chips					
	Whole plant					
	Top portion	0.6343	0.5343	0.5643	0.5443	0.6043
	Middle portion	0.6843	0.5743	0.5843	0.5243	0.5943
	Bottom	0.6543	0.5443	0.5543	0.4943	0.5743

TABLE 9.2 Proximate chemical analysis of barks, the whole plant materials of roselle and few other selected plant species. Result on oven dry basis.

Plant species	Solubility in								Ash content (%)		α-Cellulose (%)		Lignin (%)		Cellulose (%)		Pentosan (%)		Moisture (%)	
	Cold water (%)		Hot water (%)		1% NaOH (%)		Alcohol-benzene (%)													
	Plant	Bark	Plant	Bark	Plant	Bark	Plant	Bark	Plant	Bark	Plant	Bark	Plant	Bark	Plant	Bark	Plant	Bark	Plant	Bark
Hibiscus cannabinus	4.41	5.87	10.00	8.68	11.45	20.76	6.60	7.86	2.26	1.68	53.00	66.01	17.25	21.06	83.50	38.43	18.35	17.91	7.95	6.60
Hibiscus sabdariffa	4.82	5.03	10.85	9.40	11.90	30.10	6.40	4.68	1.95	1.10	54.50	66.45	17.92	16.90	84.36	39.04	18.60	17.80	8.13	7.10
Buddleja nivea	5.70	13.21	13.28	16.86	13.93	43.40	8.18	4.75	1.21	1.25	83.00	93.00	13.38	18.97	91.00	71.00	5.23	4.80	8.68	7.64
Alpinia allughas	8.38	9.47	9.38	13.34	11.10	39.87	7.23	7.55	5.00	3.63	51.08	67.30	12.50	13.83	58.00	38.88	15.35	14.86	8.73	7.73
Cordia dichotoma	8.65	9.15	9.50	17.12	10.70	23.05	5.33	7.86	4.93	2.38	53.50	63.00	14.23	22.03	58.25	39.75	14.50	7.75	8.03	7.00

Thereafter, the pulp is bleached and processed according to the end use of the paper that is to be produced. There are a variety of methods used for pulping nonwood pulp, but amongst those, only a few ones discussed below are used frequently for roselle pulping.

The prepared roselle stem components are cooked in a stainless steel digester or steel bomb containers. The digester or steel bomb containers are loaded with the required amount of oven-dried samples. The components are cooked with four different concentrations of active alkalis, namely, 10%, 12%, 14%, and 16%. The cooking process takes about 3.5 hours at 170°C where the first 1.5 hour is required to gradually attain the temperature of 170°C, and for the last 2 hours, the temperature is maintained at 170°C. Before the loading of the digesters, the pulping machine is preheated to ensure the proportionate increase in temperature inside it. The loading operation has to be done very fast to avoid the temperature inside the pulping machine slowed down. Following the completion of a 3 hours cooking process, the digesters are removed from the pulping machine and placed in a cooling tank until it is cooled enough to be handled. The pulp produced from the cooking process is taken out and washed thoroughly with water to remove the entire chemical from the pulp sample.

The alkali–oxygen and sulfate processes are used to prepare pulp from the roselle bark chips. In the alkali–oxygen process, the chips are soaked in hot water, 1% NaOH of raw material, or 0.5%–1% NaOH of liquid, with liquor-to-bark ratio of 10:1 at 90°C for 10 minutes. Thereafter, they are disgested with concentrated 10%–20% NaOH, having a liquor-to-bark ratio of 4:1 at an oxygen pressure of 7 kg/cm^2 at 105°C–120°C for approximately 90–120 minutes. In sulfate pulping, the chips are pretreated by soaking them in hot water in liquor-to-bark ratio of 10:1 at 90°C for 10 minutes using active alkali of 14%–18%, sulfidity of 25%, and a liquor ratio of 4:1, maintaining a maximum temperature of 165°C for 2 hours.

Pulp production from roselle bark can also be done using the sulfate, soda, and neutral sulfite, semichemical processes without any significant difference among pulp yields. In addition to pulp production by the soda process and oxygen delignification, a two-step soda oxygen process has also been developed. The first step of the soda oxygen process involves cooking the roselle bark with 18% NaOH, 10:1 liquor-to-bark ratio at 170°C for 4 hours. This is followed by the second step that involves cooking with 4% NaOH and 0.6% MgO at a 4:1 liquor-to-bark ratio, with oxygen at a pressure of 7 kg/cm^2 maintaining a temperature of 110°C for 90 minutes. The soda–oxygen pulp displays brightness as high as 45%, which facilitates its direct use as pulp for paper with low brightness such as newsprint paper.

There is another method for roselle pulping, which can also be prepared by prehydrolysis and kraft pulping process in an electrically heated, rotary, stainless steel digester. The process of hydrolysis can be carried out at different conditions. Prehydrolyzed chips are washed with deionized water and then with distilled water and finally air-dried. Prehydrolyzed chips of roselle obtained after 2.5 and 3 hours cooking at different temperatures are digested with 16% active alkali ($NaOH + Na_2S$) as Na_2O for 3 hours at 160 ± 2°C maintaining sulfidity at 20%, maintaining temperature from 45 to 60 minutes. At the end of the digestion period, the pulp is discharged from the autoclave on a 60 mesh sieve and washed thoroughly with deionized water and then distilled water and finally air-dried.

The chemical composition of roselle stem is indicated lower lignin and higher pentosans, which are usually good screened yield (47.32%) of bleachable pulp (kappa number 21) at 22% active alkali charge as NaOH (Hbs5A).

9.3.1.1 Screening process

The purpose of screening and purification of the pulp is to remove the crude fiber bundle and nonfibrous constituent in the pulp, thus delivering a better product quality. Screening is done based on distinctions in size and shape of fibers with the aid of a sieve through which fine pulp passes through and crude fiber is left behind. The screening process conditions are selected according to the size of the slurry and the desired results. A few commonly used screening instruments are Johnson Screen, centrifugal screen, and electrifier screen. Roselle pulps are screened through a 60 mesh sieve and washed thoroughly, first with deionized water followed by distilled water and hence prepared for the next step of bleaching.

9.3.1.2 Bleaching of pulp

Bleaching can be obtained by eliminating lignin in the pulp or changing the structure of the chromophoric group with the use of chemicals, such as oxidizing bleach, reductive bleach, sodium hydroxide, chelating agents, enzymes, and biobleaching from the fungal strain. One of the frequently used bleaching processes in traditional nonwood pulping is the CEH three-stage bleaching process. However, this bleaching method has been proven to be hazardous to the environment as it has large drainage capacity and heavy wastewater pollution. Presently, the preferred clean bleaching technologies are elemental chlorine-free bleaching and total chlorine-free (TCF) bleaching.

The Screened unbleached roselle pulp can be bleached by a multistage bleaching sequence involving chlorination (C), alkali extraction (E), and hypochlorite treatment (H) with intermediate washing with deionized water after each stage. Finally, a mild HCl acid treatment is given to the bleached pulp to improve brightness and to prevent color reversion. The bleaching can be done in two different stages, that is, the prebleaching stage and the brightening stage. Prebleaching is done to eliminate as much lignin from the pulp as possible to reduce the volume of the more expensive chemicals. Bleaching at the brightening stage removes less of lignin but brings out the brightness of the pulp. Bleaching using TCF sequences, the P–P bleaching sequence for the alkali–oxygen pulp provided lower brightness than the O–P–P bleaching sequence for the sulfate pulp. The beached pulps manufactured using the alkali–oxygen process had higher burst index and tensile index but are weaker in tear index and folding endurance and darker in brightness than sulfate-bleached pulp. Where the alkali–oxygen pulp is pretreated with 1% HC_3OOH, followed by bleaching with the P–P sequence, it is brighter (70.24%) than the alkali–oxygen pulp bleached by the P–P sequence and the sulfate pulp bleached by the O–P–P sequence and is suitable for printing and writing paper. The morphological characteristics of O–P–P sequence bleached pulp fibers are presented in Table 9.3.

9.3.1.3 Determination of kappa number

Kappa number is the number of milliliters of 0.1 $KMnO_4$ consumed by 1 g of pulp in 0.5 N sulfuric acids after a 10 minutes reaction and is determined by TAPPI T236. The kappa test helps to determine lignin content by the consumption of permanganate ion by lignin, possessing highly useful properties like long fiber length, low lignin, and high alpha-cellulose. The roselle bark helps to produce a pulp with a low kappa number. Different pulping processes produced different kappa numbers with the alkali–oxygen process and kappa numbers under the sulfate process are diminished when the alkali

TABLE 9.3 Morphological characteristic of O–P–P sequence bleached pulp fibers.

Properties	Maximum	Minimum	Average
Fiber length (*L*) mm	2.5	0.5	1.74
Fiber diameter (*D*) μm	24.5	12.0	21.00
Lumen width (*d*) μm	16.0	7.2	13.90
Cell wall thickness (*W*) μm	4.8	3.5	4.26
Slenderness ratio (*L*/*D*)	–	–	82.86
Runkel ratio (2*W* × *D*)	–	–	0.61
Flexibility coefficient (*d*/*D* × 100)	–	–	66.19
Wall fraction (2*W*/*D* × 100)	–	–	40.57
Ratio of cell wall thickness to fiber diameter (2*W*/*D*)	–	–	0.41
Ratio of cell wall thickness to the lumen diameter (*W*/*d*)	–	–	0.31
Shape factor ($D^2 - d^2/D^2 + d^2$)	–	–	0.39

dosage is increased. Kappa numbers with the alkali–oxygen process results reduced by pretreatment with NaOH and cooking with extra NaOH for a longer period. The pulp with a lower concentration of active alkali had a higher kappa number where the pulps produced are brown. However, the pulp with a higher concentration of active alkali had lower kappa number and is of a lighter color. This difference is due to the delignification process of the pulp caused by the concentration of active alkali used in the pulping process. As the active alkali charges increase 15%–17%, kappa number decreases. Pulp quantity can be well determined by kappa number and pulp lignin. Kappa number greater than 20 signifies the nonfeasibility of bleaching the pulp as it will require a higher concentration of bleaching chemicals.

9.3.1.4 Beating process/refining

Beating and refining are the processes by which the undesirable characteristics are changed. The refining process is executed by the desired properties required in the end use of the paper that develops different fiber properties in various ways for specific grades of paper. The different types of fiber react differently to beating owing to their differences in their morphological properties. The beating process of fibers can bring about changes that drastically affect the character of the paper produced from them. Beating is executed to remove the residual part of the outer surface that is left over after cooking, bleaching, and refinement. With continuous beating, the inner wall of the fiber starts to swell and collapse into fragments. The unbleached and bleached roselle pulps obtained using active chemicals in either process are separately beaten in laboratory beater to three different freeness levels—200, 300, and 400 degrees Canadian Standard freeness at 1.57 consistencies. In other methods, bleached roselle pulps are prepared and weighed (oven-dried weight) for the beating process. The pulps are beaten at 0, 2500, and 5000 revolutions with a PFI Mill at a consistency of 10%. The beating process is done to refine the pulp fibers where the refined or beaten pulp fibers would form a more conformed handsheet in the handsheet making process. The pulps from roselle bark obtained by using the alkali–oxygen process and sulfate process are beaten until a freeness of 300 mL is obtained. For producing a high-quality handsheet, the pulp beaten for 5000 revolutions are used. Due to the beating process, the handsheet produced has more strength as compared to the unbeaten pulp. The properties of a paper could be optimized by beating or refining the process. The beating process improved the properties of the pulp fibers and handsheets produced from unbeaten pulp had lower strength properties than the handsheet produced from the beaten pulp.

9.3.1.5 Freeness

After that, the unbeaten and beaten pulps are tested for their freeness using a freeness tester. Freeness tester is a standard method for measuring the drainage of 1 L of stock containing 3 g of pulp through a screen plate. The pulps are diluted in water with air pressure in a stock tank. The stock is poured inside a steel glass and drained through a screen plate. When the stock is drained, some of it would overflow and collect in a graduated cylinder and the amount of water collected is recorded. The freeness test is repeated three times for each pulp; unbeaten, 2500 revolutions pulp, and 5000 revolutions pulp. Roselle pulp is tested for freeness of 300 mL (T227-om-99).

9.3.1.6 Evaluation and characterization of O–P–P sequence bleached pulp fiber

IR spectra of pulp fiber obtained from *H. sabdariffa* exhibit characteristic absorption bands of O–H stretching (3360 cm^{-1}), C–H stretching (2920 cm^{-1}), C–H bending (1380 cm^{-1}), O–H bending (1320–1360 cm^{-1}), and so on. The X-ray diffraction patterns of pulp fiber indicate that the pulps are typical native cellulose, consisting of both crystalline and amorphous phases. The higher crystalline value is also responsible for the development of higher strength of the paper. The scanning electron micrographs of pulp fibers of roselle show hollow and cylindrical fibers with distinct nodes. Fibrils are regularly arranged in a horizontal plane direction thereby indicating good quality fibers of flexible nature (Figs. 9.2–9.4).

9.4 Handsheet making

For making the paper sheet, in the paper industry, the pulp from roselle is dried and pressed to manufacture paper sheets. The process of paper fabrication can be divided into two steps that comprise the fibrous raw material first being transformed into pulp, which is then changed into the paper. After the freeness test, the leftover roselle pulp stock is used to produce a handsheet where it is diluted inside the stock tank by adding water. The tank containing the pulp stock is introduced with air bubbles at a constant rate to allow the spreading of the roselle fibers inside the tank and then poured it into the handsheet making machine from British Evaluation Apparatus. Water is added inside the apparatus and then the pulp or fiber is agitated inside the apparatus. After that, the roselle pulp is left to settle down for a few seconds before the water is drained off from the apparatus. A wet handsheet is obtained on the wire mesh plate, and it is couched to remove the excess water from the handsheet. Then the handsheet is placed on a steel plate and pressed by using the pressing machine from British Evaluation Apparatus. The handsheet is pressed twice where the first pressing took 5 minutes and the second pressing took about 2.5 minutes. After pressing, the roselle handsheets are air-dried in a stack rings.

9.4.1 Properties of roselle handsheet

The fiber morphology is an important determinant of the papermaking properties of raw materials. A higher fiber length of roselle showed a higher

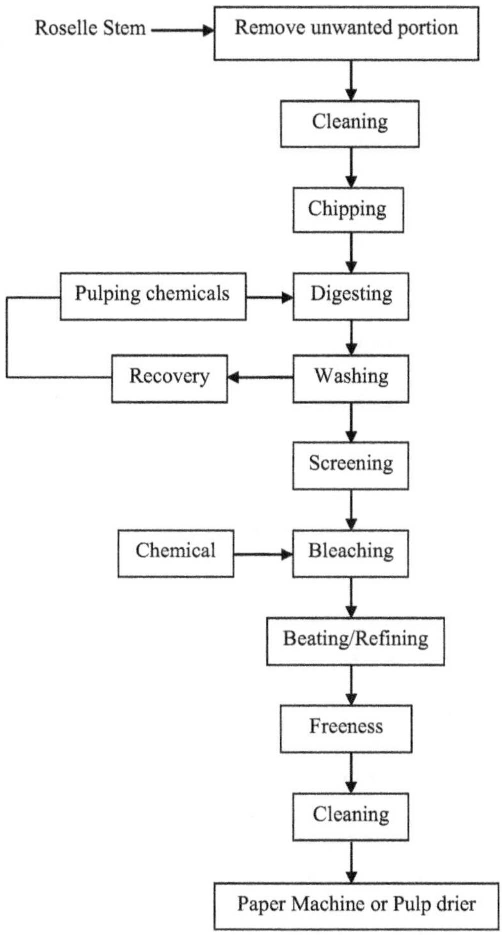

FIGURE 9.2 Pulp and paper making process of roselle.

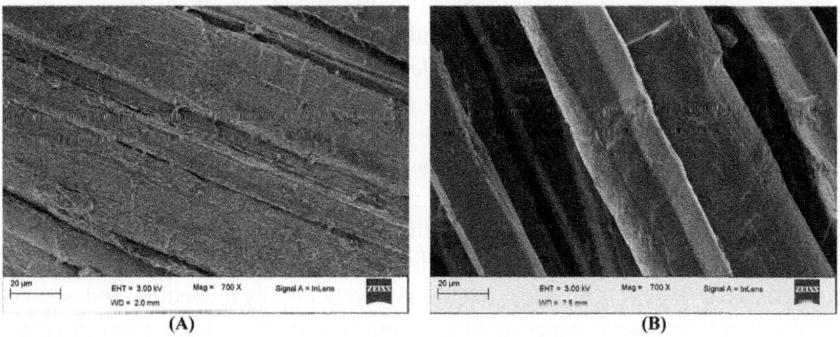

FIGURE 9.3 (A) SEM image of unbleached roselle pulp fibers and (B) bleached roselle pulp fibers. *SEM*, scanning electron microscopy.

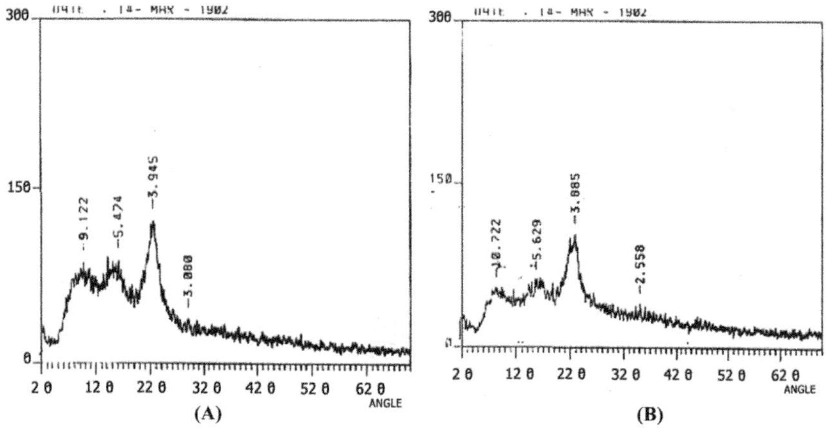

FIGURE 9.4 X-ray diffraction patterns of (A) unbleached and (B) bleached roselle pulp fibers.

tearing strength of paper. Paper manufactured from the thin-walled fiber is well formed and dense. Slenderness ratio (fiber length/fiber diameter) is also considered vital in the papermaking quality; if the slenderness ratio for the fiber is less than 70, then pulp would have poor strength.

The tearing resistance of roselle fiber is higher might be due to their fibers' morphological properties, roselle bast is determined to have the longest fibers (2.286 mm), lastly roselle core (0.849 mm). Both bast and core roselle fiber produced different handsheet properties because the bast had longer fiber, which can build up stronger fiber bonding when formed into handsheets. The tensile strength properties of all the pulps produced are different from each other, and the handsheets produced from different revolutions of beating are also considerably different, same performance for the fold and freeness testing.

The tear strength of the roselle paper sheet depends on fiber length, fiber bonding, and the total number of fibers that are involved in the sheet rupture. The flexibility of the sheet is influenced by the number of fibers involved in the tearing of the sheet. In the case of a rigid sheet the concentration of force is limited to a few fibers in a small area while in a flexible sheet distribution of force is over a much larger area with a large number of fibers. The force required to tear the roselle handsheet paper is much less than the force necessary to break a strip of the paper because the roselle handsheet is more flexible. Furthermore, in a weakly bonded sheet, as more fibers pull out than break in the tear zone, the tearing resistance is restricted more by the number of bonds that break along the length of the fibers; thus fiber length effect the tearing resistance. Tensile strength as well as burst index mainly depends on fiber bonding and showed similar trends in roselle paper making. Moreover, the beating process increased the tensile and burst indices of the roselle pulps. In identical freeness, recombined roselle pulps showed higher tensile and burst index than beaten pulps, which can be explained as due to improved fibrillation and enhancement of fiber bonding. The handsheet properties of roselle are different according to the level of active alkali used for pulp making.

It was also noticed that from the different studies, paper sheets made from unbleached and bleached roselle soda–AQ pulps produce better physical strength than that of the paper sheet from sulfate pulps, which is comparable to paper made from any other standard raw material. Alkali–oxygen-bleached pulp is higher in burst and tensile indexes, but weaker in tear index and folding endurance and darker in brightness than the sulfate-bleached pulp.

The mechanical properties (burst, tensile, and double folds) of the pulp obtained are lower, but the tear is higher compared to pulping with lower alkali charge and lower screened yield. Pulping trials with core alone at the same active alkali levels showed that the delignification is lower mechanical properties are lowered except for tear at the same degree of freeness (40°SR).

9.5 Conclusion

Roselle bark has long fiber the same as in softwood. The amount of lignin in the bark is less and high alpha cellulose, which helps to produce a pulp with high strength properties, low kappa number, and high yield. That is an important advantage in the selection of a softwood fiber substitute for the production of newsprint, paperboard, and other grades of paper products. Roselle bark was studied for pulp production using the alkali–oxygen, sulfate, soda, and neutral sulfite semichemical processes, and there is no significant difference observed among pulp yields between processes. In addition to studying pulp production by the soda process and oxygen delignification, a two-step soda oxygen process is also studied. Different pulping processes produced different pulp yields and kappa numbers with the chemical, semichemical land mechanical process. Roselle plant is potentially good to be utilized as a fiber source and also as an alternative of nonwood fiber. Roselle bark is a feasible substitute for paper manufacturers in pursuit of supreme quality and economical sources of nonwood fiber.

Chapter 10

Pulping and papermaking of sarkanda

N. Sharma, Nishi Kant Bhardwaj and P. Kaur
Avantha Centre for Industrial Research & Development (ACIRD), Paper Mill Campus, Yamuna Nagar, Haryana, India

10.1 Introduction

Pulp and paper industry plays an important role in the economic progress of any country. Different species of wood are the leading raw materials used for papermaking (Rodríguez et al., 2008). The world's total production of paperboard and paper is more than 400 million tons (Dillen et al., 2016), and this is expanding with the increasing demand for pulp and paper. Growing world population has increased the demand for paper consumption almost all over the world (López et al., 2010). The growing demand for paper has led to more deforestation for raw materials, and acute shortage of fibrous raw materials has increased the complexity of papermaking industries. Increased paper consumption over the world has forced pulp and paper manufacturers to find out alternative raw materials in response to economic and environmental demands (Gominho et al., 2001; González-García et al., 2010). Pulp and paper industry faces the scarcity of wood-based raw materials ever since the manufacturing capabilities as well as environmental awareness has increased (Liu et al., 2018). In the early 1970s for pulp and papermaking the share of wood, agricultural residue, and wastepaper was 84%, 9%, and 7%, respectively, whereas presently it is 31%, 22%, and 47%, respectively. Steady growth rate of paper demand and nonavailability of wood based raw materials have increased the utilization of nonwood fiber resources. India is an agriculture-based country, and crop production is being practiced in several states as the main occupation. But a portion of agricultural waste is burnt on-site by farmers, which is known as stubble burning. The main factors include shorter time window for the next crop, bulk volumes, and unavailability of some practical agricultural residue management means. Stubble burning initiates a chain of phenomena that cause a detrimental impact on environment. The more direct and visible issues include air

Pulping and Papermaking of Nonwood Plant Fibers. DOI: https://doi.org/10.1016/B978-0-323-91625-7.00011-4

pollution, hammering of valuable nutrients in soil, adding to greenhouse effect by increasing carbon footprint, etc. Poor air quality mainly results from the discharge of the pollutants, like CH_4, CO_2, N_2O, SO_2, CO, and NH_3, along with volatile organic substances and particulate matter causing respiratory and allergic ailments (Jain et al., 2014; Rodriguez et al., 2010). It increases the winter smog and lessens the visual ability, consequently amplifying the figure of road accidents. Therefore it is in common interest and benefit to valorize such wastes to make some useful products like paper, handicrafts, bags, and ropes instead of burning them. Plants like jute, kenaf, and hemp exhibit fine pulp and papermaking abilities and also have exceptional fiber properties (Jiménez et al., 2008) that are important for papermaking. In recent years, interest of manufacturers has increased in using nonwoods for making pulp and paper (Ververis et al., 2004). Many countries, like India, China, and Iran, are successfully utilizing these raw materials for papermaking (Agnihotri et al., 2010; Liu et al., 2018; Mousavi et al., 2013). The utilization of nonwood raw materials is a possible substitute for wood-based raw materials to manufacture pulp and paper and may sort out the problem of cutting down the forests for raw material shortage. Of the total pulp and paper production worldwide, nonwoods as raw material share about 5%–7% (Ward et al., 2008). Considering the benefits nonwoods offer to both the environment and the manufacturers, this share should be increased and more nonwood materials should be explored for pulp and papermaking.

Indian pulp and paper industry uses around 39% fibers from forest-based sources, 31% fibers from nonwood (agro-residue)-based sources, and the remnant 30% fibers are derived after recycling of wastepaper (Subrahmanyam et al., 2004). A range of nonwood fibers are available from the sources like grasses (sabai, sarkanda, kans, etc.) (Sharma et al., 2019) and agro-residues (rice straw, bagasse, wheat straw, etc.). The major sources of raw materials from nonwoods have been agro-residues from monocots specifically cereal straws (Saijonkari-Pahkala, 2001). The nonwood plant fibers can be grouped into four different types depending upon the part of the plant used for pulp or papermaking. These types can be grouped as following: (1) Stalk (culms fiber) that includes cereal straws, grasses, reeds, bamboo, and sugarcane (bagasse); (2) bast fibers such as flax, jute, kenaf, and hemp; (3) leaf fibers such as sisal and abaca; and (4) seed hull fibers, that is, cotton (Subrahmanyam et al., 2004).

Although bamboo and some grass plants have also been used for papermaking, the potential of natural fibers from abundantly available Indian grass plants have remained unharnessed. India is a country with the richness and variety of natural fibers available from the innate abundant flora (Singh et al., 2014). While agro-residues have been extensively studied for the fiber extraction and subsequent application, scarce literature is available for fiber extraction from Indian grasses. Out of all subcontinents, India lags behind in studying grasses for potential applications (Kellogg et al., 2020). The country has been chiefly inaccessible to overseas botanists for the past several decades, and the revolution in molecular systematics has mainly failed to take account of Indian grasses because of the challenges of obtaining material (Kellogg et al., 2020).

In monograph on Indian papermaking fibers, Subrahmanyam et al. (2004) reported papermaking potential of major nonwood plants commonly used by Indian paper mills along with wood-based raw material. A few grass species that have attracted attention for their application in papermaking recently include sarkanda (Sharma et al., 2020) and another much similar kans grass (Sharma et al., 2019). While this chapter will deal with papermaking from sarkanda in detail, a brief note on closely related kans grass is mentioned here. Pulp from kans grass constitutes of around 30% long fibers along with high fraction of primary fines. The long fiber contributes to good strength properties by showing zero-span tensile strength, while the large proportion of fines leads to poor drainage issues during papermaking. Pulp of kans grass has been shown comparable in terms of its quality to that of bagasse and wheat straw and therefore can be utilized for the production of good-quality writing and printing-grade paper. Besides, higher (ISO) brightness of up to 85% can be possibly attained by using modern bleaching sequences for kans grass pulp (Sharma et al., 2019). Similar potential for papermaking has also been shown by sarkanda in other studies (Sharma et al., 2020; Subrahmanyam et al., 2004).

10.2 Global scenario of nonwood materials for pulp and paper industry

The term "nonwood" plant refers to a variety of plants that possess remarkably different characteristics, although the basic chemical constituents, like cellulose, hemicelluloses, and lignin, remain the same except for the quantitative changes. Moreover, nonwood plants have been found to show high annual yields per hectare (Ashori, 2006). For this reason, nonwood fibers are also referred to as "alternate fibers" to wood-based fibers for providing cellulosic pulp to make paper. The majority of the nonwoods used for papermaking belong to monocotyledons specifically cereal straws or agro-residues, including straws, sugarcane bagasse, and abaca, and also from a few grass families such as bamboo and reeds, and bast fibers, including jute, hemp, sisal, and kenaf (Kozlowski et al., 2005). Cotton linters are mostly used for the production of high-quality cellulosic pulp. Similarly, several nonwood plant fibers are used to produce specialty pulp or different grades of paper. Since most of the nonwood plants are annual plants, they are able to grow full fiber in a single growing season. Based on their chemical composition, availability, and ease of fiber extraction, a vast range of nonwood plant fibers are available as a potential raw material for papermaking (Ashori, 2006; Touzinsky, 1993). All over the world, several nonwoods, such as wheat, bagasse, rice straws, kenaf, and bamboo, are being utilized for the production of pulp and paper (Ashori, 2006; Pande and Roy, 1996). Nonwood raw material shares only a small fraction of 6.5% for papermaking worldwide (Ashori, 2006). For pulp and papermaking all over the world, nonwoods include annual agricultural waste and natural growing plants or artificial cultivation. In past, China contributed 77% to the total pulp production from nonwood, which is turned down to 44% in recent years. India also utilizes nonwood raw materials in major parts for pulp and paper production. Utilization of nonwood resources has been comparatively limited in Europe, North America, Africa, and the Russian republics. In recent years, extensive research on nonwood materials has been carried out in North America and Europe due to scarcity of raw material and surplus of agro-residues. Plant families like Gramineae/Poaceae, Malvaceae, and Leguminosae have been of particular interest as a source of nonwood fiber for pulp and papermaking. Northern Europe witnessed specific interest in the utilization of grasses and different monocotyledons for fiber production (Olsson, 1993; Saijonkari-Pahkala, 2001). In Finland and Sweden, reed canary grass has shown promising results for fine paper production as compared to other field crops (Saijonkari-Pahkala, 2001). Elephant grass as a raw material has been explored for the production of both paper and energy in Central Europe (Scurlock, 1999). Some other grasses that have been specifically explored over the globe for paper production include *Festuca arundinacea* Schreb. or tall fescue (Janson et al., 1996), *Panicum virgatum* L. or switch grass (Radiotis et al., 1999), and different cereal straws (Liu et al., 2018; Subrahmanyam et al., 2004). Another perennial and leguminous grass tagasaste (*Chamaecytisus proliferus*) that was endemic to Canary Islands and later introduced to Australia has also been found to produce good-quality fiber for papermaking (Jiménez et al., 2008). It is now widely distributed to African countries, Portugal, California, and Hawaii with a production capacity of 15–20 t/ha in plantations of 1000 bushes and reaching the height of 7–8 m. Along with the fodder for cattle and livestock, it has been a promising nonwood source as fuel and as raw material for the production of pulps (Jiménez et al., 2008).

10.2.1 Major nonwood raw materials used for papermaking

Many countries successfully utilize nonwood resources for papermaking at commercial scale. Different nonwoods have been studied owing to their abundance, availability, and comparable lignocellulosic content, which include agro-residues, wild grasses, and industrial by-product. Kaur et al. (2018) and Rodriguez et al. (2010) have reported the utilization of agro-residue rice straw that is available in abundance after paddy harvesting. Another agro-residue banana stem has been successfully utilized as furnish for papermaking by Tripathi et al. (2017). Sharma et al. (2020) has studied the wild grass sarkanda for its potential in papermaking. Wheat straw has shown promising results as an alternative raw material for papermaking (Ghosh, 2006). Sharma et al. (2015) reported the use of industrial by-product (empty fruit bunches) of palm oil industry for papermaking. Other nonwoods include bagasse, rapeseed straw, hemp, kenaf, karkadeh, jute, kans grass, and sesame that have been studied for their prospective utilization in pulp and papermaking (Jiménez et al., 2008; Mousavi et al., 2013; Sharma et al., 2019). Study on the use of wheat straw and bagasse for pulp and papermaking has been explored extensively (Jiménez et al., 2000; Garg and Singh, 2004a).

10.2.2 Characteristics

Nonwoods are an important resource for pulp and papermaking because of availability in abundant quantity, compatibility with existing pulp and paper manufacturing infrastructure, comparable pulp and paper qualities, and economic viability in the marketplace. Most of the nonwoods contain less amount of lignin content in comparison to wood and therefore can be cooked with mild pulping conditions (Sharma et al., 2020) and involve uniform cooking due to the thin structure. There is no requirement of chipping and debarking of the nonwood raw material in comparison to wood, and also lower energy is required to refine the pulp to achieve the same freeness level comparable to that of wood. In addition, nonwoods are also required in specialty-grade papermaking due to some unique properties they confer. Cotton linters are used in making dissolving grade pulp and currency paper (Ateş et al., 2015). Hemp is used in making specialty-quality papers, for example, cigarette, bible, currency paper, and insulating paper (Bambach, 2017). Wheat straw and bagasse have the potential for the production of good-quality writing and printing-grade paper and therefore can be a substitute for wood. Maximum nonwood pulp fibers are comparable with the hardwood short fibers, except for some fraction of long fibers that require processing before undergoing papermaking process. Generally, the diameter of the nonwood fibers is small, and they consequently involve the property of lower coarseness. These fiber proportions give a thought of the probable usefulness for papermaking. Actually, from the technological standpoint, several varieties of paper can be manufactured by using the suitable blend of nonwood and wood pulps. Several nonwood plants are used for the specialty fibers, which they possess. The special properties of few nonwood fibers, such as abaca, make them in demand for specialty pulp or papermaking since the wood pulp or fiber has a limited variety to offer. Thin lightweight papers can be easily made from high-strength and long abaca fibers. These properties make abaca fiber a superior and unique material to manufacture high-porosity paper with excellent tensile strengths and tear burst (Peralta, 1996). Furthermore, specialty paper produced from abaca fiber can be used to make products like tea bags, currency paper, large sausage casings, filter paper, and cigarette and other special products that need high porosity along with high wet strength (Laftah and Abdul Rahman, 2016). Good mechanical properties and high cellulose content offered by the nonwood fibers such as straw make them feasible to be utilized as bleached pulp for making writing and printing-grade papers (Laftah and Rahman, 2016; Wahit et al., 2012). It has been demonstrated that, by proper selection of appropriate nonwood fibers and their pulping process, almost all grades of paperboard and paper can be manufactured. There are already several grades of paper being manufactured with 100% of nonwood-based fibers (Atchison, 1996). Nonwood raw materials are mainly dominated by straws, like wheat straw and rice straw, or grasses, such as reed and sugarcane bagasse. These few nonwood plant fibers account for the major proportion of total nonwood-based pulping capacity. Wheat straw fiber exhibits better quality for papermaking as compared to the other nonwood plant fibers due to its superior breaking length or other strength properties of handsheets (Laftah and Rahman, 2016).

Another nonwood plant fiber that has gained considerable attention, especially in Asian countries, is bamboo from grass family that is widely used for manufacturing paper and paperboard. Bamboo fibers exhibit long fiber length with an average range between 1.5 and 4.4 mm that is comparable to that of softwood fibers. Attributing to the high-strength properties of bamboo pulps, these pulps found the application in the production of more versatile paper products as compared to most of the other nonwood-based pulps. Sugarcane bagasse is also of a great significance for its fiber characteristics, which demonstrated that the bagasse fiber was better than those of hardwoods used for pulp production at commercial scale. Bagasse has lower lignin content in comparison to usual quantity present in moderate hardwoods. Furthermore, chemical composition of the bagasse reveals a similarity to the tropical hardwoods, which makes it suitable alternative raw material (Khristova et al., 2006). Another nonwood plant that has shown good strength properties and pulp yield is kenaf. The soda pulping of kenaf's stalk resulted in pulps suitable for papermaking. Kenaf pulp can also be used to blend with other nonwood pulps to improve mechanical properties of papers. Blending of around 30% of kenaf pulp with sunflower pulp makes the pulp suitable for the production of printing-grade papers (Khristova et al., 1998). The chemical composition of rice straw material exhibited the qualities required for papermaking. Its utilization to produce pulp and paper may prove beneficial to paper industry, farmers and environment. Pulps produced from the pulping of rice straw using soda and soda–anthraquinone (AQ) processes showed fine pulp properties; but improved delignification, reduced kappa number, and high pulp yield were obtained in soda–AQ pulping process than do the soda pulping (Kaur et al., 2018). Overall, the pulp was found suitable for its application in papermaking.

Some of the agro-residue-based paper mills utilize *Saccharum spontaneous* (sarkanda) along with other nonwood plants like rice straw, wheat straw, jute/rags, and bagasse in the furnish mix. Utilization of sarkanda not only fulfills the fiber requirement of the pulping unit but also enhances the textual quality of the final product (http://www.dcpulppaper.org/gifs/report31.pdf).

10.2.3 Availability

Nonwoods are available in abundance in many countries either free or at cheap price and have provided an alternative source of fiber for pulp and paper in developing countries like India and China (Ashori, 2006). Availability of raw material is a key factor for setting up the process or industry for its utilization. Worldwide, around 2.5 billion tons per annum of nonwood plants are available as raw materials that largely remain underutilized for pulp and papermaking (Ashori, 2006). Among different nonwood plant materials, *Hibiscus cannabinus* or kenaf had the production of 0.51 million tons in 1999–2000. Different countries contributed to the total production of kenaf with the highest 44% from China, followed by India (39%), Thailand (12%), and the remaining part of produce coming from countries, including Vietnam and Indonesia (Ashori, 2006; Fu-Wang et al., 1996). Wide geographic distribution and easy availability of the raw material along with desired chemical composition have been the factors for kenaf to be explored as an alternative raw material for the extraction of fiber for papermaking in developed and developing countries.

10.3 Problems with utilizing nonwood raw materials

10.3.1 Technical

Use of nonwood for papermaking involves some problems like collection, seasonal availability, transportation, washing, poor bleachability, chemical recovery, and properties of end product. The utilization of some nonwood annual plants poses a real culture challenge for the pulp and paper industry. This includes the issues with significant implications for setup for the specific requirements that include capital and operating costs, products uniformity, quality, and reliability. The primary concern for all the pulp and paper mills based on nonwood fibers remains the availability of a regular, year-round fiber supply. Since most of the nonwood plants are annual plants, it is a prerequisite to install large storage capacity that can ensure a constant supply of the raw material. Further complication arises from the fact that most of the nonwood plant sources are with high bulk and low density as compared to wood.

Major disadvantage of nonwood is its high ash and silica content in raw materials. High silica content generates great problem in recovery system of pulp mill. Most of the nonwood-based paper mills either have small capacity or inadequate chemical recovery system to cope with the challenge of high silica and ash content in some nonwood raw materials. Another disadvantage of using specific nonwood fibers like kenaf lies in the more necessities required for their development and subsequent harvesting of such nonwoods (Ashori, 2006).

10.3.2 Economic

Nonwoods are not available throughout the year; these are available seasonally. This requires large storage spaces for uninterrupted supply of raw material throughout the year. Since nonwood fibers are known to be bulky with low density, transportation and storage increase the costs of the process (Ashori, 2006). Use of nonwood depends to a large extent on the economics of transport and has a direct relation to suitable storage to ensure minute or no degradation during the storage period. High silica content in nonwoods also reduces the water drainage in paper machine by choking it and reduces the production, consequently creating great economic losses to the industry. But with innovations, problems are being resolved, and this makes the nonwoods a reasonable alternative raw material for pulp and papermaking.

10.3.3 Advantages of using nonwoods for papermaking

Nonwood raw materials are favorable as compared to wood-based raw materials in numerous ways. They grow fast and have shorter life cycle in comparison to wood. The average yield per annum per hectare of nonwood plant kenaf is around two folds to that of commonly used fast-growing softwoods. Another advantage is the lower lignin content in nonwoods than woods, which makes easy cooking of the former. Besides, they show lower activation energies. In addition to easier delignification, nonwoods offer benefits of cost-effectiveness with lower or nearly no costs involved (for wheat straw and bagasse). Nonwood plantations need little care and consume less water and fertilizers. Manufacturing of paper from nonwood fibers would assist in reducing the requirement of wood and put a check on deforestation. The utilization of nonwood plants can reduce the release of carbon dioxide and carbon monoxide that increase from the burning of these agricultural residues. Use of nonwoods also has a constructive impact on societal arrangement as well as employment in less populated areas.

10.4 Sarkanda

Sarkanda is available in large quantities and found along the river banks and in arid regions. Sarkanda is also known by other names such as sarpat, kanda, munja, and pin red grass. Botanical name of sarkanda is *Saccharum munja* Roxb., and other related names are *Erianthus munja* Jesw., *Saccharum bengalense* Retz., and *Saccharum sara* Roxb. (Khare, 2007). It is

FIGURE 10.1 Sarkanda grass in the fields (https://www.flowersofindia.net/catalog/slides/Sarkanda.html as accessed on 31 August 2022).

mostly found in the sandy soils alongside the rivers and dry streambeds and grows up to the height of 10–12 ft. As shown in Fig. 10.1, its leaves are sharp and long, with a length of around 3–4 ft and a width of 3–10 mm. In juvenile stage, it can be used as cattle feed, but once matured, it becomes unpalatable as the leaf blades can wound the grazing animal. Flowering in sarkanda starts immediately after rainy season; flowers are generally found in bunches and its flowers are used for decoration. Dried stalks of sarkanda are used in a variety of products; stems are used to make mudda, baskets, etc. Leafy material is dried, beaten, and converted into fiber and twisted into rope, which is used for making a cot. Due to its lightweight, furniture made from sarkanda is very popular. In villages, sarkanda is also used for making roofs. It is also a good source of substrate for conversion to furfural, reducing sugars, and alcohol. Xylose, rhamnose, and galactose are present in good amount in hydrolysate after digestion with sulfuric acid. Roots of this plant are syrupy, diuretic, aphrodisiac, cool emollient, and stimulant. Rahar et al. (2011) studied preliminary phytochemical properties of root powder of *S. munja* Roxb. The investigation was carried out on aqueous, chloroform, methanol, and petroleum ether extracts of sarkanda root powder to find the presence of pharmacologically important compounds, including carbohydrates, glycosides, proteins, flavonoids, alkaloids, terpenoid, phytosterols, saponins, and tannins. Furthermore, prismatic crystals of calcium oxalate were found to be abundantly present in the parenchymatous fibers.

In Ayurveda also, sarkanda has a great value in curing blood impurity, thirst, burning sensation, etc. It belongs to Gramineae family and resembles sugarcane. Sarkanda has the capability of growing in barren lands and is utilized to restore lands because of its inherent properties. It has widespread deep root structure that helps in the development of the barren area to make it fertile. Due to this property, sarkanda has been found to help in reestablishment of land containing fly ash lagoons (Pandey et al., 2012). Thus this is identified as a promising grass species for many benefits, including rehabilitation and restoration of land also. Besides these benefits, its chemical constituent and morphological properties make it a potential furnish for pulp and papermaking. Singh et al. (2014) reported the use of sarkanda fiber for making composite material. The extracted fibers were further treated using sodium hydroxide to improve its performance in composites. Sarkanda fiber has shown its potential as a natural fiber for reinforcement in composite. Important properties for reinforcement, like tensile properties, impact, and flexural strength of sarkanda fibers, showed comparable values with the present reinforcing fibers in use. The performance of fibers in composite was further improved by chemical modifications by alkali treatment and has shown promising results in composites.

Sarkanda, along with other perennial grasses, has also been studied for its potentials in bioenergy production (Singh et al., 2014). On the basis of Scopus-based survey, these grasses have been kept in the category of neglected and underutilized, especially in Indian perspective.

10.4.1 Characteristics of sarkanda

Sarkanda is an appropriate candidate for making pulp and paper and can be utilized as a commercial raw material. It grows speedily with growth rate of about 180 t/ha/year and generate a plentiful source of biomass with small contribution (Vasudevan et al., 1984). Sarkanda culms are biennial, pale in color, having smooth pith with undistinguished growth ring and root region. Leaves can be up to 1–2 m in length and around 3 cm in breadth. Subrahmanyam et al. (2004) reported that fibers of sarkanda pulp are straight, long, and narrow with broad walls. Fiber length of sarkanda pulp varies from 0.2 to 3.5 mm and fiber width in the range of 7–40 μm (Sharma et al., 2020). Sarkanda pulp also contains epidermal cells, parenchyma, and vessels. The pulp consists of fibers, parenchyma, vessel, and epidermal cells. Sarkanda pulp fibers are lengthy, straight, thin, sharp, and thick-walled and are pointed with occasional transverse markings. The fibers are narrower than bagasse and bamboo. The parenchyma cells are abundant in quantity, narrow, rectangular, and small to medium in size. The vessels are very long and thin, and the epidermal cells are abundant, rectangular in shape, and noticeable with margins. Sarkanda can be utilized successfully in small- and medium-size agro-based paper mills. Mills can also use this raw material in blends with other nonwood species in different proportions to make good-quality paper. Sarkanda pulp was found to contain a good amount of desired polymers that include nearly 70% holocellulose with 39.9% alpha cellulose, 20.7% pentosans, and around 23% lignin (Subrahmanyam et al., 2004).

Apart from the abovementioned major constituents, sarkanda root powder extract has been found to have different trace metals and pharmacologically important compounds that can have different physiological effects (Mishra et al., 2017; Saleem et al., 2017). Sarkanda has shown to possess strong metal absorption potential in its rhizosphere and subsequent translocation of adsorbed metals to various parts of the plant. This endures the phytostabilization and restoration capability of sarkanda of land areas with high alkalinity or heavy metal contamination (Mishra et al., 2017). The order of translocation factor has been shown in the following trend: Fe > N >Cd > N > Zn > N >Pb > N > Cr > N > Cu > N > Mn. Depending upon the value of bioconcentration factor and translocation factor, sarkanda may be used for phytoremediation of Cu and Cr and also for phytoextraction of certain metals (Mishra et al., 2017). In spite of the fact that sarkanda has been used in Ayurveda for treating several ailments, the acute toxicity study states otherwise. In the study by Saleem et al. (2017), ethanol and water extract (70:30) of sarkanda root had toxic effects on mice and found to elevate lipid profile parameters along with blood dyscrasias at the tested dose.

10.4.2 Geographic distribution

Sarkanda or *S. munja* is a wild perennial grass, widely distributed to arid regions of Europe, South Asia, and America owing to its tolerance toward drought, high temperature, and biotic stress. This makes wild grasses to be called "nurse species" since extreme habitats can also be colonized by them (Rau et al., 2009; Sharma et al., 2011; Shu et al., 2005). Naturally growing sarkanda provides a good green belt as well as ecologically and sociologically significant vegetative cover on the arid areas and riversides of India, Afghanistan, and Pakistan. In India, it mainly covers plains and lower hilly areas of north and northwest area, including Jammu and Kashmir, Punjab, Himachal Pradesh, Haryana, Uttarakhand, Uttar Pradesh, Delhi, Assam, Gujarat, Bihar, Madhya Pradesh, Rajasthan, and West Bengal (Kellogg et al., 2020; Vasudeva, 2019). It is also planted in the area dividing crop fields since it is known to form extensive root network and thus prevents the soil erosion by binding the soil/pebbles within the fields (Singh et al., 2014). This shows the benefit of using sarkanda as a noncompetitive, economic raw material for papermaking, because unlike wood and nonwoods that require fertile land and optimum growth conditions, sarkanda can grow by self-propagation over the barren, abandoned, and otherwise inhabitable land. It can also withstand different biotic and abiotic stress (Sharma et al., 2011).

Sarkanda has been specifically found useful in restoration and rehabilitation of barren or abandoned harsh ecological locations, such as Bhatti mines of Delhi–Haryana Aravalli regions (Sharma et al., 2011) and fly ash lagoons of India (Pandey et al., 2012). Sarkanda has many desirable characteristics for its suitability to grow in arid, semiarid, and impoverished sites. Major factors for its succession at such harsh environment include its soil binding ability due to widespread rooting, mechanisms to combat cyanide toxicity like other grasses, tolerance to abiotic stress of high temperature and drought, and self-propagation along with harboring a vast range of microbes in its rhizosphere (Sharma et al., 2011; Pandey et al., 2015). Regardless of its geographical location or distribution, *S. munja* or sarkanda is the species that can be grown for vegetation, stabilization, and restoration of erosion-prone rugged slopes. As a consequence, these sites can be converted into biologically productive regions, which can provide dual benefits of high socioeconomic value and land restoration (Sharma et al., 2011).

10.5 Utilization of sarkanda for papermaking

10.5.1 Chemical components of sarkanda

Proportions of chemical constituents of any raw material for papermaking give a thought for their appropriateness in pulp and paper industry. Sharma et al. (2020) revealed that raw material of sarkanda contains 36.2% cellulose, 35.2% hemicelluloses, 21.2% lignin, and 8.3% ash; Subrahmanyam et al. (2004) also reported approximately the same values for hemicelluloses: 30.5%, cellulose: 39.9%, lignin: 21.2%, and ash content: 8.2%. Chemical constituents of sarkanda and other nonwoods are presented in Table 10.1. Chemical constituents of sarkanda make it suitable for pulp and papermaking. Cellulose is principally responsible for the strength of plants; its structure consists of linkage between D-glucan polymers. Cellulose is not easily degradable and considered an important component for papermaking being directly related to pulp yields (Philip, 1992; Wood, 1981). Hemicelluloses contain heterogeneous groups of branched structures of polysaccharides that mainly consist of xylose, glucose, and mannose. Saeed et al. (2017) reported that raw materials having cellulose proportion more than 34% make them suitable for pulp and papermaking. The ash content in sarkanda is high as compared to wood and can be reduced by acid treatment. However, wheat straw also contains ash content of approximately 7% (Tripathi et al., 2018a) and is utilized successfully in Indian agro-based paper mills. Sarkanda has less extractive content as compared to other agro-residues like rice straw, wheat straw, and bagasse (Sharma et al., 2020; Kaur et al., 2018; Tripathi et al., 2013).

10.5.2 FTIR–ATR analysis of sarkanda pulp

Fourier transform infrared – attenuated total reflectance (FTIR–ATR) spectrum of handsheet of unbleached sarkanda pulp has been shown in Fig. 10.2. Spectrum demonstrates an absorption band in the range of 3445–3250 cm^{-1}, which corresponds to the OH linkage of polysaccharides and lignin. Absorption peak at 3340 cm^{-1} is related to the O(3)H...O(5) intermolecular H linkage (Kaur et al., 2019b). The absorption peak near 2900 cm^{-1} corresponds to stretching vibration of the C–H or alkyl group (specifically methyl) due to structural backbone of cellulose and hemicelluloses present in the sarkanda fiber (Singh et al., 2014). Bands at 1243–1317 cm^{-1} have been attributed to the skeletal vibrations of lignin (Reis et al., 2017). High absorbance around 1030 cm^{-1} can be allotted to the extension of C–O–C bonds. The absorption peak at 1648 cm^{-1} denotes the vibration due to the presence

TABLE 10.1 Chemical constituents of sarkanda and other nonwoods.

Parameters	Extractives (%)	Cellulose (%)	Hemicellulose (%)	Lignin (%)	Ash (%)
Sarkanda (Sharma et al., 2020)	2.3 ± 0.13[a]	36.2 ± 0.5	35.2 ± 0.41	21.2 ± 0.22	8.3 ± 0.3
Sesame (Saeed et al., 2017)	–	42.3 ± 2.2	22.1 ± 2.1	20.8 ± 1.3	3.8 ± 1.0
Wheat straw (Tripathi et al., 2018a)	3.2[b]	36.6–47.8	23.6–28.1	16.4–19.9	5.2–7.1
Karkadeh (Saeed et al., 2017)	–	45.8 ± 2.5	22.8 ± 2.4	19.3 ± 1.4	3.2 ± 1.1
Bagasse (Leponiemi, 2008)	2.7[b]	32–47.2	24.7–27.0	19.2–24.0	1.5–2.6
Bamboo (Sridach, 2010)	–	26–43	–	21–31	1.7–5
Reed grass (Sridach, 2010)	–	45	–	22	3
Rice straw (Kaur et al., 2018)	2.43 ± 0.36[a]	33.3 ± 0.47	27.3 ± 0.36	13.0 ± 0.07	12.6 ± 0.11

[a] *Acetone extractives.*
[b] *Ethanol–benzene extractives.*

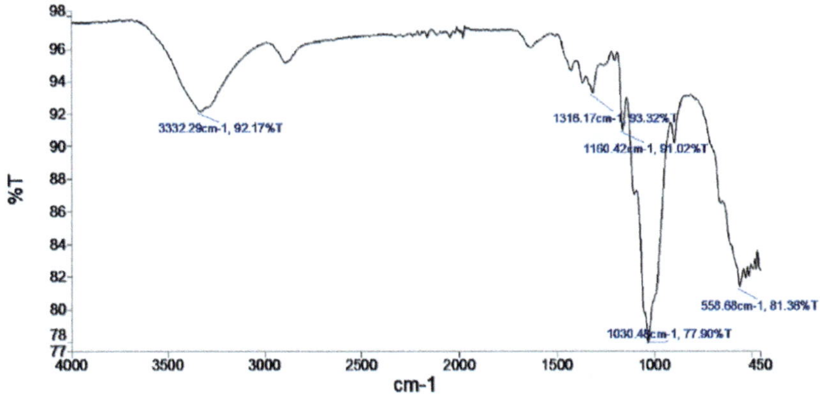

FIGURE 10.2 Fourier transform infrared – attenuated total reflectance (FTIR-ATR) spectrum of sarkanda unbleached pulp.

of lignin (Bhardwaj et al., 2006). Moreover, the peak around 1648 cm^{-1} for C–O corresponded to the ether and ester crosslink between lignin and cellulose, or hemicelluloses and cellulose. Sarkanda fiber showed a peak at 1740 cm^{-1} attributed to C = O bond in carbonyl or carboxyl group (Singh et al., 2014).

10.5.3 Morphological properties

Fiber morphology of any raw material is a significant property that depicts the feasibility or possibility of its application in papermaking. Long fiber length provides the benefit of high tear strength property (Dwivedi et al., 2010; Monje et al., 2010). On the other hand, pulp containing very long fibers causes problem in paper formation and makes nonuniform paper sheet. Kamoga et al. (2016) reported that mechanical properties of paper are affected by fiber length. Fiber length of sarkanda pulp was higher than many agro-residues like rice straw and wheat straw. Sharma et al. (2020) reported the fiber dimensions to be 0.939-mm length and 16.1-μm width (Table 10.2). Karkadeh, sesame, and okra pulp fibers also have fiber length in the range of around 0.9, 0.8, and 0.7 mm. The fiber length of these nonwood fibers is lower than that of softwoods, that is, 2.7–4.6 mm, but within the range of hardwood pulp fibers, that is, 0.7–3.0 mm (Jahan et al., 2006).

Like other nonwoods, sarkanda pulp is also having heterogeneous nature as compared to wood. Pulp of Poaceae family plants contains fiber cells of about 65%–70% by weight; pulp also contains fines, parenchyma cells, vessel elements, epidermis, etc. This makes the nonwood pulp different from wood pulp. Increment in fine content decreases the drainage time and consequently decreases the machine speed in papermaking. The microscopic view of sarkanda pulp is presented in Fig. 10.3.

TABLE 10.2 Comparison of morphological properties of sarkanda with other agro-residues.

Parameters	Sarkanda (Sharma et al., 2020)	Sesame (Saeed et al., 2017)	Wheat straw (Tripathi et al., 2019)	Karkadeh (Saeed et al., 2017)	Bagasse (Subrahmanyam et al., 2004)	Rice straw (Kiaei, 2014)
Fiber length (mm)	0.939	0.8 ± 0.2	0.818	0.9 ± 0.3	0.85–1.52	0.843
Width (μm)	16.1	16.3 ± 1.2	17.5	14.0 ± 1.5	21.8–22.8	10.3
Coarseness (μg/m)	85.0	–	71.8	–	129.4	–
0.2–0.5 mm length (%)	33.9		–	–	–	–
0.5–1.5 mm length (%)	48.3	–	–	–	–	–
1.5–3 mm length (%)	15.3	–	–	–	–	–
3.0–4.5 mm length (%)	2.5	–	–	–	–	–
Fines (length–length) (%)	2.7	–	–	–	–	–
Kink index	1.304	0.21	–	0.14	–	–
Number of fibers per gram	107,970	–	–	–	–	–

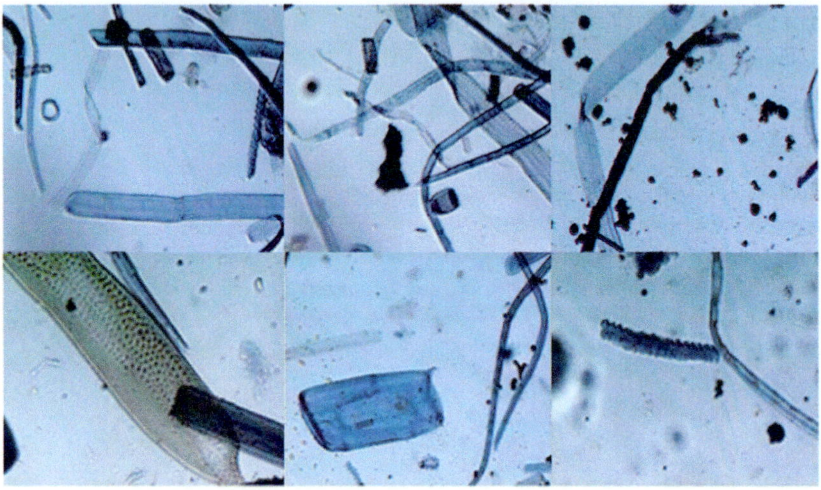

FIGURE 10.3 Microscopic view of sarkanda pulp.

10.5.4 Cooking of sarkanda

Before pulping of sarkanda, long grass culms are cut into short pieces of around 2 – 3 cm, which can be used without problems in the cooking study of sarkanda. Chopping off the raw material into shorter pieces provides larger accessible area to facilitate the movement of cooking liquor into the furnish and subsequent reaction of chemicals with the polymers for consistent pulping. After cutting, wet cleaning of furnish is carried out to remove any undesired impurities like nonfibrous materials and dust particles. Following the wet cleaning, the raw material is air dried to attain nearly 85% dryness to store in bags for subsequent use. This practice helped to keep the moisture uniform throughout the study. For pulping the alkali concentration, AQ dose is calculated on oven dry (o.d.) weight basis of raw material, and other pulping conditions, like temperature and cooking time, were optimized to get a pulp of kappa number 15. After pulping, the material was washed to remove residual cooking liquor, and after that, pulp was squeezed to remove extra water and stored for analysis, like pulp yield, kappa number, brightness, and viscosity.

Sarkanda pulp yield was reported from 58.5% to 52.3% with different dose of alkali from 13.0% to 18.0%. Degradation of cellulose content occurred with increased dose of active alkali, which was also found for other agro-residues like rice straw (Bhardwaj et al., 2006; Kaur et al., 2018). Increase in cooking time and alkali dose reduced the kappa number, reject content, and viscosity of pulp. Approximately 15 kappa number was obtained with 15% active alkali and 25-min cooking time (Sharma et al., 2020). The black liquor properties were also similar to other agro-residues; active alkali, total solids, and pH in black liquor increased with increment in cooking chemical charge.

10.5.5 Physical strength characteristics of sarkanda pulp

Sarkanda pulp has comparable physical strength properties in comparison to different agro-residues. Physical strength possessions, like tear, burst indices, and double fold, were higher as compared to those of wheat straw, rice straw, and bagasse pulps, whereas tensile index of sarkanda pulp was lower than that of bagasse pulp but higher than that of rice straw and wheat straw pulps.

Table 10.3 depicts the physical strength qualities of elemental chlorine-free (ECF) ($ODE_{OP}D$) and total chlorine-free (TCF) ($Z_{EOP}P_{O}P_{O}$)-bleached sarkanda pulp and its comparison with other pulp samples. Strength properties of ECF-bleached sarkanda pulp were at higher side in comparison to sarkanda pulp bleached by TCF sequence. Scanning electron microscopy (SEM) analysis reveals that bleached pulp shows the damaged surface in comparison to an unbleached pulp. Breaking length, burst, and tensile index of ECF-bleached pulp were better in comparison to previously reported values for rice straw and bagasse pulps but lesser than that of wheat straw pulp. Also, sarkanda pulp had higher tear index of 5.7 mN (m^2/g), as compared to rice straw pulp, while lesser than the wheat straw and bagasse pulps. Double-fold value for sarkanda is also higher than other nonwood pulp samples.

10.5.6 Bleaching of sarkanda pulp

Sarkanda pulp has the potential to be bleached with environment-friendly bleaching sequences. Sharma et al. (2020) revealed that higher brightness can be obtained by bleaching the sarkanda pulp with $DE_{OP}D$ (ECF) sequence as compared to the other straws, like wheat straw and rice straw pulps. Although the brightness of bleached sarkanda pulp was better than the straw pulps, it was inferior to that of bagasse pulp (Table 10.4). Another researcher, Ghosh (2006), also reported that pulp of wheat straw bleached through $DE_{P}D$ sequence obtains 85.0% ISO brightness. Sarkanda pulp achieved higher optical properties when bleached through ECF bleaching in comparison to TCF bleaching. Viscosity value was lower of TCF pulp, 7.5 cP, in comparison to viscosity value of ECF pulp, 10.5 cP. Many previous studies claimed that TCF bleaching sequence degraded the carbohydrate structure more in comparison to ECF bleaching sequence. TCF bleaching sequences involve the ozone and pressurized hydrogen peroxide stage; these bleaching chemicals attack the carbohydrates in the presence of metal ions and therefore degraded the pulp viscosity.

10.5.7 Surface morphological characteristics of pulps by SEM

Surface morphology of sarkanda unbleached and bleached pulps has been demonstrated in Figs. 10.4 and 10.5. The unbleached pulp has smooth surface and is less fibrillated. Following by bleaching, fiber surface gets

TABLE 10.3 Physical strength properties of unbleached and bleached pulps.

Pulp	Bleaching sequence	°SR	Breaking length (m)	Double fold	Burst index [kN (m^2/g)]	Tear index [mN (m^2/g)]	References
Sarkanda	Unbleached	32.0	5192	78	4.30	6.00	Sharma et al. (2020)
	$ODE_{OP}D$	32.5	5039	222	4.00	5.70	Sharma et al. (2020)
	$OZ_{EOP}P_OP_O$	33.0	3488	188	2.80	5.50	Sharma et al. (2020)
Rice straw	Unbleached	34.0	4304	76	2.46	–	Kaur et al. (2018)
	$DE_{OP}D$	46.0	4784	11	3.34	5.42	Kaur et al. (2018)
Wheat straw	Unbleached	29.0	4498	62	3.51	5.91	Kaur et al. (2019a,b)
	$ODE_{OP}D$	38.5	5294	18	4.90	6.70	Tripathi et al. (2018b)
Bagasse	Unbleached	37.0	6191	51	3.65	3.93	Garg and Singh (2004b)
	$ODE_{OP}D$	19.0	4447	51	3.13	6.06	Tripathi et al. (2013)

TABLE 10.4 Comparison of optical properties of sarkanda with other raw materials.

Particulars	Bleaching sequence	Brightness (% ISO)	CIE whiteness	ASTM yellowness	References
Sarkanda	$ODE_{OP}D$	85.7	77.7	4.58	Sharma et al. (2020)
	$OZE_{OP}P_OP_O$	81.9	73.7	4.90	Sharma et al. (2020)
Rice straw	$DE_{OP}D$	82.2	74.1	6.01	Kaur et al. (2018)
Wheat straw	$ODE_{OP}D$	84.5	79.4	–	Tripathi et al. (2018b)
Bagasse	$ODE_{OP}D$	89.3	83.3	–	Tripathi et al. (2013)

CIE, Commission Internationale de l'Elcairage; *ASTM*, American Society for Testing and Materials.

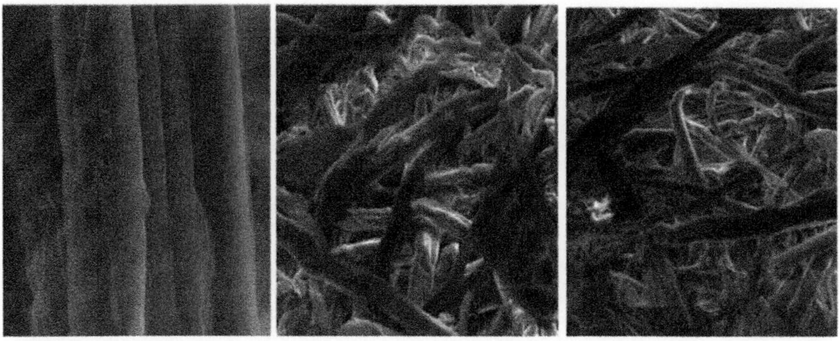

FIGURE 10.4 Scanning electron micrograph images at 5000, 1000, and 500 × magnification of unbleached pulp.

damaged due to reaction of bleaching chemicals. Bleaching process involves the bleaching chemicals that mainly react with lignin, and as lignin is attached to carbohydrates, therefore, carbohydrates are also exposed to bleaching chemicals. Extent of fiber surface degradation depends on the nature of bleaching chemicals.

10.5.8 Bleaching effluents characteristics

The characteristics of bleaching effluent of sarkanda pulp generated from both the bleaching sequences (ECF and TCF) have been shown in Table 10.5.

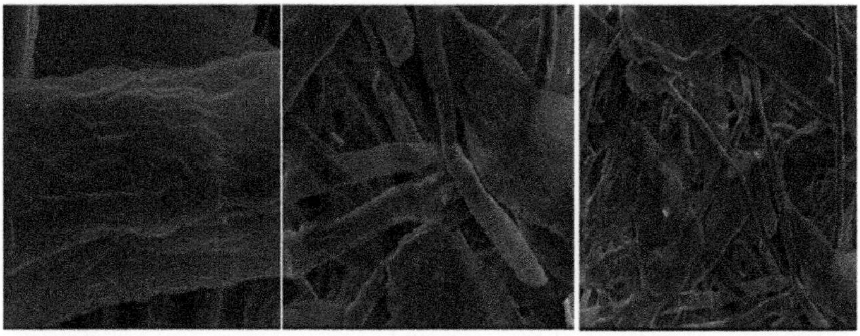

FIGURE 10.5 Scanning electron micrograph images at 5000, 1000, and 500 × magnification of bleached pulp.

TABLE 10.5 Characteristics of bleaching effluent of sarkanda.

Particulars	Bleaching sequence	AOX (mg/L)	COD (mg/L)	TDS (mg/L)	Color (PCU)	References
Sarkanda	$ODE_{OP}D$	21.0	745 ± 10	2100	522	Sharma et al. (2020)
	$OZE_{OP}P_OP_O$	–	2022	3312	677	Sharma et al. (2020)
Rice straw	$DE_{OP}D$	35.5	1422	3120	648	Kaur et al. (2018)
Wheat straw	$ODE_{OP}D$	11.0	1120	2550	406	Tripathi et al. (2018b)
Bagasse	$DE_{OP}D$	–	1976	–	810	Sudarshan et al. (2018)

AOX, adsorbable organic halogens; *COD*, chemical oxygen demand; *TDS*, total dissolved solids.

Collective effluent produced from different bleaching stages of ECF bleaching sequence has 522 PCU color, 21.0 mg/L adsorbable organic halogens (AOX), and 745 mg/L chemical oxygen demand (COD). The COD value of the bleaching effluent from rice straw and wheat straw was higher in comparison to bleaching effluent of sarkanda. Sharma et al. (2020) revealed that rice straw and wheat straw pulps had more fines and parenchyma cells responsible for high COD in comparison to sarkanda. Other environmental parameters like AOX, total dissolved solids (TDS), and color of sarkanda bleaching effluent were similar to reported values of other nonwood materials.

ECF-bleached pulp effluent had lower color, COD content, and TDS content as compared to those with TCF-bleached pulp effluent. During TCF bleaching sequence, the degradation of polysaccharides with ozone and peroxide stages was more, and this may be responsible for higher values. According to the study by Tripathi et al. (2018b), greater extent of carbohydrate depolymerization occurred in wheat straw pulp after bleaching with ozone in the presence of metal ions. As a consequence of considerable amount of degraded carbohydrates due to those bleaching sequences, the pollutant load was also found to be increased in the effluent. On the other hand, chemical recovery system can recycle the effluent generated during TCF bleaching as the chloride compounds are absent. This may provide a way to set up a bleaching process with zero discharge.

10.6 Conclusion

Sankanda has the capability to be utilized commercially for papermaking. Sarkanda could be a proficient raw material for pulp and papermaking. Chemical composition and other properties make the sarkanda suitable for papermaking. It required ordinary cooking chemical charge and time to produce bleachable pulp of 15.0 kappa number that is an appropriate kappa number for bleaching of pulp. Sarkanda pulp has superior physical strength and morphological characteristics in contrast to rice straw and wheat straw. Good optical properties can be achieved for sarkanda with environment-friendly bleaching techniques. Based on the provided information, it can be stated that sarkanda is a suitable raw material to produce excellent writing and printing-grade paper. The production of pulp and paper from sarkanda will be favorable to paper manufacturing units, villagers, as well as useful for waste management. Utilization of sarkanda for pulp and papermaking can meet the requirement of paper mills looking for a substitute to wood-based raw material.

Acknowledgments

Authors are thankful to family members as a major portion of this chapter was written during lockdown period of pandemic Covid-19.

References

Agnihotri, S., Dutt, D., Tyagi, C.H., 2010. Complete characterization of bagasse of early species of *Saccharum officinarum*-Co 89003 for pulp and paper making. Bioresources 5 (2), 1197–1214.

Ashori, A., 2006. Non-wood fibers—a potential source of raw material in papermaking. Polym. Technol. Eng. 45 (10), 1133–1136. Available from: https://doi.org/10.1080/03602550600728976.

Atchison, J.E., 1996. Twenty five years of global progress in non-wood plant fiber repulping. Tappi J. 79 (10), 87–95.

Ateş, S., Deniz, I., Kirci, H., Atik, C., Okan, O.T., 2015. Comparison of pulping and bleaching behaviors of some agricultural residues. Turk. J. Agric. For. 39 (1), 144–153.

Bambach, M.R., 2017. Compression strength of natural fibre composite plates and sections of flax, jute and hemp. Thin Walled Struct. 119, 103–113.

Bhardwaj, N.K., Vinh, Q.D., Kien, L.N., 2006. Determination of carboxyl content in high-yield kraft pulps using photoacoustic rapid-scan Fourier transform infrared spectroscopy. Anal. Chem. 78 (19), 6818–6825.

Dillen, J.R., Dillen, S., Hamza, M.F., 2016. Pulp and paper: wood sources. Reference Module in Materials Science and Materials Engineering. Elsevier. Available from: http://doi.org/10.1016/B978-0-12-803581-8.09802-7.

Dwivedi, P., Vivekanand, V., Pareek, N., Sharma, A., Rajesh, P.S., 2010. Bleach enhancement of mixed wood pulp by xylanase–laccase concoction derived through co-culture strategy. Appl. Biochem. Biotechnol. 160 (1), 255–268.

Available from: https://www.flowersofindia.net/catalog/slides/Sarkanda.html as accessed on 31 August 2022.

Fu-Wang, H., Chin, H., Zhi-bin, H.E., 1996. The characteristics of tobacco stem. In: Proc. 3rd Int. Non-wood Fiber Pulping Papermaking Conf., Beijing, China, pp. 81–90.

Garg, M., Singh, S.P., 2004a. Response of bagasse and wheat straw recycled pulps to refining. Tappi J. 3 (10), 11–17.

Garg, M., Singh, S.P., 2004b. Recycling potential of bagasse and wheat straw pulps. Tappi J. 3 (9), 25–31.

Ghosh, U.K., 2006. Short sequence environment friendly bleaching of wheat straw pulp. J. Sci. Ind. Res. 65 (1), 68–71.

Gominho, J., Fernandez, J., Pereira, H., 2001. *Cynara cardunculus* L.—a new fibre crop for pulp and paper production. Ind. Crops Prod. 13 (1), 1–10.

González-García, S., Teresa Moreira, M., Artal, G., Maldonado, L., Feijoo, G., 2010. Environmental impact assessment of non-wood based pulp production by soda-anthraquinone pulping process. J. Cleaner Prod. 18 (2), 137–145.

Jahan, M., Sarwar, D.A., Nasima, Chowdhury, Khalidul Islam, M., 2006. Characterization and evaluation of golpata fronds as pulping raw materials. Bioresour. Technol. 97 (3), 401–406.

Jain, N., Bhatia, A., Pathak, H., 2014. Emission of air pollutants from crop residue burning in India. Aerosol Air Qual. Res. 14 (1), 422–430.

Janson, J., Jousimaa, T., Backman, R., 1996. Phosphate-based pulping of agrofibre including papermaking and spent liquor recovery. Nord. Pulp Pap. Res. J. 11 (1), 4–9.

Jiménez, L., Pérez, I., De la Torre, M.J., López, F., Ariza, J., 2000. Use of formaldehyde for making wheat straw cellulose pulp. Bioresour. Technol. 72 (3), 283–288.

Jiménez, L., Rodríguez, A., Pérez, A., Moral, A., Serrano, L., 2008. Alternative raw materials and pulping process using clean technologies. Ind. Crops Prod. 28 (1), 11–16.

Kamoga, O.L., Kirabira, J.B., Byaruhanga, J.K., Godiyal, R.D., Anupam, K., 2016. Characterisation and evaluation of pulp and paper from selected Ugandan grasses for paper industry. Cellul. Chem. Technol. 50 (2), 275–284.

Kaur, D., Nishi, K.B., Lohchab, R.K., 2018. A study on pulping of rice straw and impact of incorporation of chlorine dioxide during bleaching on pulp properties and effluents characteristics. J. Cleaner Prod. 170, 174–182.

Kaur, D., Nishi, K.B., Lohchab, R.K., 2019a. Environmental aspect of using chlorine dioxide to improve effluent and pulp quality during wheat straw bleaching. Waste Biomass Valoriz. 10 (5), 1231–1239.

Kaur, D., Nishi, K.B., Lohchab, R.K., 2019b. Effect of incorporation of ozone prior to ECF bleaching on pulp, paper and effluent quality. J. Environ. Manage. 236, 134–145.

Kellogg, E., Richard Abbott, J., Bawa, K., et al., 2020. Checklist of the grasses of India. PhytoKeys 163, 1. Available from: https://doi.org/10.3897/phytokeys.163.38393.

Khare, C.P., 2007. Indian medicinal plants. In: Khare, C. (Ed.), *Saccharum munja* Rox. Springer, New York, pp. 52–53. Available from: http://doi.org/10.1007/978-0-387-70638_1394.

Khristova, P., Gabir, S., Bentcheva, S., Dafalla, S., 1998. Soda-anthraquinone pulping of sunflower stalks. Ind. Crops Prod. 9 (1), 9–17.

Khristova, P., Kordsachia, O., Patt, R., Karar, I., Khider, T., 2006. Environmentally friendly pulping and bleaching of bagasse. Ind. Crops Prod. 23 (2), 131–139.

Kiaei, M., 2014. Investigation on biometrical properties and mineral content of rice residues and its application in pulp and paper production.". Adv. Environ. Biol. 8 (13), 952–959.

Kozlowski, R., Baraniecki, P., Barriga-Bedoya, J., 2005. Bast fibres (flax, hemp, jute, ramie, kenaf, abaca). *Biodegradable and Sustainable Fibres*. Woodhead Publishing Limited and CRC Press LLC, pp. 36–88. Available from: http://doi.org/10.1533/9781845690991.36.

Laftah, W.A., Abdul Rahman, W.A.W., 2016. Pulping process and the potential of using non-wood pineapple leaves fiber for pulp and paper production: a review. J. Nat. Fibers 13 (1), 85–102. Available from: https://doi.org/10.1080/15440478.2014.984060.

Leponiemi, A., 2008. Non-wood pulping possibilities-a challenge for the chemical pulping industry. Appita: Technol. Innov. Manuf. Environ. 61 (3), 234.

Liu, Z., Wang, H., Hui, L., 2018. Pulping and papermaking of non-wood fibers. In: Newaz Kazi, S. (Ed.), Pulp and Paper Processing, vol. 1. IntechOpen, pp. 4–31Available from: http://doi.org/10.5772/intechopen.79017. Available from: https://www.intechopen.com/books/pulp-and-paper-processing/pulping-and-papermaking-of-non-wood-fiber.

López, F., García, J.C., Pérez, A., García, M.M., Feria, M.J., Tapias, R., 2010. *Leucaena diversifolia* a new raw material for paper production by soda-ethanol pulping process. Chem. Eng. Res. Des. 88 (1), 1–9. Available from: https://doi.org/10.1016/j,cherd.2009.06.016.

Mishra, T., Pandey, V.C., Singh, P., Singh, N.B., Singh, N., 2017. Assessment of phytoremediation potential of native grass species growing on red mud deposits. J. Geochem. Explor. 182, 206–209.

Monje, P.G., González-García, S., Moldes, D., et al., 2010. Biodegradability of kraft mill TCF biobleaching effluents: application of enzymatic laccase-mediator system. Water Res. 44 (no. 7), 2211–2220.

Mousavi, S.M., Mazhari, S.Z., Hosseini, H., Resalati, S., Mahdavi, Garmaroody, E.R., 2013. Papermaking potential of rapeseed straw, a new agricultural-based fiber source. J. Cleaner Prod. 52, 420–424.

Olsson, R., 1993. Production methods and costs for reed canary grass as an energy crops. Bioenergy Res. Programme, Publ. 2, 14–15.

Pandey, V., Chandra, O., Bajpai, D.N., Pandey, Singh, N., 2015. *Saccharum spontaneum*: an underutilized tall grass for revegetation and restoration programs. Genet. Resour. Crop. Evol. 62 (no. 3), 443–450. Available from: https://doi.org/10.1007/s10722-014-0208-0.

Pandey, V.C., Singh, K., Singh, R.P., Singh, B., 2012. Naturally growing *Saccharum munja* L. on the fly ash lagoons: a potential ecological engineer for the revegetation and stabilization. Ecol. Eng. 40, 95–99.

Pande, H., Roy, D.N., 1996. Delignification kinetics of soda pulping of kenaf. J. Wood Chem. Technol. 16 (3), 311–325.

Peralta, A.G., 1996. Pulp produced from decorticated abaca fiber. Tappi J. 79 (3), 263–266.

Philip, J., 1992. Biosynthesis of Major Crop Products. John Wiley and Sons, Chichester, UK, p. 154.

Radiotis, T., Li, J., Goel, K., Eisner, R., 1999. Fiber characteristics, pulpability, and bleachability of switchgrass. Tappi J. 82 (7).

Rahar, S., Nagpal, N., Swami, G., Nagpal, M.A., Kapoor, R., 2011. Pharmacognostical studies of *Saccharum munja* Roxb. Root. Int. J. PharmTech Res. 3 (2), 792–800.

Rau, N., Mishra, V., Sharma, M., Das, M.K., Ahaluwalia, K., Sharma, R.S., 2009. Evaluation of functional diversity in rhizobacterial taxa of a wild grass (*Saccharum ravennae*) colonizing abandoned fly ash dumps in Delhi urban ecosystem. Soil. Biol. Biochem. 41 (4), 813–821.

Reis, C.L.B., Alexandre e Silva, L.M., Soares Rodrigues, T.H., Nogueira Félix, A.K., de Santiago-Aguiar, R.S., Marques Canuto, K., et al., 2017. Pretreatment of cashew apple bagasse using protic ionic liquids: enhanced enzymatic hydrolysis. Bioresour. Technol. 224, 694–701.

Rodríguez, A., Moral, A., Serrano, L., Labidi, J., Jiménez, L., 2008. Rice straw pulp obtained by using various methods. Bioresour. Technol. 99 (8), 2881–2886.

Rodríguez, A., Sánchez, R., Requejo, A., Ferrer, A., 2010. Feasibility of rice straw as a raw material for the production of soda cellulose pulp. J. Cleaner Prod. 18 (10–11), 1084–1091.

Saeed, H.A.M., Liu, Y., Chen, H., Lucia, L.A., 2017. Suitable approach using agricultural residues for pulp and paper manufacturing. Nord. Pulp Pap. Res. J. 32 (4), 674–682.

Saijonkari-Pahkala, K., 2001. Non-Wood Plants Raw Material for Pulp and Paper.

Saleem, U., Amin, S., Ahmad, B., Azeem, H., Anwar, F., Mary, S., 2017. Acute oral toxicity evaluation of aqueous ethanolic extract of *Saccharum munja* Roxb. roots in albino mice as per OECD 425 TG. Toxicol. Rep. 4, 580–585.

Scurlock, J.M.O., 1999. Miscanthus: A Review of European Experience With a Novel Energy Crop. United States. https://doi.org/10.2172/4538. https://www.osti.gov/servlets/purl/4538.

Sharma, A., Kumar, K., Anupam, V., Swaroop, P.S., Lal, Bist, V., 2015. Pilot scale soda-anthraquinone pulping of palm oil empty fruit bunches and elemental chlorine free bleaching of resulting pulp. J. Cleaner Prod. 106, 422–429.

Sharma, A., Kumar, R., Godiyal, Thapliyal, B.P., 2019. Kans grass – a promising raw material for papermaking. Cellul. Chem. Technol. 53 (7–8), 747–753.

Sharma, M., Mishra, V., Rau, N., Sharma, R.S., 2011. Functionally diverse rhizobacteria of *Saccharum munja* (a native wild grass) colonizing abandoned morrum mine in Aravalli hills (Delhi). Plant Soil 341 (1), 447–459.

Sharma, N., Tripathi, S.K., Bhardwaj, N.K., 2020. Utilization of sarkanda for making pulp and paper using elemental chlorine free and total chlorine free bleaching processes. Ind. Crops Prod. 149. Available from: https://doi.org/10.1016/j.indcrop.2020.112316.

Shu, W.S., Ye, Z.H., Zhang, Z.Q., Lan, C.Y., Wong, M.H., 2005. Natural colonization of plants on five lead/zinc mine tailings in Southern China. Restor. Ecol. 13 (1), 49–60.

Singh, G.P., Madiwale, P.V., Jagtap, R.N., Adivarekar, R.V., 2014. Extraction of fibers from *Saccharum munja* grass and its application in composites. J. Appl. Polym. Sci. 131 (19). Available from: https://doi.org/10.1002/APP.40829.

Sridach, W., 2010. The environmentally benign pulping process of non-wood fibers. Suranaree J. Sci. Technol. 17 (2), 105–123.

Subrahmanyam, S.V., Godiyal, R., Janbade, R., Sharma, A., Kulkarni, A.G., 2004. Preparation of monograph of different fibrous raw materials used by Indian paper industry. In: CESS. Proj. Rep. CPPRI.

Sudarshan, K., Venkateshwaran, R., Kotteeswaran, P., Murugan, A., 2018. Treatment of bagasse based pulp mill bleach plant effluent by coagulation method. Res. J. Chem. Environ. 22 (9), 7–14.

Touzinsky, G.F., 1993. Kenaf. In: third ed Hamilton, F.; Leopold, B. (Eds.), Secondary Fibers and Non-wood Pulping, vol. 3. Pulp and Paper Manufacture; Joint Textbook Committee of the Paper Industry, Atlanta, GA; Montreal, pp. 106–109.

Tripathi, S., Alam, I., Bhardwaj, N.K., 2017. Effect of blending banana stem and hardwood pulps on sizing, ash retention, physical strength and optical properties of paper. Appita: Technol. Innov. Manuf. Environ. 70 (4), 378.

Tripathi, S.K., Bhardwaj, N.K., Ghatak, H.R., 2019. Effect of introducing ozone prior to elemental chlorine free bleaching of wheat straw pulp on pulp, paper and effluent characteristics. Cellul. Chem. Technol. 53 (1–2), 105–112.

Tripathi, S., Kumar, N.K., Bhardwaj, Ghatak, H.R., 2018a. Determination of main parameters affecting ozone bleaching of wheat straw pulp using Plackett–Burman design. Ozone: Sci. Eng. 40 (2), 148–156.

Tripathi, S., Kumar, N.K., Bhardwaj, Ghatak, H.R., 2018b. Improvement in pulp quality and effluent properties using methanol as carbohydrate protector during ozone bleaching of wheat straw pulp. Appita: Technol. Innov. Manuf. Environ. 71 (4), 338.

Tripathi, S., Singh, S., Gangwar, A., Mishra, O.P., Chakrabarti, S.K., Bhardwaj, N.K., et al., 2013. Blending of banana stem with wheat straw and bagasse to enhance physical strength properties of paper. IPPTA J. 25 (2), 121–125.

Vasudeva, R., 2019. Vanishing realm of basketry in Punjab. Res. Rev. Int. J. Multidiscip. 4 (6), 1435–1438.

Vasudevan, P., Gujral, G.S., Madan, M., 1984. *Saccharum munja* Roxb., an underexploited weed. Biomass 4 (2), 143–149.

Ververis, C., Georghiou, K., Christodoulakis, N., Santas, P., Santas, R., 2004. Fiber dimensions, lignin and cellulose content of various plant materials and their suitability for paper production. Ind. Crops Prod. 19 (3), 245–254.

Wahit, M., Uzir, N.I., Akos, Laftah, W.A., 2012. Influence of natural fibers on the mechanical properties and biodegradation of poly (lactic acid) and poly (ε-caprolactone) composites: a review. Polym. Compos. 33 (7), 1045–1053.

Ward, A.J., Phil, J.H., Holliman, P.J., David, L.J., 2008. Optimisation of the anaerobic digestion of agricultural resources. Bioresour. Technol. 99 (17), 7928–7940.

Wood, I.M., 1981. The utilization of field crops and crop residues for paper pulp production. Field Crop Abstr. 34, 557–568.

Chapter 11

Pulping and papermaking of reed bagasse

Widya Fatriasari[1], Azizatul Karimah[1,2] and Nyoman J. Wistara[2]
[1]Research Center for Biomass and Bioproducts, National Research and Innovation Agency (BRIN), Cibinong, Bogor, West Java, Indonesia, [2]Department of Forest Products, Faculty of Forestry, IPB University, Dramaga, Bogor, West Java, Indonesia

11.1 Introduction

The process of removing cellulose is known as pulping to produce pulp for papermaking. Currently, wood (hardwood or softwood) is still the primary source of raw material in the pulp and paper industry, numbering almost 91% of all types of raw materials (Bajpai et al., 2004). From 2023 to 2027, the size of global paper and paperboard market is predicted to grow with a 4.14% yearly compound growth rate, reaching US$409.50 billion by 2027 (Mordor Intelligence, 2023). Paper and paperboard demand is forecast to expand at a rate of 2.1% each year until 2020 with developing regions also predicted to increase up to 4%. In the future paper market, China will play the most important role (Szabó et al., 2009). While paper consumption increases significantly in North America, Europe and Asia were reported to be the dominant consumers, contributing more than 90% to global paper and paperboard consumption (Szabó et al., 2009). Thus the consumption of wood fiber to make pulp and paper experiences a tendency of constant growth (Kassim et al., 2015a).

Nonwood plants are interesting candidates for alternative resources in the industry of pulp and paper. Such consideration is proposed because nonwood plants wield abundant potential at lesser costs. The continuity and abundance of resources in the pulp and paper industry are important factors that need to be addressed. The idea of utilizing in the pulp and paper industry, nonwood plants are being used as an alternative resource is strengthened by increasing concern for environmental issues caused by the utilization of wood in pulp and papermaking, mainly deforestation. El-Sakhawy et al. (1995) noted that approximately 10% of total pulp and paper production in the world used nonwood raw materials, which included straw (44%), bagasse (18%), reeds

Pulping and Papermaking of Nonwood Plant Fibers. DOI: https://doi.org/10.1016/B978-0-323-91625-7.00003-5

(14%), bamboo (13%), and others (11%) (Sridach, 2010). China led the use of nonwood-based pulp production around the world (Hammett et al., 2001). On the other hand, Indonesia is 1 of the top 20 global pulp and paper producers. In an industrial plantation forest (HTI), fast-growing wood species are planted, such as *Acacia* and *Eucalyptus*, are primary sources of virgin fiber for pulp and paper industry in Indonesia (Sari et al., 2018). Indonesia's capacity for pulp production is estimated to be roughly 11.6 million tons will be produced in 2020 [Asosiasi Pulp dan Paper Indonesia. (2016)]. According to Ahmad et al. (2018), Indonesia's pulp and paper industry contributes significantly to the country's economy.

Nonwood plants have been developed for various applications in composite, textile, fuels, enzymes, chemicals, animal feed, and pulp and paper production (Reddy and Yang, 2005; Khalil et al., 2006). Several nonwood biomasses, namely, pineapple leaves (Asim et al., 2015), bamboo (Fatriasari et al., 2010; Fatriasari and Risanto, 2011; Fatriasari et al., 2009), *Musa* stem (Ramdhonee and Jeetah, 2017), oil palm empty fruit bunches (Ibrahim, 2003), rice straw (Kaur et al., 2017), switchgrass (Madakadze et al., 2010), giant grass (Madakadze et al., 2010), and reed grass (Sutiya et al., 2012; Kassim et al., 2015a; 2015b; Kassim et al., 2016), have been reported to be potential alternative sources of virgin fiber in pulp and paper production.

Reed grass or cogon grass (*Imperata cylindrica*) is a perennial grass species found in open areas in tropical and subtropical regions (Hubbard et al., 1944). Cogon grass can be used in soil stabilization, though it has aggressively invaded all continents except Antarctica from its original habitat in Southeast Asia (MacDonald, 2004) and caused problems for plantations (Sutiya et al., 2012; MacDonald, 2004). Cogon grass has rough, pointed leaves that make it unsuitable for grazing or commercial use. As such, this species is proposed to become an alternative virgin fiber source in the pulp and paper industry (Kassim et al., 2016). Furthermore, based on its chemical properties, reed grass is suitable to produce pulp and paper sheets because it is low in lignin (Kassim et al., 2016; Sutiya et al., 2012), hot water, and 1% NaOH solubility, as well as high in cellulose and hemicellulose (Sutiya et al., 2012; Kassim et al., 2016). Evaluation of the physical properties of cogon grass showed that its fiber is of moderate length, diameter, and felting rate compared to the fibers of cocoa pod husk and oil palm leaf. Cogon grass fiber also has long fiber with thin cell walls (Sari et al., 2018; Sutiya et al., 2012). Thus cogon grass is prospective for active packaging material (Zulfiana et al., 2020). Long fiber facilitates strong interfiber bonding that allows the production of tight and strong paper sheets (Zulfiana et al., 2020). Using cogon grass for pulp and paper production will help to alleviate the pulp and paper industry's negative environmental effects as well as increase the economic value of an invasive plant species. In this chapter, the potential of reed grass as a future raw material of high-quality pulp and paper was discussed.

11.2 The potential of reed grass (*Imperata cylindrica*) as pulp and paper

The most widely known potential of reed grass is being a main ingredient in traditional medicine, ranking sixth among plants most widely used as herbal medication. Reed grass is commonly used to treat fever, heartburn, malaria, cancer, diabetes, kidney stones, jaundice, and headaches, as well as urine smoothing (Subositi and Widodo, 2018). As a weedy species, reed grass is among the 10 worst weeds in the world, noted as a hazard for 35 crops in 73 countries (Holm et al., 1977; MacDonald, 2004) due to its quick growth and tolerance to nutrient-poor soil (Zulfiana et al., 2020). Based on Lucia (2005), only 14.3% of reed grass growing in China is used as alternative material in the local papermaking industry (Fig. 11.1). In addition, the utilization of reed grass other than that of its rhizomes to make herbal medication needs to be developed to solve the reed grass pest problem.

Table 11.1 shows some known utilizations of reed grass extract has been used as an antihypertensive and antibacterial substance (Prisdiany and Levita, 2019; Parkavi et al., 2012; Mulyadi et al., 2017). Reed grass pulp has been effectively employed in the pulp and paper industry as a nonwood alternative raw material, reinforcement for recycled paper, and the production of antimicrobial paper for active packaging (Zulfiana et al., 2020; Sutiya et al., 2012; Wibisono et al., 2012; Kassim et al., 2016; Kassim et al., 2015a; Sari et al., 2018). The high cellulose and low lignin content of reed grass also render it suitable as raw material for bioethanol production, wherein pretreated reed grass was subjected to enzymatic hydrolysis using *Pennicilium* sp. and then fermented by *Zymomonas mobilis* yeast (Kartikasari et al., 2013).

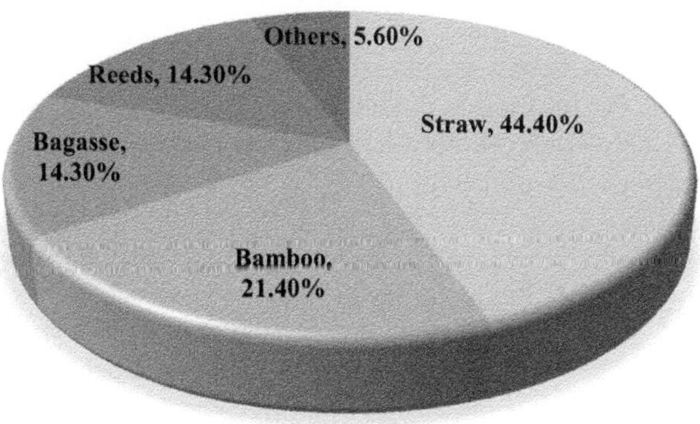

FIGURE 11.1 Ratio of the utilization of nonwood materials in papermaking (Lucia, 2005 in Liu et al., 2018). *Reprints with permission from Intech Open, License: Creative Commons Attribution License BY 3.0.*

TABLE 11.1 Utilizations of some species of reed grass.

Variation of reed grass	Utilizations	References
Reed grass (*Imperata cylindrica* L.)	Antimicrobial paper	Zulfiana et al. (2020)
	Raw material for pulp and paper	Sutiya et al. (2012) and Wibisono et al. (2012)
	Alternative material for pulp and paper	Kassim et al. (2016) and Kassim et al. (2015a)
	Antihypertensive substance	Prisdiany and Levita (2019)
	Antibacterial substance	Parkavi et al. (2012) and Mulyadi et al. (2017)
	Bioethanol production	Kartikasari et al. (2013)
	Reinforcement for recycled paper	Sari et al. (2018)
Giant reed grass (*Arundo donax* L.)	The pulp and paper substitutes	Shatalov et al. (2001)

11.3 The pulp and paper characteristics of reed grass

Fig. 11.2 depicts the reed grass plant and fresh reed grass stem. A plant with a high cellulose content, low lignin content, and long fiber must have the best quality as an alternative supply of fiber for the pulp and paper industry (Kassim et al., 2016; Sari et al., 2018). Cellulose, the most abundant component in lignocellulosic materials (Bertella and Luterbacher, 2020) with a crystalline phase that gives it a stable structure (Lavoine et al., 2012), can form bundled fibrils bound together by hemicellulose and lignin (Bertella and Luterbacher, 2020). Lignin, the third most abundant component in lignocellulosic materials after cellulose and hemicellulose, has a complex, rigid structure (Bertella and Luterbacher, 2020), which makes it capable of inhibiting the interaction between fiber bundles rich in cellulose, thus allowing the production of noncompact paper with low strength. Each fibril bundle will interact to form paper. Based on Table 11.2, compared to other nonwood materials, reed grass has relatively higher cellulose and lower lignin contents. Those major chemical components significantly impact the characteristics of pulp and paper, thus the cost of pulping and papermaking processes in the industry. The use of materials with a higher cellulose content and a lower lignin percentage allows for the production of stronger paper sheets.

Reed grass has long fiber at a relatively higher amount than switchgrass and elephant grass and at a similar amount to giant reed grass, kenaf, wheat,

FIGURE 11.2 (A) Reed grass plant and (B) fresh reed grass stem (Pictures from author collections).

and canola straw. It falls in the range of good alternative fibers suitable for utilization in the pulp and paper industry (Kassim et al., 2016). This physical property is supported by morphological analysis as performed by Kassim et al. (2015a). Because of the compact interfiber bonding during the production of paper sheets, the morphology of reed grass reveals an abundance of long fibers that can make high-strength paper. Furthermore, a reed grass fiber has a rough surface that protects its fiber (Kassim et al., 2015a).

Fiber length, diameter, and cell wall thickness all correlate with the chemical components of lignocellulosic material, which are cellulose and lignin. Lignin is present on the outside of bundled α-cellulose (Bertella and Luterbacher, 2020) and affects fiber diameter. Reed grass has lower fiber diameter compared to other nonwood materials. Fibers with low lignin content tend to have thinner cell walls. Furthermore, the thin cell walls of a fiber make it easier to soften and flatten to provide wider surface area and increase interfiber interaction, thus producing paper with higher tensile strength (Fatriasari and Hermiati, 2008). The long and regular shape of reed grass fibers makes it possible to produce compact paper sheets from reed grass fibers. Reed grass has long and regularly shaped fibers that lead to a considerably good arrangement of fibers and an increase in tensile strength (Kassim et al., 2015b).

TABLE 11.2 Characteristics of nonwood materials.

Nonwood material	Cellulose (%)	Lignin (%)	Length (mm)	Diameter (μm)	Cell wall thickness (μm)	References
Reed grass (*Imperata cylindrica*)	37.13	5.67	1.04	7.49	–	Kassim et al. (2016) and Kassim et al. (2015a)
Giant reed grass (*Arundo donax* L.)	29.18–32.93	19.03–19.60	1.2	14.6–16.9	5.3–4.6	Shatalov et al. (2001)
Switchgrass	41.2	23.9	0.76	13.9	–	Madakadze et al. (2010)
Elephant grass	45.6	17.7	0.75	15.1	–	Madakadze et al. (2010)
Kenaf	40.2–53.8	13.21	1.29	22.1	4.3	Khalil et al. (2010) and Ververis et al. (2004)
Cotton	43.8	17.6	–	–	–	Ververis et al. (2004)
Wheat straw	43.2	21.1	1.18	13.6	3.96	Singh et al. (2011)
Canola straw	41.1	17.2	1.21	28	7.43	Yousefi (2009)

Based on previous findings on its properties, reed grass is the best alternative source of fiber compared to other nonwood materials because it has higher cellulose content, lower lignin content, and good fiber length. It also has a lower hot water solubility of 3.84%, indicating low tannins, gums, sugar, coloring matter, and starch content, all of which could have an impact on the quality of paper sheets. Compared to other nonwood plants, reed grass has the lowest 1% NaOH solubility at 19.6%, which indicates that only a low amount of fiber will disintegrate during the pulping process; thus the pulp yield would be higher compared to the rest of the nonwood plants (Kassim et al., 2016). Fig. 11.3 shows scanning electron microscopy (Scanning Electron Microscope SEM) images of reed grass. Based on the images, reed grass has a long, abundance fiber and has a rough surface that can protect the lignocellulosic content (Kassim et al., 2015a).

11.4 Recent development of pulping method of reed grass

The sequence of steps in producing pulp and paper are the preparation of the raw material, pulping process based on the method used, bleaching for bleached pulp, and the making of paper sheets. Pulping is a method to delignify the structure of biomass to extract cellulose. As previously stated, reed grass consists of the main chemical components of a plant cell wall, which include cellulose, hemicellulose, and lignin. Low lignin content and high cellulose content is targeted in this pulping process before further processes in papermaking. Papermaking is a process to make paper sheets by drying a diluted suspension of fibers in water to form interwoven fibers in the mat. To obtain maximum water removal, the fiber mat is then pressed and dried.

Generally, nonwood plants have lower lignin content, shorter growing cycles, annual renewability, and a higher annual yield of cellulose than wood

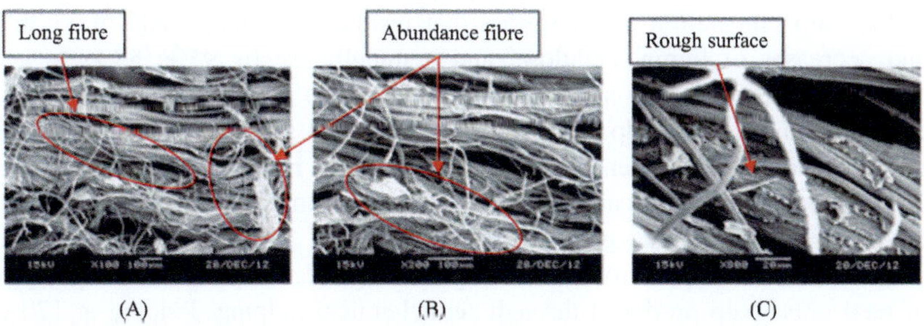

FIGURE 11.3 Scanning electron microscopy (SEM) images of reed grass on magnification level at (A) 100 ×, (B) 200 ×, and (C) 800 × (Kassim et al., 2015a). *Reprints with permission from Scientific.net, License: Creative Commons CC BY 4.0.*

plants (Alila et al., 2013). Therefore, the pulping of nonwood plants required moderate chemical system and lower alkaline charge in the same degree of delignification. As such, the pulping of nonwood plants comes at a lower production cost than that of wood plants (Hammett et al., 2001). Several pulping methods have been developed for pulp extraction, that is, mechanical, thermal, semichemical, biopulping, and chemical methods, with chemical pulping being the most used method (74.1%), followed by mechanical pulping process (21.4%) and other processes (4.5%) (Sabharwal et al., 1994). Most pulping industries use kraft pulping as a part of chemical pulping to extract higher amount of cellulose and degrade lignin and hemicellulose from the biomass into liquor. These substances in the black liquor are recovered in the form of steam as the mean to recycle the energy needed for in mill processes. Intermolecular rearrangements of carbohydrate in kraft pulping allow stabilization toward alkali, which yields superior kraft pulp. However, rich condensed lignin brings about a darker colored pulp. To achieve a high brightness level, this pulp needs to be bleached (Johansson et al., 1987).

The pulping method, chip size, chemical charges, bleaching procedure, and additives used in the papermaking process, all have an impact on the qualities of pulp and paper. Some methods have been used to prepare pulp from reed grass, which were reported in previous studies. The comparison of pulping methods of reed grass is summarized in Table 11.3.

Zulfiana et al. (2020) used a semichemical method for 1 hour and soda pulping for 1 and 2 hours to pulp reed grass. The yield obtained for semichemical pulping, 1 hour soda pulping, and 2 hours soda pulping was of 48.96%, 30.32%, and 28.32%, respectively. The lowest kappa number was achieved by soda pulping for 1 hour (20.66), which is relatively similar to that achieved by soda pulping for 2 hours (20.86), while semichemical pulping achieved the opposite (78.63). It was also reported that kappa number is directly proportional to acid insoluble lignin (AIL) content. Improvement of cellulose content was observed after longer cooking time and soda pulping is more effective than semichemical pulping at reducing lignin and increasing cellulose content. Sari et al. (2018) used semichemical pulping with lower NaOH concentration in the pulp preparation of reed grass that resulted in a pulp yield of 31.26% and decreases of lignin, cellulose, and hemicellulose by 95%, 57.19%, and 36.44%, respectively. As such, it is shown that the use of lower NaOH concentration and shorter pulping time resulted in lower pulp yield. The utilization of the alkaline pretreatment of reed grass was also reported by (Lin and Lee, 2011). There is a significant increase of glucan from 38.5% in the control to 60.8% in pretreated reed grass. On the other hand, almost 50% of xylan and arabinan were removed from the reed grass. Fig. 11.4 depicts the morphology of reed grass pulp produced through semichemical pulping. Pulping at 170°C for 1 hour with a liquor-to-sample ratio of 15 and a 20% alkali charge produced a more homogeneous pulp because amorphous material such as lignin and hemicellulose were effectively removed (Zulfiana et al., 2020).

TABLE 11.3 Recent developments in the pulping of reed grass.

No	Pulping method	Pulping condition	References
1	Semichemical pulping	10% w/v NaOH concentration, liquid-to-sample ratio of 10, 2 h at 100°C	Sari et al. (2018)
2	Alkaline pulping	15% active alkali with liquid-to-sample ratio of 7 for 170°C for 90 min. After the cooking temperature reaches 170°C, for a further 120 min, the sample was isothermally cooked	Kassim et al. (2016)
3	1. Semichemical pulping 2. Soda chemical pulping	Pulping temperature of 100°C for 1 h, liquor-to-sample ratio of 15 with 20% alkali charge Pulping temperature of 170°C for 1 h with liquor-to-sample ratio of 20 and 2 h with 20% alkali charge using a rotary digester	Zulfiana et al. (2020)
4	Semichemical pulping	Alkaline pretreatment with NaOH 2% (w/v) with liquid-to-sample ratio of 20 at room temperature for 24 h, followed by heating using autoclave at a temperature of 121°C for 1 h	Lin and Lee (2011); Kartikasari et al. (2013)
5	Acetosolv pulping	Pulping using acetic acid with liquid-to-sample ratio of 10, 150 rpm. The concentration of acetic acid was 60%, 75%, and 90% at temperatures of 70, 85 and 100°C for 30–150 min	Wibisono et al. (2012)

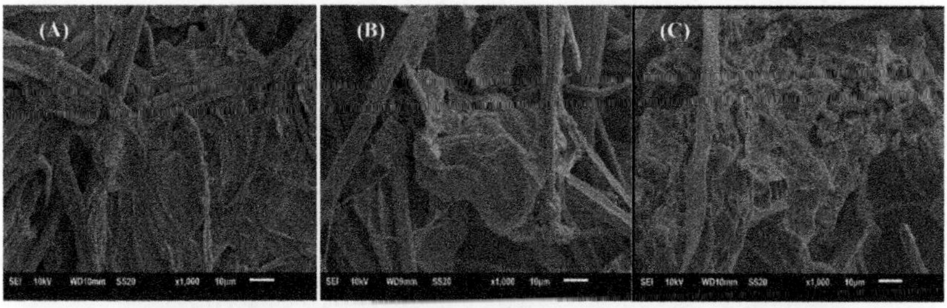

FIGURE 11.4 Scanning electron microscopy (SEM) images of pulps (A) PA-100°C/1 h pulp, (B) PA-170°C/1 h pulp, (C) PA-170°C/2 h pulp (Zulfiana et al., 2020). *Reprints with permission from Elsevier, License number: 5098321127215.*

Kassim et al. (2016) reported that observation on the morphological surface of reed grass paper sheet showed that it has abundant, straight, and smooth fibers with higher tensile strength compared to other published nonwood fibers. Fig. 11.5 depicts the morphology of reed grass paper. The long fibers of reed grass pulp can be seen in the SEM image (*red arrow*), and the ones created in the lengthy lump revealed that the reed grass pulp had not been fibrillated previously (*yellow arrows*).

Soda pulping of reed grass was reported to result in a holocellulose content of 64.9%, with a pulp yield of 38.2%, tensile strength of 45.1 Nm/g, burst index of 3.90 kPa m^2, and tear index of 2.17 mN (m^2/g) (Kassim et al., 2016). Cellulose content is the primary parameter of pulp quality besides kappa number; the higher the cellulose content, the higher the pulp quality. Cellulose content is affected by pulping condition. The aforementioned yield is higher than the yield reported by Zulfiana et al. (2020) with higher alkali charge. The most important physical property of paper and paperboard is tensile strength, which is defined as the greatest stress required to break a strip of paper sheet. It correlates with fiber length. Longer fiber allows stronger interfiber bonding in a paper sheet, which then results in higher tensile strength of the sheet. In addition, fiber refining degree and fiber direction also have significant effects on this value. In industries, additive substances such as filler are usually added. Compared to other nonwood fibers, the tensile strength of reed grass sheet is higher than that of nonwood crops, such as kenaf bast, palmyra plant fruit (Ashuvila, 2014), canola straw, vine shoot, and rice straw (Kassim et al., 2016). However, this strength is lower than that of annual plants such as elephant grass and switchgrass and residues from the utilization of agricultural

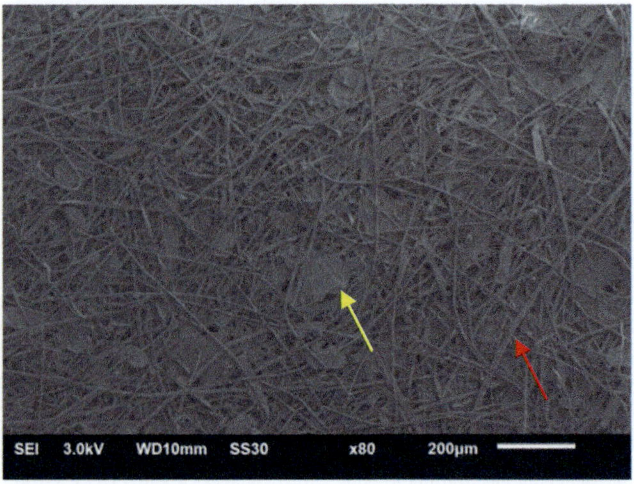

FIGURE 11.5 Scanning electron microscopy (SEM) images of reed grass paper (Zulfiana et al., 2020). *Reprints with permission from Elsevier, License number: 5101340588729.*

plants, such as banana stem, sugarcane bagasse, and wheat straw (Ashuvila, 2014). Compared to agricultural wastes and annual plants, for example, are nonwoody plants, reed grass has a moderate burst index. The tear index of reed grass sheet is lower than those of nonwood crops and annual plants (bamboo, elephant grass, and switchgrass) (Ashuvila, 2014).

Kappa number represents residual lignin content in pulp, thus indicating the cooking degree of a pulping process. The volume of strong oxidizing chemical potassium permanganate consumed by 1 g of dry pulp is the basic principle for determining kappa number (TAPPI, 2003). Lower kappa number of pulps indicates a better degree of delignification that leads to the requirement of a smaller quantity of bleaching chemicals (Fatriasari et al., 2015). Furthermore, a kappa number reaching up to 20 shows that a higher amount of bleaching chemicals is needed (Casey, 1980).

Wibisono et al. (2012) pulped reed grass using the acetosolv pulping method. They reported that the lowest kappa number was 20.41, meaning more amount of bleaching chemicals is required to bleach reed grass pulp. At the best pulping condition (100°C for 60 minutes with acetic acid concentration of 60%), the α-cellulose content of the pulp is 84.6%. Such pulp yield is higher than that of commercial chemical pulping, which numbers around 40%–55% (Karlsson et al., 2006). The highest pulp yield of reed grass using the acetosolv pulping method was 88.2% at a temperature of 70°C for 30 minutes Wibisono et al., (2012). That indicates this process to be more effective in preserving cellulose as represented by the higher pulp yield compared to that of the chemical pulping method. The effectivity of a pulping process is also indicated by pulp yield; the pulping process is more efficient if the pulp yield is higher (Fatriasari and Risanto, 2011).

11.5 Opportunities and future challenges

The growing demand for pulp and paper, combined with environmental concerns, encourages the development of nonwood plants as pulp and paper raw materials. Nonwood plants have a higher ash content than wood plants, although having lower lignin content (Azeez, 2018), which should be taken into consideration. To succeed in utilizing nonwood biomass, including reed grass, for pulp and papermaking with better product quality, the following factors should be noted (Azeez, 2018):

1. Availability. Most nonwood biomass is bulky and has a low yield per hectare of land; thus densification can be addressed to facilitate handling, storage, and transportation.
2. Storage and handling. Nonwood plants are frequently harvested after shorter periods of time and are more easily deteriorated, resulting in low pulp yield and limiting paper production. As a result, appropriate storage and handling methods must be devised.

3. Pulping. Washing nonwood pulp is more difficult due to higher silica content; thus the washer should be set to twice the normal size.
4. Bleaching. Bleaching is more difficult due to discoloration that occurs during storage.
5. Paper production. The speed of the paper machine needs to be adjusted to lower drainage speed.
6. Chemical recovery. Black liquor of nonwood plants has 10 times higher solid content than that of black liquor of wood; thus it is required more treatment in evaporation, boiler recovery, causticizing equipment, and lime kiln.

Based on its physical and chemical qualities, reed grass has been stated to have the ability to make pulp and paper throughout this chapter, both in pulp and paper sheets. Other than that, pretreated reed grass is also used for bioethanol production (Kartikasari et al., 2013). Reed grass pulp has contributed in the improvement of recycled paper strength (Sari et al., 2018). Reed grass pulp and paper may be utilized for a variety of uses in the future, namely, food packaging, bio-composite, and textile. Reducing the size of reed grass fiber from micro (in pulp) to nano (nanocellulose) can open a possibility for wider range of applications. Nanocellulose has high tensile strength and rigidity, and surface area, with an aspect ratio of more than 100 with uniform structure (Fatriasari et al., 2019). These properties are important for nanocomposite that needs improvement in strength, toughness, and barrier properties (Favier et al., 1997; Bondeson et al., 2006). Nanocellulose is also more hydrophilic, as such it is easier to modify further (Fatriasari et al., 2019). Various fields in which biomaterials containing nanocellulose can be applied include biomedicine, nanocomposite, smart or active packaging, drug delivery, biosensors, superabsorbent, heavy metal absorber, renewable energy and energy device, and nano paper.

11.6 Conclusions

The increasing demand on pulp and paper in which mainly from wood as a fiber source can increase forest exploitation. As a result, it is necessary to seek out new fiber sources from nonwood plants. Reed grass is one of them, and it has the potential to be used as a pulp and paper raw material. Using this grass increases the economic worth of underutilized plants, such as weeds, on the land. This grass has a long fiber with a high cellulose content and a low lignin content, resulting in good pulp quality. Thus the physical paper sheet includes good properties. Some pulping methods have been used for processing reed grass, and the acetosolv pulping provides high pulping yield. Soda pulping for 1 hour at 170°C results in lower kappa number and high cellulose content. Pulp of reed grass is prospective to be developed advanced biomaterials by converting in nanocellulose.

Acknowledgments

The authors are grateful for the facilities, as well as the scientific and technical assistance from the Integrated Laboratory of Bioproducts at the Indonesian Institute of Science. Authors gratefully acknowledge all publishers (Elsevier, Scientific.net, Intech Open) for all permissions.

References

Alila, S., Besbes, I., Rei Vilar, M., Mutjé, P., Boufi, S., 2013. Non-wood plants as raw materials for production of microfibrillated cellulose (MFC): a comparative study. Ind. Crops Prod. 41 (1), 250–259. Available from: https://doi.org/10.1016/j.indcrop.2012.04.028.

Ahmad, T., Arief Daryanto, R.O., Priyarsono, D.S., 2018. Global value chain of Indonesian pulp and paper industry. Jurnal Manajemen Dan Agribisnis 15 (2), 118–128. Available from: https://doi.org/10.17358/jma.15.2.118.

Ashuvila, M.A., 2014. Potential of Non-Wood Fibres for Pulp and Paper-Based Industries. 63. https://books.google.com.my/books?id = Jyt2swEACAAJ.

Asim, M., Abdan, Khalina, Jawaid, M., Nasir, M., Dashtizadeh, Zahra, Ishak, M.R., Hoque, M. Enamul, 2015. A review on pineapple leaves fibre and its composites. International Journal of Polymer Science 1–16, 950567. Available from: https://doi.org/10.1155/2015/950567.

Asosiasi Pulp dan Paper Indonesia, 2016. Opportunities and challenges of Indonesian pulp and paper industry. Available from: http://www.pptgroup.se/wp-content/uploads/2016/02/Indonesian-Pulp-and-Paper-Industry-APKI-1.pdf. [Accessed: 4 Mei 2020].

Azeez, M.A., 2018. Pulping of non-wood biomass. Pulp Paper Process. Available from: https://doi.org/10.5772/intechopen.79749.

Bajpai, P., Mishra, S.P., Mishra, O.P., Kumar, S., Bajpai, P.K., Singh, S., 2004. Biochemical pulping of bagasse. Biotechnol. Prog. 20 (4), 1270–1272. Available from: https://doi.org/10.1021/bp040005q.

Bertella, S., Jeremy, S.L., 2020. Lignin functionalization for the production of novel materials. Trends Chem. 2 (5), 1–14. Available from: https://doi.org/10.1016/j.trechm.2020.03.001.

Bondeson, D., Mathew, A., Oksman, K., 2006. Optimization of the isolation of nanocrystals from microcrystalline cellulose by acid hydrolysis. Cellulose 13 (2), 171–180. Available from: https://doi.org/10.1007/s10570-006-9061-4.

Casey, J.P., 1980. 3rd ed. Pulp and Paper Chemistry and Chemical Technology.", 1. Interscience Publishers, Inc, New York.

El-Sakhawy, M., Fahmy, Y., Ibrahim, A., Lönnberg, B., 1995. Organosolv pulping. I. Alcohol pulping of bagasse. Cellulose Chem. Technol. 29 (5), 615–629.

Fatriasari, W., Hermiati, E., 2008. Analisis Morfologi Serat Dan Sifat Fisis-Kimia Pada Enam Jenis Bambu Sebagai Bahan Baku Pulp Dan Kertas. Jurnal Ilmu Dan Teknologi Hasil Hutan 2 (1), 67–72.

Fatriasari, W., Ermawar, R.A., Falah, F., Heri, D., Yanto, Y., Hermiati, E., 2009. Pulping Soda Panas Terbuka Bambu Betung Dengan Praperlakuan Fungi Pelapuk Putih (Pleurotus Ostreatus Dan Trametes Versicolor) Open hot soda pulping of betung bamboo using white rot pretreatment (pleurotus ostreatus and trametes versicolor). Jurnal Ilmu dan Teknologi Hasil Hutan 2 (2), 45–50.

Fatriasari, W., Anita, S.H., Falah, F., Adi, T.N., Hermiati, E., 2010. Biopulping Bambu Betung Menggunakan Kultur Campur Jamur Pelapuk Putih (Trametes Versicolor, Pleurotus Ostreatus Dan Phanerochaete Crysosporium). Jurnal Selulosa 45 (02), 44–56.

Fatriasari, W., Risanto, L., 2011. The properties kraft pulp sengon wood (*Paraserianthes falcataria*): differences of cooking liquor concentration and bleaching sequence. Widyariset 14 (3), 589–598.

Fatriasari, W., Supriyanto, Iswanto, A.H., 2015. The kraft pulp and paper properties of sweet sorghum bagasse (*Sorghum bicolor* L Moench. J. Eng. Technol. Sci. 47 (2), 149–159. Available from: https://doi.org/10.5614/j.eng.technol.sci.2015.47.2.4.

Fatriasari, W., Masruchin, N., Hermiati, E., 2019. Selulosa: Karakteristik dan Pemanfaatannya. LIPI PRESS.

Favier, V., Canova, G.R., Shrivastava, S.C., Cavaillé, J.Y., 1997. Mechanical percolation in cellulose whisker nanocomposites. Polym. Eng. Sci. 37 (10), 1732–1739. Available from: https://doi.org/10.1002/pen.11821.

Hammett, A.L., Youngs, R.L., Sun, X., Chandra, M., 2001. Non-wood fibre as an alternative to wood fibre in china's pulp and paper industry. Holzforschung 55 (2), 219–224. Available from: https://doi.org/10.1515/HF.2001.036.

Holm, L.G., Pucknett, D.L., Pancho, J.B., Herberger, J.P., 1977. The World's Worst Weeds: Distribution and Biology. Univ. Press of Hawaii, Honolulu.

Hubbard, C.E., Whyte, R.O., Brown, D.C., Gray, A.P., 1944. Imperata cylindrica. Taxonomy, distribution, economic significance and control. Imperial Agricultural Bureaux, Aberystwyth. UK, p. 53.

Ibrahim, R., 2003. Structural, mechanical and optical properties of recycled paper blended with oil palm empty fruit bunch pulp.". J. Oil Palm Res. 15 (2), 28–34.

Johansson, A., Aaltonen, O., Ylinen, P., 1987. Organosolv pulping - methods and pulp properties. Biomass 13 (1), 45–65. Available from: https://doi.org/10.1016/0144-4565(87)90071-0.

Karlsson, H., Lorentzen dan Wettre, A.B., 2006. Fibre Guide: Fibre Analysis and Process Applications in the Pulp and Paper Industry. AB Lorentzen & Wettre, Box 4, SE-164 93, Kista, Sweden.

Kartikasari, S.D., Nurhatika, S., dan Muhibuddin, A., 2013. Potensi Alang-Alang (*Imperata cylindrica* (L.) Beauv.) Dalam Produksi Etanol Menggunakan Bakteri *Zymomonas mobilis*. Jurnal Sains Dan Seni Pomits 2 (2), 1–11.

Kassim, A.S.Mohd, Ashuvila, M.A., Ishak, N., Zainulabidin, M.H., 2015a. Cogon grass as an alternative fibre for pulp and paper-based industry: on chemical and surface morphological properties. Appl. Mech. Mater. 773–774 (July), 1242–1245. Available from: https://doi.org/10.4028/www.scientific.net/amm.773-774.1242.

Kassim, A.S.M., Aripin, A.M., Hatta, Z., Daud, Z., 2015b. Exploring non-wood plant as alternative pulp: from the physical and chemical perspectives. ICGSCE 19–24. Available from: https://doi.org/10.1007/978-981-287-505-1.

Kassim, A.S.M., Aripin, A.M., Ishak, N., Hairom, N.H.H., Fauzi, N.A., Razali, N.F., et al., 2016. Potential of cogon grass (*Imperata cylindrica*) as an alternative fibre in paper-based industry. ARPN J. Eng. Appl. Sci. 11 (4), 2681–2686.

Kaur, D., Kant, N., Kumar, R., 2017. Prospects of rice straw as a raw material for paper making. Waste Manage. 60, 127–139.

Khalil, H.P.S.A., Siti Alwani, M., Mohd Omar, A.K., 2006. Chemical composition, anatomy, lignin distribution, and cell wall structure of Malaysian plant waste fibres. BioResources 1 (2), 220–232. Available from: https://doi.org/10.15376/biores.1.2.220-232.

Khalil, H.P.S., Ireana Yusra, A.F., Bhat, A.H., Jawaid, M., 2010. Cell wall ultrastructure, anatomy, lignin distribution, and chemical composition of Malaysian cultivated kenaf fiber. Ind. Crops Prod. 31 (1), 113–121. Available from: https://doi.org/10.1016/j.indcrop.2009.09.008.

Lavoine, N., Desloges, I., Dufresne, A., Bras, J., 2012. Microfibrillated cellulose – its barrier properties and applications in cellulosic materials: a review. Carbohydr. Polym. 90 (2), 735–764. Available from: https://doi.org/10.1016/j.carbpol.2012.05.026.

Lin, Y.S., Lee, W.C., 2011. Simultaneous saccharification and fermentation of alkali-pretreated cogon grass for bioethanol production. BioResources 6 (3), 2744–2756. Available from: https://doi.org/10.15376/biores.6.3.2744-2756.

Liu, Z., Wang, H., Hui, L., 2018. Pulping and papermaking of non-wood fibers. Pulp and Paper Processing. pp. 3–32. Available from: https://doi.org/10.5772/intechopen.79017.

Lucia, L.A., 2005. Novel Pulping Technology: Directed Green Liquor Utilization (D-GLU) Pulping. https://doi.org/10.2172/859901.

MacDonald, G.E., 2004. Cogon grass (*Imperata cylindrica*)-biology, ecology, and management. Plant Sci. 23 (5), 367–380.

Madakadze, I.C., Masamvu, T.M., Radiotis, T., Li, J., Smith, D.L., 2010. Evaluation of pulp and paper making characteristics of elephant grass (*Pennisetum purpureum* Schum) and switchgrass (*Panicum virgatum* L.). Afr. J. Environ. Sci. Technol. 4 (July), 465–470. Available from: https://doi.org/10.5897/AJEST10.097.

Mordor Intelligence, 2023. Global Paper and Paperboard Packaging Market-Growth, Trends, Covid-19 Impact, and Forecasts (2023 - 2028) Available from: https://www.mordorintelligence.com/industry-reports/global-paper-and-paperboard-packaging-market (2023).

Mulyadi, M., Wuryanti, W., Sarjono, P.R., 2017. Konsentrasi Hambat Minimum (KHM) kadar sampel alang-alang (*Imperata cylindrica*) dalam etanol melalui metode difusi cakram. Jurnal Kimia Sains Dan Aplikasi 20 (3), 130–135. Available from: https://doi.org/10.14710/jksa.20.3.130-135.

Parkavi, V., Vignesh, M., Selvakumar, K., Muthu Mohamed, J., Ruby, J.J., 2012. Antibacterial activity of aerial parts of *Imperata cylindrica* (L) beauv. Int. J. Pharm. Sci. Drug Res. 4 (3), 209–212. Available from: http://www.ijpsdr.com.

Prisdiany, Y., Levita, J., 2019. Aktivitas Antihipertensi Tanaman Genus Imperata. Farmaka 17 (2), 306–314. Available from: http://jurnal.unpad.ac.id/farmaka/article/view/22145/pdf.

Ramdhonee, A., Jeetah, P., 2017. Production of wrapping paper from banana fibres. J. Environ. Chem. Eng. 5 (5), 4298–4306.

Reddy, N., Yang, Y., 2005. Biofibres from agricultural byproducts for industrial applications. Trends Biotechnol. 23 (1), 22–27. Available from: https://doi.org/10.1016/j.tibtech.2004.11.002.

Sabharwal, H.S., Akhtar, M., Blanchette, R.A., Young, R.A., 1994. Biomechanical pulping of kenaf. Tappi J. 77 (12), 105–112.

Sari, F.P., Ghozali, M., Damayanti, R., Fatriasari, W., 2018. Peranan Serat Alang-Alang (*Imperata cylindrica*) Sebagai Penguat Kertas Daur Ulang. Majalah Polimer Indonesi 21 (1), 1–19.

Shatalov, A.A., Quilho, T., Pereira, H., 2001. *Arundo donax* L. reed: new perspectives for pulping and bleaching 1. Raw material characterisation. *Tappi J.* 84(1), 1–12. https://doi.org/10.1016/j.biortech.2007.01.014.

Singh, S., Dutt, D., Tyagi, C.H., 2011. Complete characterization of wheat straw (*Triticum aestivum* PBW-343 L. Emend. Fiori & PAOL.) – a renewable source of fibres for pulp and paper making. BioResources 6 (1), 154–177. Available from: https://doi.org/10.15376/biores.6.1.154-177.

Sridach, W., 2010. The environmentally benign pulping process of non-wood fibres. Suranaree J. Sci. Technol. 17 (2), 105–123.

Subositi, D., Widodo, H., 2018. Genetic diversity of cogon grass (*Imperata cylindrica* (L.) beauv) based on the intersimpel marking of the sequence repeat (ISSR). Jurnal Ilmu-Ilmu Hayati 17 (2), 115–122.

Sutiya, B., Istikowati, W.T., Rahmadi, A., 2012. Kandungan Kimia Dan Sifat Serat Alang-Alang (*Imperata cylindrica*) Sebagai Gambaran. Bioscientiae 9 (May 2014), 8–19.

Szabó, L., Soria, A., Forsström, J., Keränen, J.T., Hytönen, E., 2009. A world model of the pulp and paper industry: demand, energy consumption and emission scenarios to 2030. Environ. Sci. Policy 12 (3), 257–269. Available from: https://doi.org/10.1016/j.envsci.2009.01.011.

TAPPI, 2003. Kappa Number of Pulp TAPPI/ANSI T 236 om-13. TAPPI.

Ververis, C., Georghiou, K., Christodoulakis, N., Santas, P., Santas, R., 2004. Fibre dimensions, lignin and cellulose content of various plant materials and their suitability for paper production. Ind. Crops Prod. 19 (3), 245–254. Available from: https://doi.org/10.1016/j.indcrop.2003.10.006.

Wibisono, I., Hugo Leonardo, A., Aylianawati, 2012. Pembuatan Pulp Dari Alang-Alang. Widya Tenik 10 (1), 11–20.

Yousefi, H., 2009. Canola straw as a bio-waste resource for medium density fibreboard (MDF) manufacture. Waste Manage. 29 (10), 2644–2648. Available from: https://doi.org/10.1016/j.wasman.2009.06.018.

Zulfiana, D., Karimah, A., Anita, S.H., Masruchin, N., Wijaya, K., Suryanegara, L., et al., 2020. Antimicrobial *Imperata cylindrica* paper coated with anionic nanocellulose crosslinked with cationic ions. Int. J. Biol. Macromol. 164, 892–901. Available from: https://doi.org/10.1016/j.ijbiomac.2020.07.102.

Chapter 12

Pulping and papermaking of sorghum bagasse

Nyoman J. Wistara[1] and Widya Fatriasari[2]
[1]*Department of Forest Products, Faculty of Forestry, IPB University, Dramaga, Bogor, West Java, Indonesia,* [2]*Research Center for Biomass and Bioproducts, National Research and Innovation Agency (BRIN), Cibinong, Bogor, West Java, Indonesia*

12.1 Introduction

Increasing demand for pulp and paper products increases pulpwood plantation that could threaten the existence of natural forests. Therefore, environmental concern has been among the critical factors that lead to the exploration of nonwood fibers. Nonwood fibers can substitute a portion of wood fibers in papermaking. Despite its potential, nonwood fibers variation is tremendous compared to wood fibers. It is highly varied in chemical compositions and fiber properties. Nonwood fibers are classified into the mean of production and plant fibers (Han, 1998). Fibers from sugarcane bagasse, sorghum bagasse, wheat straw, and alike are byproducts of biomass processing. Meanwhile, jute, flax, ramie, and cotton fibers are derived from plants purposely planted for fiber production.

Sorghum [*Sorghum bicolor* (L.) Moench] (Fig. 12.1) is among the five most popular prominent food grains in the semiarid tropics of Africa, Asia, and Latin America (Food Security Department – NRI, 1999). Sorghum is adaptable to a relatively harsh climate and, therefore, a preferred grain producer plant to be cultivated in relatively marginal lands. It has many varieties, from traditional to hybrid, with different productivity and climatic adaptability. Sorghum productivity is governed by planting site quality (soil fertility), water input, and other climatic factors. The grain productivity of traditional varieties is low compared to that of hybrid varieties. Open-pollinated traditional varieties under rain-fed conditions produce 0.3–1.0 t/ha, while hybrid varieties under ideal input of soil and water condition with a higher density of planting produce up to 12 t/ha (Food Security Department – NRI, 1999). In terms of biomass yield, hybrid sorghum planted in the marginal land of Flores (an island in Eastern Indonesia)

Pulping and Papermaking of Nonwood Plant Fibers. DOI: https://doi.org/10.1016/B978-0-323-91625-7.00013-8

FIGURE 12.1 Sorghum cultivation in Gresik, Indonesia. *Pictured by Dr. Dede Hermawan, Department of Forest Products, Faculty of Forestry and Environment, IPB University.*

produces 24 t/ha biomass (Wiloso et al., 2020), an average of 17.4–42.1 t/ha from various types of planting soil in northern China (Tang et al., 2018). Hui et al. (2014) reported a 43.0 t/ha dry matter yield from trial fields in China's Jiangsu province.

Sorghum is a versatile plant usable for the raw material of the various types of derivative products. The grain, for instance, is being developed to a staple food in East Nusa Tenggara-Indonesia (Winarti et al., 2020), considering sorghum adaptability to relatively harsh climates such as that in East Nusa Tenggara (NTT). Sorghum can be feedstock for starch-, sugar- and cellulosic-based bioethanol. Its seed, stalk juice, and bagasse are the source of starch, sucrose, and cellulose, respectively. The ethanol productivity of sorghum grain (Dyartanti et al., 2015; Barcelos et al., 2011), stalk juice (Bridgers et al., 2011; Mei et al., 2009; Liu and Shen, 2008), and cellulosic components (Cardoso et al., 2018; Tang et al., 2018; Solihat et al., 2017; Cardoso et al., 2013; McIntosh and Vancov, 2010) have been evaluated. Approximately 3000–4000 L/ha/year of ethanol can be produced from sorghum stalk (Carioca and Leal, 2011). Sorghum processing wastes are also useful raw material for the production of bio-oil and biochar via pyrolysis process (Yin et al., 2013; Piskorz et al., 1998), activated carbon (Senthilkumar et al., 2011), and biopellet (Wiloso et al., 2020; Theerarattananoon et al., 2011).

Although bioprocessing research on sorghum biomass is dominated by bioethanol production, its lignocellulosic components such as stem, leaves, and panicles are also potential sources of fiber for pulp and paper products. The use of sorghum stem for pulp and paper products is relatively promising, considering that the cell-wall content of the stem is as much as 39% (Billa et al., 1997). The fiber length of sorghum stem is comparable to that of deciduous wood, and it was thought to be suitable as a short fiber component in papermaking (Albert et al., 2011). The pulping properties of the sorghum bagasse have been evaluated employing the pulping process of soda (Gencer and Sahin, 2015), soda–anthraquinone (AQ) (Khristova and Gabir, 1990; Belayachi and Delmas, 1997; Belayachi and Delmas, 1995), and kraft (Gencer and Hatil, 2019; Fatriasari et al., 2015).

In the chapter the review emphasized on pulping and pulp properties of sorghum bagasse. The suitability of sorghum bagasse for the raw material of pulp and paper was also examined from chemical composition and fiber properties point of view.

12.2 Chemical composition and fiber properties of sorghum bagasse

Chemical composition and fiber properties are essential parameters in determining the suitability of biomass for papermaking. It has been well understood that the chemical compositions of the raw materials of pulp are essential in determining the appropriate pulping process for quality pulp. Meanwhile, fiber properties affect the mechanical properties of the paper.

12.2.1 Chemical composition

Pulp production mainly depends on the main structural chemical components, that is, cellulose, hemicelluloses, and lignin. On average, cellulose, hemicellulose, and lignin content of wood are 40%–50%, 20%–30%, and 26%–32% (softwood) and 20%–25% (hardwood), respectively (Bajpai, 2018; Sjöström, 1993). Cellulose is a linear homopolymer at a macromolecular level consisting of glucose monomer units (Fig. 12.2). The β-D-glucopyranose chain units are in the $^4C_1$ chain conformation with an equatorial orientation of the HO-2, HO-3, and CH_2OH substituents (Sjöström, 1993). Cellulose chains aggregate to form a supramolecular structure with a crystalline and amorphous area that is decisive for applying cellulose fibers. Cellulose is the most important for any pulp and its derivated products.

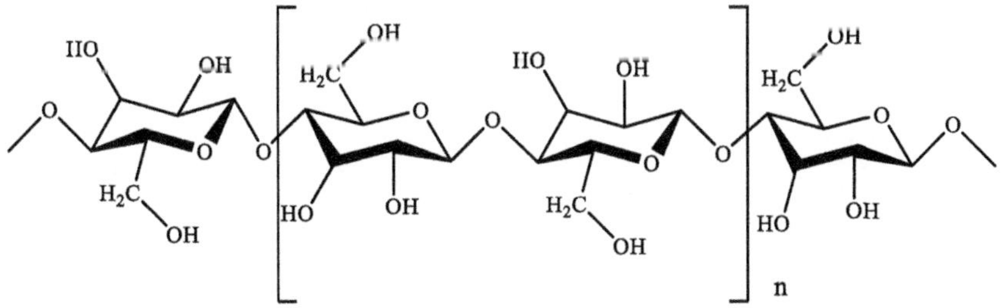

FIGURE 12.2 A structure of cellulose (Bertella and Luterbacher, 2020). *Reprints with permission from Elsevier, License number: 5115291121514.*

Hemicellulose, a branched heteropolymer, varied in structure and composition among lignocellulosic materials. Acid hydrolysis of hemicelluloses results in hexose and pentose sugar types and various acid types such as D-glucuronic acid, 4-*O*-methyl-D-glucuronic acid, and D-galacturonic acid (Sjöström, 1993). The dominant type of hemicellulose in softwood and hardwood is galactoglucomannan and glucuronoxylan, respectively. The structure of hemicellulose is depicted in Fig. 12.3. Hemicellulose is preserved in pulp for papermaking; however, it is removed when producing dissolving pulp.

Lignins are amorphous three-dimensional phenylpropane unit polymers. The precursor consists of *p*-coumaryl alcohol, coniferyl alcohol, and sinapyl alcohol (Fig. 12.4). Softwood lignin is entirely composed of coniferyl alcohol monolignol. In contrast, hardwood is composed of coniferyl and sinapyl monolignols, and *p*-coumaryl alcohol is present in grass lignins (Hatfield et al., 2009). The polymeric structure of lignin is formed by various types of intermolecular linkages of lignin precursors, where β-*O*-4 linkage is predominant in both softwood (50%) and hardwood (60%) lignin (Sjöström, 1993). Lignin is almost totally and partly preserved in mechanical and high-yield pulp production, respectively. However, most lignin is removed in the case of chemical pulping-based papermaking.

In general, the chemical compositions of plants varied according to the plant structure differentiation and the growth stages. As for the sorghum, hemicellulose, cellulose, and lignin content changed from panicles to preboot dependent on growth stages (McBee and Miller, 1993). Considering the optimization of sorghum for industrial products, Billa et al. (1997) thoroughly characterized the chemical components of sorghum stalk. They found that the pith of sorghum stalk contains a higher amount of nonstructural carbohydrate (sucrose and glucose) and lower bark structural chemical components (hemicellulose, cellulose, and lignin). Inhomogeneous distribution of structural chemical components between rind and pith of sorghum stalk was previously reported by Richie and McBee (1991). Table 12.1 summarizes the chemical compositions of sorghum stalk from several references.

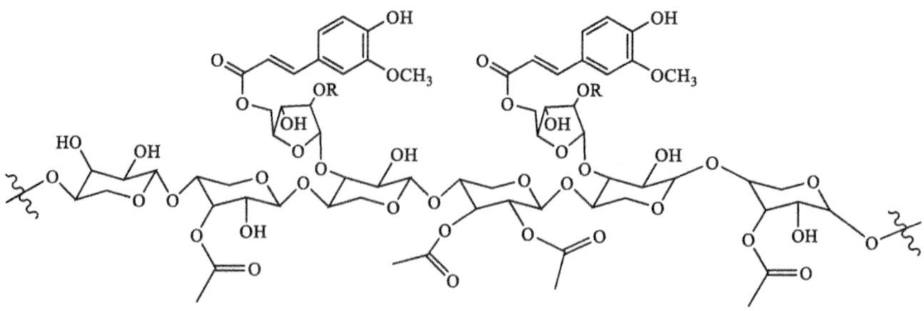

FIGURE 12.3 A structure of hemicellulose (Bertella and Luterbacher, 2020). *Reprints with permission from Elsevier, License number: 5115291121514.*

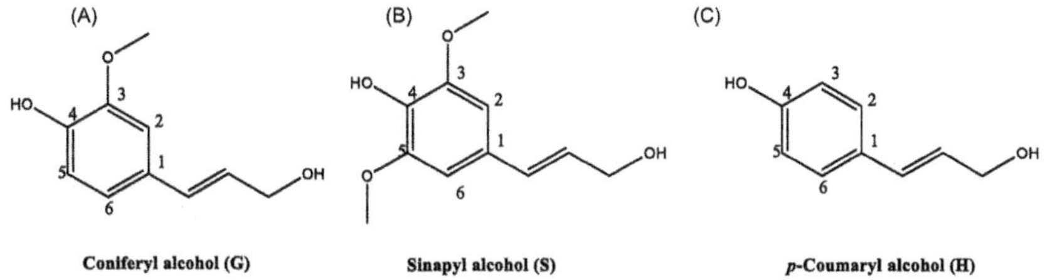

FIGURE 12.4 Lignin precursors: (A) coniferyl alcohol, (B) sinapyl alcohol, and (C) *p*-coumaryl alcohol (Chio et al., 2019). *Reprints with permission from Elsevier, License number: 5115290858023.*

TABLE 12.1 Chemical composition of sorghum bagasse.

Components (%)	Belayachi and Delmas (1997)	Gencer and Sahin (2015)	Fatriasari et al. (2015)	Alves et al. (2019)
EtOH–benzene extractives	24.0	–	2.9	–
Acid-insoluble lignin	14.9	13.0	25.0	16.7
Holocellulose	61.6	71.0	73.0	68.7
Hemicellulose	16.6	30.7	30.7	19.2
α-Cellulose	45.0	40.3	42.4	49.4
Ash	2.7	–	–	–

It can be seen from Table 12.1 that the chemical compositions of sorghum are relatively inhomogeneous. It ascertains that the planting area, plant differentiation, and analysis method influence the chemical composition. Water extraction could dissolve low molecular weight carbohydrates and increase the percentage of cellulose relative to the weight of extracted samples. It is indicated by the slightly higher cellulose content of the water-extracted stem bagasse found by Belayachi and Delmas (1997) compared to that of the untreated stem bagasse found by Fatriasari et al. (2015). The lignin content of sorghum is typical for nonwood lignocellulosic materials. Sorghum stalk could be an excellent raw material for pulp and paper based on its cellulose and lignin content. Regardless of its lignin structure, a sorghum stalk with relatively low lignin content should be relatively easy to be pulped. The extractive content of the sorghum stalk is relatively high. It should be considered when the soda pulping process is employed. Soda pulping liquor is sensitive to extractives, mainly at the residual pulping stage.

Compared to many sources of nonwood fibers, on average, sorghum could be a better raw material for pulp and paper products, considering its cellulose content ranged from 40% to 49% (Table 12.1). Nonwood fiber sources listed in Table 12.2 indicate 30%–45% cellulose content. Sorghum contains a relatively higher amount of cellulose with lower lignin content. The acid-insoluble lignin content of sorghum (determined as Klason lignin) is somewhat lower than those of the references (Table 12.2). Lignocellulosic materials with low lignin content are preferred as the raw material of pulp, mainly for chemical pulps. On average, sorghum contains lower hemicellulose than those of other nonwood fiber sources referred to, indicating that sorghum could be a better raw material for dissolving pulp. Even though most lignocellulosic materials can be used as dissolving pulp since in the production of dissolving pulp, hemicellulose can be reduced through the prehydrolysis stage before the chemical pulping process is employed (Smook, 1992).

12.2.2 Fiber properties

Nonwood fibers are produced from a wide range of plants with diverse characteristics. Rice straw, bagasse, and bamboo are leading nonwood raw materials used for industrial pulp and paper production (Atchison, 1995). Even though the potential is hampered by density, collection, transportation, and storage, its annual harvesting nature makes it promising for a sustainable source of pulp raw materials. Pulping nonwood fiber sources has been a common practice mainly in the Asia-Pacific region, with the top five

TABLE 12.2 Chemical components (%) of several nonwood lignocellulosic materials.

Raw materials	Hemicellulose	α-Cellulose	Lignin	Extractives	Ash
Cotton stalks[a]	33.1	29.74	23.79	–	5.0
Corn stalks[b]	32.0	34.5	15.7	–	4.1
Sunflower stalks[c]	25.5	41.8	20.1	2.2	3.1
Sugarcane bagasse (depithed)[d]	28.7	42.3	21.7	1.9	2.1
Bamboo[e]	18.7	44.8	28.0	3.9	1.7

[a] *Ateş et al. (2015).*
[b] *He et al. (2020): average of pith and bark.*
[c] *Barbash et al. (2016).*
[d] *Agnihotri et al. (2010).*
[e] *Fatriasari et al. (2016). Hemicellulose is the difference between α-cellulose and holocellulose.*

countries in nonwood pulping capacity, including China, India, Pakistan, Indonesia, and Thailand (McCloskey, 1995). The latest data indicate that the total capacity of nonwood papermaking pulp of the top five producing countries in the Asia-Pacific region (China, India, Pakistan, Thailand, and Vietnam) was 20,485 million tonnes (Abd El-Sayed et al., 2020). The share of each country to the total capacity is illustrated in Fig. 12.5.

Fiber properties influence the properties of the resulting paper products. Therefore, the simplest means of predicting the suitability of raw materials for pulp and papermaking is by determining the fiber properties. It includes measuring the fiber length, cell-wall thickness, and fiber diameter. The derivative value of fibers is commonly determined to evaluate the influence of combined fiber dimensions on the properties of paper. The derivative value of fiber dimensions includes slenderness, flexibility ratio, Runkle ratio, and rigidity coefficient. However, an easier and faster way of determining fiber properties is by measuring fiber length, strength, and coarseness. Coarseness measurement includes fiber thickness, lumen diameter, and cellulose density of the cell wall. Coarseness is proportional to the cross-section area of the fiber and thus to the cell diameter. Higher coarseness indicates a thicker cell wall, subsequently stiffer, inflexible, and resistant to collapse fibers (Seth, 1990).

Bearing in mind that nonwood plants are suitable alternatives for wood fiber sources, the properties of numerous nonwood fibers have been characterized. Table 12.3 indicates a wide variation in the fiber characteristics of nonwood plant fibers. It can be seen that, except for bamboos, the fiber length of nonwood raw materials is between hardwood and softwood. It is, in fact, closer to the length of hardwood fibers. Therefore, it could substitute part of hardwood fibers in the papermaking process. Shorter fiber benefits the formation during paper sheet production and, when appropriately beaten,

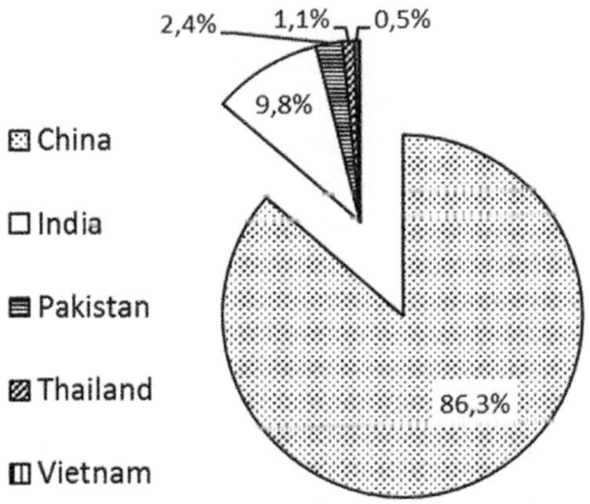

FIGURE 12.5 The top five Asia-Pacific countries in nonwood papermaking pulp capacity.

TABLE 12.3 The average fiber dimension of nonwood plant fibers.

Fiber source	Fiber dimension				References
	Length (mm)	Diameter (μm)	Lumina diameter (μm)	Cell-wall thickness (μm)	
Sorghum stalk	1.8	19.5	6.6	6.5	Albert et al. (2011)
Corn stalk	1.3	24.3	10.7	6.8	Usta et al. (1990)
Cotton stalk	1.3	29.3	23.0	3.6	Eroğlu et al. (1992)
Sago pith	1.9	34.0	25.0	5.0	Qanytah et al. (2019)
Bamboo	4.6	30.0	18.0	6.0	Qanytah et al. (2019)
Hardwood	1.2	24.3	13.7	5.3	Kiaei and Samariha (2011)
Softwood	2.6	31.7	19.2	6.2	Bektaş (2018)

could increase the tensile strength of the paper sheet. At the same time, thick cell-wall fiber reduces the interfiber bonding due to a low surface area. The fiber needs to be swollen and flattened. Flattening a thick cell-wall fiber requires higher beating energy. Internal defibrillation due to pulp beating increases the swelling capability of pulp. Internal defibrillation of the pulp can be measured by determining water retention value (WRV). Fig. 12.6 indicates an example of increasing WRV due to pulp beating. The cell-wall thickness of sorghum and all other fibers from sources listed in Table 12.3 is classified as thick based on the classification of International Association of Wood Anatomists (IAWA) (Wheelers et al., 1989). According to IAWA classification, thick fiber retains a luminal diameter of less than threefold of double cell-wall thickness.

As previously mentioned, fiber dimensions intricately influence the strength properties of a paper sheet. Therefore, simplifying fiber properties' influence on paper properties is carried out by determining the fiber derivative values. The use of fiber derivative value is a common practice employed by the Forest Products Research Institute of Indonesia (LPHH, 1976). The derivative values of fibers are scored, and the total score is used to classify the fiber source grade, in which grade one (I) is the best and grade IV is the

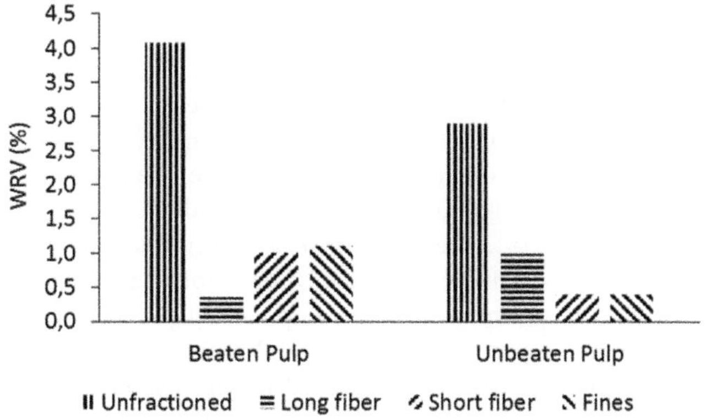

FIGURE 12.6 The pulp beating and fiber fraction effect on WRV of pulp (Wistara and Effendi, 2011). *WRV*, water retention value. *Reprints with permission from the editor-in-chief of Journal of Tropical Wood Science and Technology.*

TABLE 12.4 Scoring of fiber derivative value and grade classification of several nonwood fibers referring to softwood and hardwood fibers.

Fiber source	Fiber length (mm)	Fiber derivative value					Score	Grade
		RR	FP	MR	RC	FR		
Sorghum stalk	1.8	2.0	92.3	88.5	0.3	0.3	275	III
Corn stalk	1.3	1.3	53.5	80.6	0.3	0.4	225	III
Cotton stalk	1.3	0.3	44.4	38.4	0.1	0.8	400	II
Sago pith	1.9	0.4	55.9	45.9	0.1	0.7	400	II
Bamboo	4.6	0.7	153.3	64.0	0.2	0.6	400	II
Hardwood	1.2	0.8	49.4	68.2	0.2	0.6	275	III
Softwood	2.6	0.6	82.0	63.3	0.2	0.6	400	II

FP, Felting power; *FR*, flexibility ratio; *MR*, Muhlstep ratio; *RC*, rigidity coefficient; *RR*, Runkle ratio.

worst for the raw material of pulp and paper products. Grading is carried out following the criteria in Table 12.4. It can be seen that sorghum fiber is in the same grade as hardwood. Grade III is characterized by its thick cell wall and is thus resistant to collapse (Seth, 1990). This characteristic brings about the higher effort required so that the fiber forms a good paper quality.

12.3 Pulping of sorghum biomass

Sorghum is planted mainly for its grain and stalk juice that can be used, respectively, for food and bioethanol. Its stalk juice is also the primary raw material for food flavor, such as monosodium glutamate. In most cases, upon harvesting the grain and extraction of the stalk juice, the sorghum leaves and pinnacles are left in the ground, and extracted stalk bagasse is burned for energy (Alves et al., 2019). The waste of sorghum cultivation contains a relatively high amount of valuable cellulose (see Table 12.1) convertible into cellulose derivatives, such as various pulp and paper products. Different pulping methods have attempted to extract cellulosic materials from sorghum biomass. Sorghum stalks have been pulped by the method of soda pulping (Alves et al., 2019; Gencer and Sahin, 2015), soda–AQ pulping (Saeed et al., 2017; Ateş et al., 2015; Belayachi and Delmas, 1995; Khristova and Gabir, 1990), kraft pulping (Gencer and Hatil, 2019; Fatriasari et al., 2015), KOH pulping (Rainey and Doherty, 2013), and organosolv pulping (Kordsachia and Patt, 1991; Teramura et al., 2016, 2018). Even though it was not intended to produce pulp for paper products, thermomechanical pulping has been done as a pretreatment for bioethanol production using sorghum stalk (Gonzalez et al., 2011).

12.3.1 Alkaline pulping

Most works on pulping of sorghum biomass were done by alkaline pulping, such as soda, soda–AQ, KOH–AQ, and kraft pulping. Soda pulping might be the oldest among chemical pulping in existence. It was invented by Burgess and Watts of England in 1851 (Bajpai, 2018). The process uses sodium hydroxide as a cooking chemical, and it is not suited well for softwood and most hardwoods pulping, mainly for resinous softwood. The soda process was soon converted into a kraft process. However, the soda process is still being used for easily pulped nonwood raw materials, such as pulping of rice straw.

Gencer and Sahin (2015) identified the condition of soda pulping required to produce pulp from sorghum stalk grown in Turkey. Fig. 12.7 indicates that the tear strength of pulp increased quantitatively by increasing active alkali and decreased by increasing pulping temperature. On the other hand, burst strength tends to fall by increasing active alkali and pulping temperature. Calculated from pulping yield and the kappa number of pulps, both active alkali charge and cooking temperature increased the selectivity of lignin dissolution during pulping, as shown by the data listed in Table 12.5. Even though the delignification in pulping with 20% of active alkali at 130°C was the most selective, the scoring system involving pulping properties and strength properties of pulp used by these researchers suggested pulping at 120°C, and active alkali of 20% was the most appropriate condition

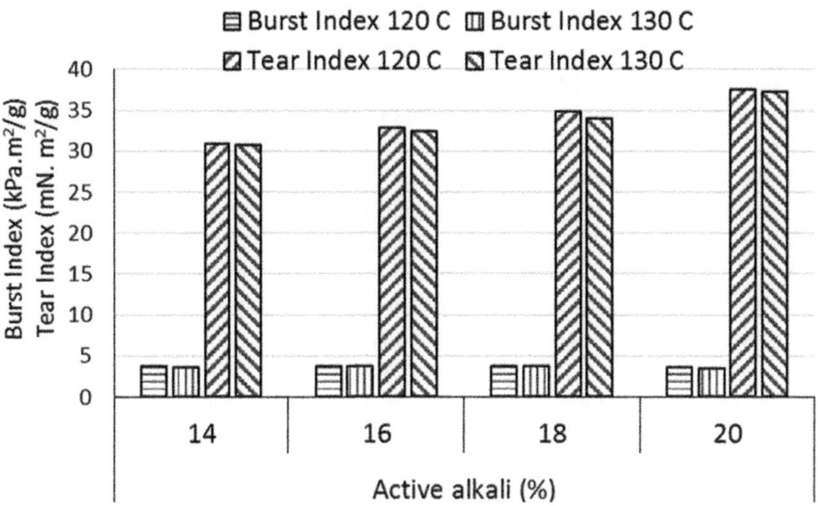

FIGURE 12.7 The effect of active alkali and pulping temperature on burst and tear indices of sorghum pulp. *Reproduced from the data of Gencer, A., Hatil, C., 2015. Identifying the conditions required for the NaOH method for producing pulp and paper from sorghum grown in Turkey. BioResources, 10(2), 2850–2858.*

TABLE 12.5 Delignification selectivity of sorghum stalk soda pulping.

Temperature (°C)	Active alkali (%)			
	14	**16**	**18**	**20**
120	1.8	1.6	1.4	1.3
130	1.7	1.5	1.3	1.0

Note: Calculation is based on yield, kappa number, and amount of oven-dried raw material cooked per batch. Delignification selectivity = (lignin in pulp/carbohydrate in pulp) × 100, lower value indicates higher selectivity.

for sorghum. Pulping was carried out at the liquor to stalk sorghum (L/W ratio) of 5/1 and cooking time of 60 minutes (maximum cooking temperature). A similar finding has been reported by Zendrato et al. (2021) on the soda pulping of torch ginger, a nonwood fiber source. They found that the highest delignification selectivity was achieved using 25% active alkali and a maximum cooking temperature of 170°C at an L/W ratio of 6/1 for 150 minutes. It seems that a better alkali profiling at higher active alkali keeps the delignification selective.

Alves et al. (2019) carried out a soda pulping on fine particles of sorghum straw. The straw was milled to a 1 mm mesh sieve. Reducing particle size eases the delignification of sorghum straw. A low alkali concentration of only 1.25% with a cooking time of 2.5 hours at 90°C, a kappa number of 2.7 was achieved. The finding also indicates that higher alkali gives a better delignification of the raw materials (alkali concentration used was 0.75%–1.25%). The pulp was intended for the production of cellulose acetate. Thus, it needs to be bleached. Bleaching was carried out with an acetate buffered–sodium chlorite. Indeed, raw material milling is not appropriate for producing pulp and paper products since the paper strength will be lost.

It is well understood that soda pulping is characterized by low yield and low pulp strength. It is brought about by the unselective delignification of soda cooking liquor, mainly at the residual cooking stage. Cellulose undergoes a peeling reaction (Fig. 12.8) that reduces its degree of polymerization (DP) and gives low pulp strength. AQ as a catalyst in soda pulping has successfully overcome the loss of strength and yield. The cyclic action of AQ (Fig. 12.9) increases pulping yield and promotes delignification. The drawback of the soda–AQ pulping method is that the AQ is expensive and cannot be recovered from the black liquor. Pulping various nonwood lignocellulosic materials with soda–AQ by Ateş et al. (2015) resulted in a bleachable pulp

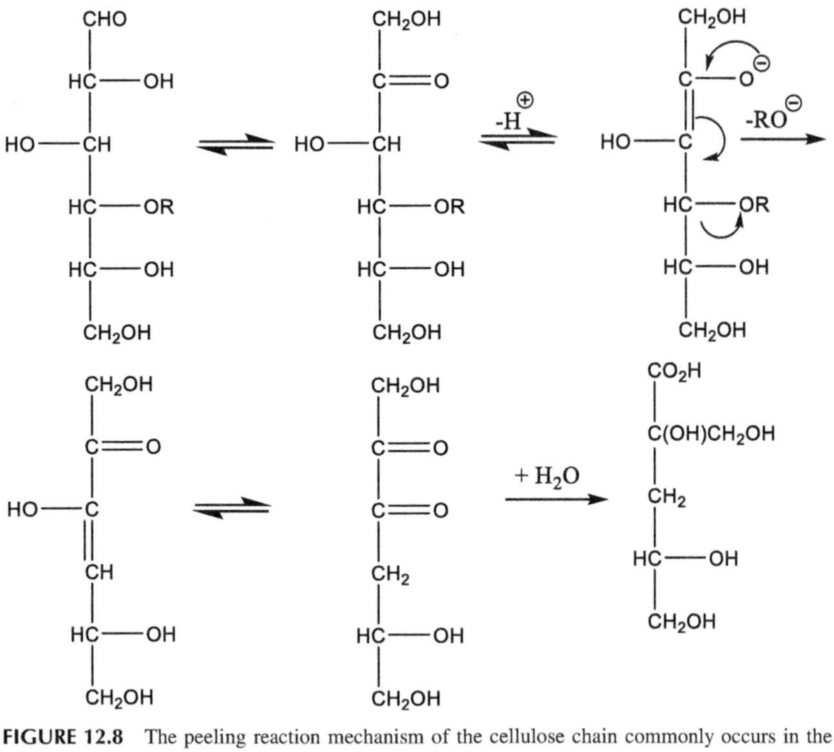

FIGURE 12.8 The peeling reaction mechanism of the cellulose chain commonly occurs in the alkaline pulping, R is a glucan (cellulose) chain (Sjöström, 1993).

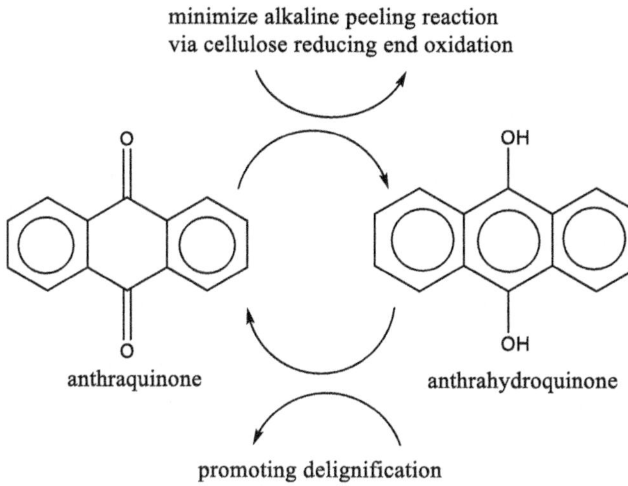

FIGURE 12.9 Cyclic action of AQ reduces the peeling reaction of cellulose, the reducing end group, and promotes delignification (Bajpai, 2018). *AQ*, Anthraquinone. *Reprints with permission from Elsevier, License number: 5115180129296.*

for most raw materials. Depending on the characteristics of nonwood fiber sources, a bleachable pulp yield as high as 44% has been achieved. However, only wheat straw gave a relatively high strength (tensile index of close to 80 N m/g). Depithing is beneficial in the case of sorghum stalk pulping with soda–AQ (Belayachi and Delmas 1995). Relatively high yield (app. 45%), low kappa number (app. 10), and exceptional strength properties of pulps were obtained by depithing before pulping with soda–AQ was carried out. Even though soda–AQ can increase delignification selectivity and strength properties of depithed sorghum stalk, the effect was less definite than tropical hardwoods (Khristova and Gabir, 1990).

The kraft pulping method is superior to the soda pulping method, mainly for wood pulping, so attempts have been made to apply kraft pulping to sorghum stalk. Fatriasari et al. (2015) showed that pulping yield was influenced by active alkali charge but not by sulfidity. However, an improvement of delignification occurred with increasing sulfidity in the white liquor. The pulp result also showed an acceptable strength property, mainly for the bursting strength. The pulping yield is dependent on the dissolution and degradation of lignin and carbohydrates influenced by alkali charge, concentration and cooking time, and temperature (Gullichsen, 1999). In contrary to the results of Fatriasari et al. (2015), which achieved a pulping yield of up to 55% and kappa number of 7 and 8, Gençer and Hatil (2019) obtained a relatively lower yield of approximately up to 38% with the lowest kappa number of 23 for sorghum stalk kraft pulping. The former uses a higher alkali charge (17%–22%) and cooking temperature (170°C) than the latter (10%–14% at

155°C). As mentioned previously, the dissolution of carbohydrates depends on the alkali charge, cooking time, and temperature. Employing low alkali charge and sulfidity will not retain a relatively constant active pulping species (OH^-) during all cooking stages, which will promote the degradation of cellulose at the final cooking stage. Interestingly, Gencer and Hatil (2019) obtained a relatively high tensile strength after beating the pulp to °SR of 50 +/−2, which was much higher than the tensile index of unbeaten pulp obtained by Fatriasari et al. (2015).

Apart from using a standard alkali such as soda, soda−AQ, and kraft pulping, AQ−potassium hydroxide (KOH−AQ) pulping has been applied for sorghum stalk (Rainey and Doherty, 2013). At a cooking temperature of 170°C, an L/W ratio of 6/1, and a cooking time of 75 minutes, a 29% KOH was appropriate to pulp sorghum stalk to result in a kappa number of bleachable pulps. Sorghum pulp is also relatively easy to be bleached with the totally chlorine-free bleaching method of chelation-peroxide-peroxide (DTPA−H_2O_2−H_2O_2) to a brightness of 68% ISO. The tensile index of unbleached pulp and bleached pulp of sorghum stalk were approximately 44 and 29 N m/g. A higher concentration is needed to pulp sorghum biomass using KOH than that required by NaOH (usually up to 25%).

12.3.2 Organosolv pulping

Even though chemical pulps are dominantly produced by the kraft pulping method, the process is associated with numerous environmental concerns. The use of sulfurous chemicals brings about the release of problematic malodorous gasses. Endeavors to invent nonsulfur environment-friendly pulping processes such as biopulping and organosolv pulping have been undertaken. Organic solvent-based-pulping methods known as organosolv processes using methanol, ethanol, and acetic acid were intensively investigated (Muurinen, 2000). Several developed organosolv processes have been given industrial names such as ASAM, Organocell (methanol-based), Alcell (ethanol-based), Milox (formic acid-based), Acetocell, Acetosolv, and Formacell (acetic acid-based) (Young and Akhtar, 1998). Organosolv can be applied to pulp hardwoods, softwood, and annual plants. In the beginning, the direct distillation of the spent pulping liquor and recovering the energy from the combustion of organic components was thought among the advantages of organosolv processes (Fig. 12.10). However, none of the processes succeeded in being industrially beneficial.

In the case of organosolv pulping of sorghum, instead of using high solvent concentration, Teramura et al. (2016) carried out sorghum bagasse pulping using ethanol, 1-propanol, 2-propanol, 1-butanol, and 1-pentanol at the solvent concentration of 12.5% catalyzed with 1% sulfuric acid. Pulping was carried out at 180°C for 45 minutes. They could recover approximately 72%−91% cellulose with about 44%−74% lignin removal. Interestingly,

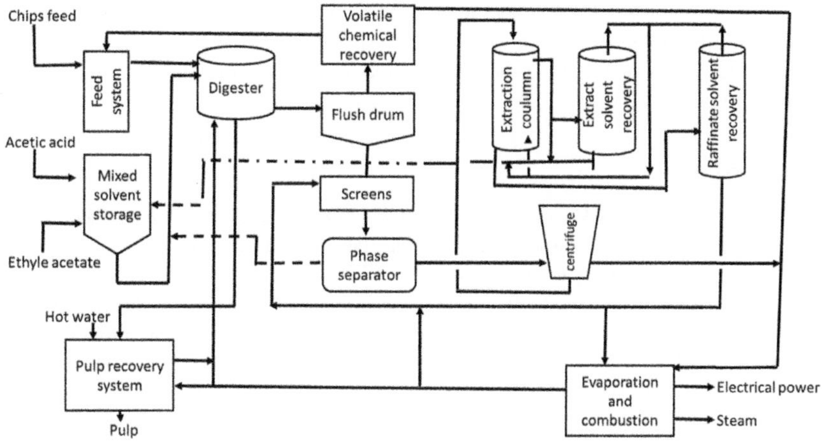

FIGURE 12.10 Ester pulping system by Professor Raymond Young of the Department of Forestry, University of Wisconsin-Madison, United States (Haggin, 1986).

three fractions were obtained using a solvent with high hydrophobicity, such as 1-butanol and 1-pentanol. It would facilitate the recovery of the solvent and organic component of the spent liquor. Increasing the butanol concentration to 25% and pulping temperature to 200°C resulted in a cellulose recovery of about 85% with a residual lignin of 15% (Teramura et al., 2018).

12.4 Conclusions

Sorghum biomass can be a source of food, energy, and cellulosic fibers. It is an annual plant adaptable to the relatively harsh climatic growing site. Even though most research on the utilization of sorghum biomass emphasized its utilization as bioenergy (bioethanol), investigations of its potential for the raw material of pulp production have been done. The relatively high cellulose and low lignin content of its stalk indicate that sorghum stalk is suitable for the raw material of pulp and paper production. The fiber length of sorghum is between hardwoods and softwoods and could substitute part of short hardwood fiber in papermaking. An effort is needed to prepare sorghum fiber for good paper quality since sorghum is classified as grade III fiber quality, indicating a thick fiber wall. Sorghum stalk can be converted into a satisfying pulp quality using soda, soda–AQ, KOH–AQ, kraft, and organosolv pulping methods.

Acknowledgments

The authors gratefully acknowledge the Division of Forest Products Chemistry—Department of Forest Products, Faculty of Forestry and Environment of the IPB University, and the Research Center for Biomaterial, Indonesian Institute of Sciences

(LIPI) for the facilities given during the preparation of this manuscript. Special thanks are addressed to Dr. Dede Hermawan for his permission to use the sorghum picture displayed in the manuscript and also to Azizatul Karimah, S.Si for processing the license number of figures.

References

Abd El-Sayed, E.S., El-Sakhawy, M., El-Sakhawy, M.A.M., 2020. Non-wood fibers as raw material for pulp and paper industry. Nordic Pulp Pap. Res. J. 1–12. Available from: https://doi.org/10.1515/npprj-2019-0064.

Agnihotri, S., Dutt, D., Tyagi H., C., 2010. Complete Characterization of Bagasse of Early Species of Saccharum officinerum-CO 89003 for Pulp and Paper Making. BioResources 5 (2), 1197–1214.

Albert, S., Padhiar, A., Gandhi, D., 2011. Fiber properties of *Sorghum halepense* and its suitability for paper production. J. Nat. Fibers 8 (4), 263–271. Available from: https://doi.org/10.1080/15440478.2011.626236.

Alves, J.A.A., dos Santos, M.D.L., Morais, C.C., Ascheri, J.L.R., Signini, R., dos Santos, D.M., et al., 2019. Sorghum straw: pulping and bleaching process optimization and synthesis of cellulose acetate. Int. J. Biol. Macromol. 135, 877–886. Available from: https://doi.org/10.1016/j.ijbiomac.2019.05.014.

Atchison, J.E., 1995. Twenty-five years of global progress in non-wood plant fibre pulping-historical highlights, present status and future prospects. *Proceedings, Tappi pulping conference* (Book 1). Tappi Press, Atlanta, GA, pp. 91–101.

Ateş, S., Deniz, İ., Kirci, H., Atik, C., Okan, O.T., 2015. Comparison of pulping and bleaching behaviors of some agricultural residues. Turkish J. Agric. For. 39, 144–153. Available from: https://doi.org/10.3906/tar-1403-41.

Bajpai, P., 2018. Biermann's Handbook of Pulp and Paper: Raw Material and Pulp Making, 3rd ed. Elsevier, Oxford, pp. 295–351.

Barbash, V., Trembus, I., Alushkin, S., Yashchenko, O., 2016. Comparative pulping of sunflower stalks. ScienceRise 3/2 (20), 71–78. Available from: https://doi.org/10.15587/2313-8416.2016.63098.

Barcelos, C.A., Maeda, R.N., Betancur, G.J.V., Pereira Jr., N., 2011. Ethanol production from sorghum grains [*Sorghum bicolor* (L.) MOENCH]: evaluation of the enzymatic hydrolysis and the hydrolysate fermentability. Braz. J. Chem. Eng. 28 (04), 597–604.

Bektaş, İ., 2018. The suitability of radiata pine wood for papermaking. Pro Ligno 14 (2), 38–44.

Belayachi, L., Delmas, M., 1997. Sweet sorghum bagasse: a raw material for the production of chemical paper pulp: effect of depithing. Ind. Crop. Products 6, 229–232.

Belayachi, L., Delmas, M., 1995. Sweet sorghum: a quality raw material for the manufacturing of chemical paper pulp. Biomass Bioenergy 8 (6), 411–417.

Bertella, S., Luterbacher, J.S., 2020. Lignin functionalization for the production of novel materials. Trends Chem. 1–14.

Billa, E., Koullas, D.P., Monties, B., Koukios, G.G., 1997. Structure and composition of sweet sorghum stalk components. Ind. Crop. Products 6, 297–302.

Bridgers, E.N., Chinn, M.S., Veal, M.W., Stikeleather, L.F., 2011. Influence of juice preparations on the fermentability of sweet sorghum. Biol. Eng. Trans. 4 (2), 57–67.

Cardoso, W.S., Queiroz, P.V., Tavares, G.P., Santos, F.A., de Freitas Soares, F.E., Kasuya, M.C. M., et al., 2018. Multi-enzyme complex of white rot fungi in saccharification of lignocellulosic material. Braz. J. Microbiol. Available from: https://doi.org/10.1016/j.bjm.2018.05.006.

Cardoso, W.S., Tardin, F.D., Tavares, G.P., Queiroz, P.V., Mota, S.S., Kasuya, M.C.M., et al., 2013. Use of sorghum straw (*Sorghum bicolor*) for second generation ethanol production: pretreatment and enzymatic hydrolysis. Quim. Nova 36 (5), 623–627.

Carioca, J.O.B., Leal, M.R.L.V., 2011. Ethanol production from sugar-based feedstocks. Compr. Biotechnol. 27–35. Available from: https://doi.org/10.1016/b978-0-08-088504-9.00184-7.

Chio, C., Sain, M., Qin, W., 2019. Lignin utilization: a review of lignin depolymerization from various aspects. Renew. Sustain. Energy Rev. 107, 232–249.

Dyartanti, E.R., Margono, S.H., Pranolo, B., Setiani, Nurhayati, A., 2015. Bioethanol from sorghum grain (*Sorghum bicolor*) with SSF reaction using biocatalyst co-immobilization method of glucoamylase and yeast. Energy Procedia 68, 132–137. Available from: https://doi.org/10.1016/j.egypro.2015.03.241.

Eroğlu, H., Usta, M., Kırcı, H., 1992. A review of oxygen pulping conditions of some non wood plants growing in Turkey. In: Tappi Pulping Conference Proceedings, Book:1, November 1–5, Boston, U.S.A., pp 215–222.

Fatriasari, W., Syafii, W., Wistara, N., Syamsu, K., Prasetya, B., 2016. Lignin and cellulose changes of Betung bamboo (*Dendrocalamus asper*) pretreated microwave heating. Int. J. Adv. Sci. Eng. Inf. Technol. 6 (2), 186–195.

Fatriasari, W., Supriyanto, Iswanto, A.H., 2015. The kraft pulp and paper properties of sweet sorghum bagasse (*Sorghum bicolor* L Moench). J. Eng. Technol. Sci. 47 (2), 149–159. Available from: https://doi.org/10.5614/j.eng.technol.sci.2015.47.2.4.

Food Security Department – NRI, 1999. Sorghum: Post-Harvest Operations. Food and Agriculture Organization of the United Nations.

Gencer, A., Hatil, C., 2019. Determination of alkaline and sulfite ratio in paper pulp produced from sorghum stalks via the kraft method. BioResources 14 (1), 922–930.

Gencer, A., Sahin, M., 2015. Identifying the conditions required for the NaOH method for producing pulp and paper from sorghum grown in Turkey. BioResources 10 (2), 2850–2858.

Gonzalez, R., Jameel, H., Chang, H., Treasure, T., Pirraglia, A., Saloni, D., 2011. Thermomechanical pulping as a pretreatment for agricultural biomass for biochemical conversion. BioResources 6 (2), 1599–1614.

Gullichsen, J., 1999. Papermaking Science and Technology. Book 6A. Chemical Pulping. Helsinki, Fapet Oy, p. A51.

Haggin, J., 1986. Ester pulping process avoids use of sulfur compounds. Chem. Eng. News 64 (4), 25–26. Available from: https://doi.org/10.1021/cen-v064n004.p025.

Han, J.S. 1998. Properties of non-wood fibers. In: Proc. of the 1998 Annual Meeting of the Korean Society of Wood Science and Technology, pp. 3–12.

Hatfield, R.D., Marita, J.M., Frost, K., Grabber, J., Ralph, J., Lu, F., et al., 2009. Grass lignin acylation: *p*-coumaroyl transferase activity and cell wall characteristics of C3 and C4 grasses. Planta 229 (6), 1253–1267. Available from: https://doi.org/10.1007/s00425-009-0900-z.

He, L., Yang, S., Chen, D., Peng, L., Liu, Y., Guan, Q., et al., 2020. Hemicellulose transportation from different tissues of corn stalk to alkaline hydrogen peroxide solution. Cellulose . Available from: https://doi.org/10.1007/s10570-020-03088-8.

Hui, Qu, Bao Liu, Xin, Fei Dong, Chen, Yan Lu, Xiao, Xin Shen, Yi, 2014. Review: Field performance and nutritive value of sweet sorghum in eastern China. Field Crop. Res. 157, 84–88. Available from: https://doi.org/10.1016/j.fcr.2013.12.010.

Khristova, P., Gabir, S., 1990. Soda-anthraquinone pulping of sorghum stalks. Biol. Wastes 33, 243–250.

Kiaei, M., Samariha, A., 2011. Fiber dimensions, physical and mechanical properties of five important hardwood plants. Indian. J. Sci. Technol. 4 (11), 1460–1463.

Kordsachia, O., Patt, R., 1991. Suitability of different hardwoods and non-wood plants for non-polluting pulp production. Biomass Bioenergy 1 (4), 225–231.

Liu, R., Shen, F., 2008. Impacts of main factors on bioethanol fermentation from stalk juice of sweet sorghum by immobilized *Saccharomyces cerevisiae* (CICC 1308). Bioresour. Technol. 99, 847–854. Available from: https://doi.org/10.1016/j.biortech.2007.01.009.

LPHH, 1976. Dimensi Serat Jenis Kayu di Indonesia. Laporan LPHH 75.

McBee, G.G., Miller, F.R., 1993. Stem carbohydrate and lignin concentrations in sorghum hybrids at seven growth stages. Crop. Sci. 33, 530–534.

McCloskey, J.T., 1995. What about non-woods? Tappi Global Fibre Supply Symposium. Tappi Press, Atlanta, GA, pp. 95–106.

McIntosh, S., Vancov, T., 2010. Enhanced enzyme saccharification of *Sorghum bicolor* straw using dilute alkali pretreatment. Bioresour. Technol. 101, 6718–6727. Available from: https://doi.org/10.1016/j.biortech.2010.03.116.

Mei, X., Liu, R., Shen, F., Wu, H., 2009. Optimization of fermentation conditions for the production of ethanol from stalk juice of sweet sorghum by immobilized yeast using response surface methodology. Energy Fuels 23, 487–491. 3.

Muurinen, E., 2000. Organosolv pulping. Dissertation, Faculty of Technology. University of Oulu, Linnanmaa.

Piskorz, J., Majerski, P., Radlein, D., Scott, D.S., Bridgwater, A.V., 1998. Fast pyrolysis of sweet sorghum and sweet sorghum bagasse. J. Anal. Appl. Pyrolysis 46, 15–29.

Qanytah, K., Syamsu, F.Fahma, Pari, G., 2019. Structure analysis of three non-wood materials for liner paper. Nordic Pulp Pap. Res. J. 34 (4), 453–466.

Rainey, T., Doherty, W., 2013. Caustic potash pulping study of water efficient *Arundo donax* and *Sorghum bicolor*: a comparison to sugarcane bagasse. In: Clark, T. (Ed.), Proceedings of the 2013 Appita Conference and Exhibition. Appita Inc, Australia, pp. 78–84.

Richie, W.R., McBee, G.G., 1991. Structural components in sorghum stem biomass. Bioresour. Technol. 38, 15–22.

Saeed, H.A.M., Liu, Y., Lucia, L.A., Chen, H., 2017. Evaluation of Sudanese sorghum and bagasse as a pulp and paper feedstock. BioResources 12 (3), 5212–5222.

Senthilkumar, S.T., Senthilkumar, B., Balaji, S., Sanjeeviraja, C., Selvan, R.K., 2011. Preparation of activated carbon from sorghum pith and its structural and electrochemical properties. Mater. Res. Bull. 46, 413–419. Available from: https://doi.org/10.1016/j.materresbull.2010.12.002.

Seth, R.S., 1990. Fibre quality factors in papermaking – II: the importance of fibre coarseness. Mat. Res. Soc. Symp. Proc. 197, 143–161.

Sjöström, E., 1993. Wood Chemistry: Fundamentals and Applications, 2nd ed. Academic Press, Inc, New York, p. 48.

Smook, G.A., 1992. Handbook For Pulp and Paper Technologists: Chapter 7 Kraft Pulping, 2nd ed. Angus Wilde Publications, Vancouver, pp. 74–83.

Solihat, N.N., Fajriutami, T., Adi, D.T.N., Fatriasari, W., Hermiati, E., 2017. Reducing Sugar Production of Sweet Sorghum Bagasse Kraft Pulp. In: AIP Conf. Proc. 1803, 020012-1–020012-8; Available from: https://doi.org/10.1063/1.4973139.

Tang, C., Li, S., Li, M., Xie, G.H., 2018. Bioethanol potential of energy sorghum grown on marginal and arable lands. Front. Plant. Sci. 9, 440. Available from: https://doi.org/10.3389/fpls.2018.00440.

Teramura, H., Sasaki, K., Oshima, T., Kawaguchi, H., Ogino, C., Sazuka, T., et al., 2018. Effective usage of sorghum bagasse: optimization of organosolv pretreatment using 25% 1-butanol and subsequent nanofiltration membrane separation. Bioresour. Technol. 252, 157–164. Available from: https://doi.org/10.1016/j.biortech.2017.12.100.

Teramura, H., Sasaki, K., Oshima, T., Matsuda, F., Okamoto, M., Shirai, T., Kawaguchi, H., Ogino, C., Hirano, K., Sazuka, T., Kitano, H., Kikuchi, J., Kondo, A., 2016. Organosolv pretreatment of sorghum bagasse using a low concentration of hydrophobic solvents such as 1-butanol or 1-pentanol. Biotechnol. Biofuels 9, 27. Available from: https://doi.org/10.1186/s13068-016-0427-z.

Theerarattananoon, K., Xu, F., Wilson, J., Ballard, R., Mckinney, L., Staggenborg, S., et al., 2011. Physical properties of pellets made from sorghum stalk, corn stover, wheat straw, and big bluestem. Ind. Crop. Products 33, 325–332. Available from: https://doi.org/10.1016/j.indcrop.2010.11.014.

Usta, M., Erolu, H., Kırcı, H., 1990. Soda-oxygen pulping of corn stalks (*Zea mays* indurata sturt.). In: Pulping Conference, Sheraton Centre, Toronto, Ontario, October 14–17, Proceedings Books 1, pp. 307–312.

Wheeler, E.A., Baas, P., Gasson, P.E., 1989. IAWA List of microscopic features for hardwood identification. IAWA Bull. n. s. 10 (3), 219–332.

Wiloso, E.I., Setiawan, A.A.R., Prasetia, H., Muryanto, Wiloso, A.R., Subyakto, et al., 2020. Production of sorghum pellets for electricity generation in Indonesia: a life cycle assessment. Biofuel Res. J. 27, 1178–1194. Available from: https://doi.org/10.18331/BRJ2020.7.3.2.

Winarti, C., Arif, A.B., Budiyanto, A., Richana, N., 2020. Sorghum development for staple food and industrial raw materials in East Nusa Tenggara, Indonesia: a review. IOP Conf. Series: Earth Environ. Sci. 443 (2020), 012055. Available from: https://doi.org/10.1088/1755-1315/443/1/012055.

Wistara, N.J., Effendi, H., 2011. The characteristics of beaten and unbeaten mixed tropical hardwood kraft pulp. J. Ilmu dan. Teknologi Kayu Trop. 9 (1), 34–42.

Yin, R., Liu, R., Mei, Y., Fei, W., Sun, X., 2013. Characterization of bio-oil and bio-char obtained from sweet sorghum bagasse fast pyrolysis with fractional condensers. Fuel 112, 96–104. Available from: https://doi.org/10.1016/j.fuel.2013.04.090.

Young, R.A., Akhtar, M. (Eds.), 1998. Environmentally Friendly Technologies for the Pulp and Paper Industry. John Wiley & Sons, Inc, New York, ISBN 0-471-15770-8.

Zendrato, H.M., Devi, Y.S., Masruchin, N., Wistara, N.J., 2021. Soda pulping of torch ginger stem: promising source of nonwood-based cellulose. J. Korean Wood Sci. Technol. 49 (4). Available from: https://doi.org/10.5658/WOOD.2021.49.4.1.

Chapter 13

Pulping and papermaking using pineapple leaves

Vinod Kumar Nathan
School of Chemical and Biotechnology, SASTRA Deemed to be University, Thanjavur, Tamil Nadu, India

13.1 Introduction

Natural fiber-based composites are undergoing extensive research because of their environmentally benign nature and unique characteristics. Natural fibers have the advantage of being readily available, easy to handle, and biodegradable. Natural fibers have excellent physical and mechanical qualities, although they vary depending on the source, species, locality, and many more (Asim et al., 2015). After bananas and other citrus fruits, pineapple (*Ananas comosus*) is considered the world's most popular fruit seen in tropical countries (Arib et al., 2004). Fruits of pineapple are commercially valuable, and the leaves are neglected as waste products of the fruit that are utilized to extract natural fibers. Pineapple leaf fiber (PALF) has exceptional mechanical characteristics, which could be used to create reinforced polymer composites (Pavithran et al., 1987; Mishra et al., 2001). PALF, like other natural fibers, offers a spectrum of applications, including plastic reinforcing, sound, and thermal insulation (Kengkhetkit and Amornsakchai, 2012).

The main disadvantage of PALF is that it is hydrophilic, making it difficult to bond with a matrix material, especially at high temperatures (George et al., 1998). Chemical treatments, such as NaOH treatment, dewaxing, cyanoethylation, and grafting of acrylonitrile monomer over dewaxed PALF, could all help to improve the interaction quality between PALF and polymer. Furthermore, compounds, such as NaOH, benzoyl peroxide (BPO), 2,4-dinitrochlorobenzene, and BPO/acetylation, can reduce the absorption of water along with increased mechanical characteristics (Samal and Mukul, 1997). Modification of PALF-reinforced composites chemically can lower the moisture content of the composite when compared with unmodified PALF (George et al., 1995).

Pulping and Papermaking of Nonwood Plant Fibers. DOI: https://doi.org/10.1016/B978-0-323-91625-7.00016-3

The PFs can be recycled for a variety of applications as they have high biodegradable cellulose, including paper (Yusof et al., 2012), textile industry, food packaging (Sasikala and Umapathy, 2018), acoustic barrier (Girish et al., 2018), medical applications (Kengkhetkit and Amornsakchai, 2012), automotive and electrical applications (Ramnath et al., 2014; Hoang et al., 2019). The chemical constitution of PALF is shown in Table 13.1. These values, however, will vary depending on the plant's kind and the environment in which it is produced (Mahardika et al., 2018).

This chapter mainly focuses on the PALF characteristics and their properties for papermaking. This also provides an overview of various pretreatment methods as well as the extraction of cellulose from PALF. These fibers being an ideal source for papermaking and composite preparation requires extensive research. Chemical or physical modification of PALF enables them to be utilized for hybrid composite preparation for various applications. Owing to their mechanical and physical superiority over other natural fibers, PALF finds application in many spheres of material sciences.

13.2 Properties of pineapple leaf fiber

PALF is with good mechanical qualities, as well as low density, high stiffness, and low cost (Devi et al., 2011). The thermomechanical qualities of the extracted fibers were shown to be directly related to the concentration of holocellulose and cellulose, along with the cellulose crystallinity index (CI). The mechanical characteristics of the fiber bundle were found to be inversely proportional to the lignin content and diameter (Neto et al., 2015). The PALF content of the paper affected the tensile modulus. When compared to PLA, the material containing 40% PALF increased its tensile modulus along with elongation at the break by roughly 48% and 111%, respectively (Kaewpirom and Worrarat, 2014). Young's modulus as well as the tensile

TABLE 13.1 Pineapple leaf fiber composition (Asim et al., 2015).

Constituent	Composition (w/w%)
Ash content	4.50
Cellulose	66.2
Holocellulose	85.7
Hemicellulose	19.5
1% NaOH solubility	39.8
Lignin content	4.28
Moisture content	81.6

strength of the resultant material improved when increased the fiber content, according to the rule of mixtures. By adding fiber to the paper material, the elongation of the fiber increases until the composites break. The optimum fiber length of 30 mm exhibited its best mechanical properties (Devi et al., 1997). In comparison to pure pineapple fiber, the materials' effectual thermal conductivity and diffusivity diminish as the fraction of fiber loading increases (Mangal et al., 2003). Furthermore, as the fiber content increases, all mechanical qualities improve dramatically (Motaleb et al., 2018). The storage modulus was observed to rise with the addition of PALF of 30 mm length, reaching a maximum value of 40 wt.% (Devi et al., 2011). PALF's hydrophilic characteristic causes it to absorb more moisture, causing its overall properties to deteriorate. Engineering the fiber surface can resolve this issue. As a result, for the PALF and its PALF-fortified materials to function better, a full perception of the impact of modification of fiber surface on various characteristics and polymer adhesion is required (Rajeshkumar et al., 2020). Because of the elimination of surface contaminants, breakdown of the middle lamella, and/or interfibrillar swelling, mercerization treatments reduced the diameter of fibers. Mercerized fibers' mechanical characteristics were altered (Jaramillo-Quiceno et al., 2018).

13.3 Pretreatment of pineapple leaf fiber

Pretreatment of PALF is an essential step when it comes to reducing the energy and time for pulping. The pretreated fiber pulping is easier than its untreated counterparts. Moreover, various pretreatment steps help in the exposure of fiber materials in order to remove the lignin and hemicellulose. The steam explosion of fibers improved the thermal properties along with crystallinity by reducing hemicellulose and lignin content while boosting cellulose content. Changes in fiber shape depended on the steam pressure and treatment cycles. The modified fibers could have smaller widths but shorter lengths. In addition, the steam explosion approach may cause cellulose nanofibers (CNFs) to fibrillate. More repetitions of the steam explosion treatment may reveal high quantities of CNFs (Tanpichai et al., 2019). Because PL is primarily made up of cellulose, it is ideal for nanocellulose extraction. Acid-hydrolysis and pressured acid-hydrolysis procedures were used for the extraction process. In one of the studies, cellulose extraction was performed using alkali treatment and bleaching. Acid hydrolysis and pressured acid hydrolysis were used to hydrolyze cellulose nanocrystals (CNC). CNC has been scaled down to the nanoscale. The CI was discovered to be higher. As a result, CNC could be used as an agent for reinforcing the material like natural rubber (Chawalitsakunchai et al., 2019).

CNFs were initially created with an acid-base treatment and TEMPO-mediated oxidation, followed by surface modification with polymerized methyl methacrylate (MMA) in an aqueous system using an environmentally friendly method. The MMA-modified CNF's surface hydrophobicity increased interfacial compatibility with the polymethyl methacrylate (PMMA) matrix, increasing the contact angle from 12.02 to 57.45 degrees. When 1–3 wt.% modified CNF was added to the nanocomposites, the impact strength rose by 2.7%–22.9% (Shih et al., 2018).

13.4 Cellulose extraction

At times, the paper manufacturing strategies vary depending on the demand for properties of the final paper. For papermaking, sometimes, nanocellulose-based additives are used as mentioned in the above section. Hence, there is a need for cellulose extraction from natural fibers like PALF. According to chemical analyses, pineapple leaves (PL) possess 11.8% lignin along with 55.1% of holocellulose, among which α-cellulose constitutes about 32.6%. The cellulose extracted ranges in color from bright yellow to pure white. Cellulose is known as cellulose-1 in its natural condition and is classified into either two crystalline phases or as its allomorphs, cellulose I (either α or β form) (Atalla and Vanderhart, 1984). The proportion of such allomorphs in plants may vary in different species of plants. Because of the extensive hydrogen bonding, the structure of such allomorphs was characterized with high precision using crystallographic techniques (Nishiyama et al., 2002). This intra as well as intermolecular network generates a stable and sophisticated cellulose structural hierarchy offering a high axial rigidity (Moon et al., 2011).

The method of cellulose fiber extraction used in this investigation was a multistep process that included alkaline treatment followed by bleaching (Abraham et al., 2011; Cherian et al., 2011). The chemical extraction procedure has been shown to have an impact on the chemical constituents of PALF. The raw fibers are first hydrolyzed in an alkaline environment to eliminate hemicellulose, silica, and minerals, along with ash (Deepa et al., 2015). The residual lignin is removed in the final stage of bleaching, which takes place at a low pH as well as a high temperature (Acharjee et al., 2017). The bleaching process caused cellulose to degrade unfavorably. When compared to bleaching of 1 hour duration, the content of lignin in treated PALF was reduced by 2.4% after 4 hours of bleaching. The successive alkali treatment as well as bleaching diminished the proportion of lignin in the CMF and raised the proportion of the α-cellulose but had no effect on the hemicellulose content. The lignin content deteriorated as a result of the reduction in the cleavage components in lignin that is ester-linked (Deepa et al., 2015). It was also discovered that the rate of removal of lignin was proportional to the time spent bleaching. However, the cellulose content of PALF varies greatly. Some of the differences might be because of the variation in sample preparation and extraction processes, along with geographic and climatic factors (Chandrasekhar et al., 2003). Meanwhile, as it undergoes steam explosion and bleaching, Cherian et al. (2010) discovered that PALF has a high percentage of cellulose content (98.63%) and 0.8% lignin. Due to the existence of nonselective bleaching agents, the change and degradation of cellulose produced by excessive or insufficient bleaching necessarily diminishes its strength and chemical resistance, and hence its value, due to its inferior physical properties (Stephen Pointing et al., 2003).

Amorphous cellulose and other pollutants were discovered to be extensively polluted on the surface of raw cellulose. The surface of the fiber is rough and uneven, with waxes and other encrusting compounds, such as hemicelluloses, lignin, and cellulose, coating it. The alkaline and bleaching treatments, on the other hand, caused substantial morphological alterations in CMF samples, with no significant differences in CMF samples across different bleaching periods. The production of micro-sized cellulose fibrillations was also a result of the NaOH treatment of cellulose. After 4 hours of bleaching, the CMF had a clean and smoothed surface, indicating that contaminants and amorphous cellulose had been removed. As a result, holes formed between the PALF, causing the fibers to breakdown into smaller sizes.

In alkali-bleaching process, there are increases in the 2θ values, and no peaks were observed corresponding to cellulose II at 12.2 degrees in the case of other plant biomass (French, 2014). Similarly, a higher concentration of NaOH may trigger disruption of the lattice of cellulose with a high lateral order mode as in the case of cotton fibers (Yue et al., 2013). As reported earlier, cellulose source as well as extraction procedure influence the cellulose CI (Park et al., 2010; Wada et al., 1997), and successive alkali treatments might improve the CI of cellulose of other plant biomasses like banana pseudo stem (Azraaie et al., 2016). Elevated crystallinity can be attained by removing hemicellulose along with lignin from amorphous regions, which causes cellulose molecules to realign (Mtibe et al., 2015; Rambabu et al., 2016).

PALF has a significant potential for mechanical and thermal strength due to the exorbitant amount of cellulose. Fibers were first removed using the retting and scraping procedures and then treated with varying amounts of alkali. Characterization was carried out to determine its qualities, and it was discovered that alkaline treatment improves mechanical as well as thermal properties, however, up to a concentration of 7% alkali only (Jain et al., 2021). Maceration pretreatment was used to extract the cellulose. The heating timeframes ranged from a few minutes to several hours. When compared to traditional chemical pulping and bleaching techniques, this method offers a simpler and more effective way to delignify PALFs (Kassim et al., 2019).

13.5 Pulping of pineapple leaf fibers

In one of the studies, for pulping PL, acetone in the range of 1%–10% was used. The impact of soaking as well as the duration of delignification on the characteristics of the paper was studied. Thermogravimetric analysis and tearing resistance tests were used to investigate the thermal as well as physical properties of paper sheets. The morphological qualities were examined under a microscope at a magnification of 200 times. The mechanical qualities of the paper sheet manufactured from pulp with a 3% acetone content are the best. The strength of the paper was improved by extending the delignification period. Cooking the PL at 118°C under 80 kPa pressure shortened the delignification period, resulting in a significant increase in paper strength. Laftah and Wan (2015) stated that solvent pulping may be beneficial over other processes considering the scale of production, as well as the consumption of energy, water, and chemicals along with the final quality of pulp obtained from PL waste. In countries like Nigeria where there are a lot of agricultural waste products like PL, researchers tried to prospect them for pulp and papermaking (Aremu et al., 2015). Following the chemical pulping method (soda pulping), PL resulted in a pulp yield of 15% with 65% of cellulose and 11.5% of lignin content. These papers made from pineapple pulps also possess a tensile index of 0.19 Nm/g, tear index of 17.19 Nm^2/g, and elongation at break of 1.14 mm. Scanning electron microscopy (SEM) revealed the condensed and packed arrangement of fibers in the papers produced (Aremu et al., 2015). In a similar work, researchers tried the combination of pineapple leaf fiber samples with recycled newspapers. The tensile and tearing tests observed better values compared to papers made from pineapple fiber or newspaper alone. The findings from these experiments could help in the formulation of pineapple leaf fiber-based paper manufacture (Yusof et al., 2012).

13.6 Papermaking using pineapple leaf fibers

Employing acetone as one of the pulping agents for pulping PL was investigated by Laftah and Rahaman (2015). The influence of delignification time as well as soaking on the paper characteristics was examined. The mechanical qualities of the paper sheet fabricated from pulping by adding a 3% acetone were found to be the best. The strength of the paper was improved by extending the delignification period. In a similar study, the mechanical qualities of the paper sheet produced from pulping with a 3% acetone content are the best. The strength of the paper was improved by extending the delignification period. Clay, titanium dioxide, and calcium carbonate are the most common fillers. To ensure that the ferrite magnet, an alternate filler material, had the same qualities as the commercial filler, it was employed as the filler. The magnetic PL paper, like other commercial papers, has good qualities, and this will provide numerous benefits in terms of reducing the use of wood for pulp, as well as being one of the efforts toward green technology and innovations in papermaking (Sufian and Othman, 2021).

To enhance the rate of utilization of pineapple leaf as well as other natural fibers like crop straw and to further provide technological support for creating biodegradable mulch paper through organic cultivation, the procedure and features of degradable fiber mulch paper fabricated using pineapple leaf and rice straw were investigated. The degradable fiber mulch paper was created by combining raw components such as PALF and rice straw fiber, as well as other eco-friendly agents. The tensile strength and bursting strength parameters were reached by the degradable fiber mulch paper manufactured using pineapple leaf along with rice straw (Ming et al., 2021).

Some studies attempted to recycle leaves of pineapple that usually form an agro-waste mixed with waste paper to form new paper using facile processing procedures, such as pulping, mixing, sifting, compression, and drying. The tensile strength of the pineapple leaf combined with the waste paper sample was much better than that of the PALF sample. Furthermore, this composite made had a higher density than PALF alone. To enhance the qualities of the PALF-based recycled paper in the future, more changes to the body formulation will be required (Ter Teo et al., 2020).

However, the rich content of PALF may tend to develop flocs as well as agglomerates fibers, which degraded the composite paper's performance. The organization of the fibers in the composite paper should be modified to overcome the agglomeration effect. The mechanical and moisture properties of Bagasse sugarcane (BSC) BSC/PALF with various hybrid ratios were investigated, and the outcome revealed that hybrid composite paper of multiple layers produced a hybrid composite paper with superior properties to random hybrid composite paper. When compared to random hybrid fiber composite, multilayer hybrid fiber composite created stronger interfiber bonding and overcome the agglomeration effect between BSC/PALF (Salleh et al., 2021).

In another study, using a semimechanical soda pulping procedure, the paper was fabricated using a combination of PL along with sugarcane bagasse (SB) pulp at various cooking liquor or NaOH concentrations. The maximum tensile strength observed was about 6709 kN using a 100:0% PL and SB pulp composition with 3% NaOH, whereas the highest tearing strength (1730.13 mN) was obtained using a 40:60% of PL and SB with a comparable concentration of NaOH (Adegustias et al., 2019). Similarly, at varying concentrations of pineapple juice, a continuous level of 3% sodium hydroxide (NaOH) was added to the pineapple juice/water during the pulping of PALF. The PALF was soaked in pineapple juice and NaOH thrice to achieve the best outcomes for the paper sheet during the process of pulping and bleaching. In comparison, paper sheets made from NaOH-mediated pulping have superior mechanical as well as physical qualities than paper sheets made from pulping followed by NaOH and pineapple juice treatment. Furthermore, pulping with a greater proportion of pineapple juice produces long fibers (Jamal et al., 2018).

13.7 Conclusions

Among the natural fibers, PALF is one of the potential ones. The right usage of PALF in composite preparation is possible by combination with other fiber sources. The mechanical as well as the physical properties of hybrid composites made from PALF are very superior compared with others. Hence, the sustainable utilization of PALF is needed. PALF-based paper materials are giving higher performance than other papers made of natural fibers. However, extensive research is needed to improvise functional and surface properties of these papers.

References

Abraham, E., Deepa, B., Pothan, L.A., Jacob, M., Thomas, S., Cvelbar, U., et al., 2011. Extraction of nanocellulose fibrils from lignocellulosic fibers: a novel approach. Carbohydr. Polym. 86 (4), 1468–1475.

Acharjee, T.C., Jiang, Z., Daniel Haynes, R., Lee, Y.Y., 2017. Evaluation of chlorine dioxide as a supplementary pretreatment reagent for lignocellulosic biomass. Bioresour. Technol. 244, 1049–1054.

Adegustias, R.A., Margaretha, E.A., Simanjuntak, M.S., 2019. Production and characterization of paper from pineapple leaves and sugarcane bagasse pulp. J. Phys.: Conf. Ser. 1351 (1), 012117. IOP Publishing.

Aremu, M.O., Rafiu, M.A., Adedeji, K.K., 2015. Pulp and paper production from nigerian pineapple leaves and corn straw as substitute to wood source. Int. Res. J. Eng. Technol. 2 (4), 1180–1188.

Arib, R.M.N., Sapuan, S.M., Hamdan, M.A.M.M., Paridah, M.T., Zaman, H.M.D.K., 2004. A literature review of pineapple fiber reinforced polymer composites. Polym. Polym. Compos. 12 (4), 341–348.

Asim, M., Abdan, K., Jawaid, M., Nasir, M., Dashtizadeh, Z., Ishak, M.R., et al., 2015. A review on pineapple leaves fiber and its composites. Int. J. Polym. Sci. 2015.

Atalla, R.H., Vanderhart, D.L., 1984. Native cellulose: a composite of two distinct crystalline forms. Science 223 (4633), 283–285.

Azraaie, N., Abidin, N.A.M.Z., Aziz, F.A., Radiman, S., Ismail, A., Zin, W.Y.W.M., et al., 2016. Cellulose microfibrils/nanofibrils (CMNF) produced from banana (*Musa acuminata*) pseudo-stem wastes: isolation and characterization. Mater. Sci. Forum 846, 448–453. Trans Tech Publications Ltd.

Chandrasekhar, S.A.T.H.Y., Satyanarayana, K.G., Pramada, P.N., Raghavan, P., Gupta, T.N., 2003. Review processing, properties and applications of reactive silica from rice husk—an overview. J. Mater. Sci. 38 (15), 3159–3168.

Chawalitsakunchai, W., Dittanet, P., Loykulnunt, S., Tanpichai, S., Prapainainar, P., 2019. Extraction of nanocellulose from pineapple leaves by acid-hydrolysis and pressurized acid hydrolysis for reinforcement in natural rubber composites, IOP Conference Series: Materials Science and Engineering, vol. 526. IOP Publishing, p. 012019, no. 1.

Cherian, B.M., Leão, A.L., De Souza, S.F., Thomas, S., Pothan, L.A., Kottaisamy, M., 2010. Isolation of nanocellulose from pineapple leaf fibres by steam explosion. Carbohydr. Polym. 81 (3), 720–725.

Cherian, B.M., Leão, A.L., de Souza, S.F., Costa, L.M.M., de Olyveira, G.M., Kottaisamy, M., et al., 2011. Cellulose nanocomposites with nanofibres isolated from pineapple leaf fibers for medical applications. Carbohydr. Polym. 86 (4), 1790–1798.

Deepa, B., Abraham, E., Cordeiro, N., Mozetic, M., Mathew, A.P., Oksman, K., et al., 2015. Utilization of various lignocellulosic biomass for the production of nanocellulose: a comparative study. Cellulose 22 (2), 1075–1090.

Devi, L.U., Bhagawan, S.S., Thomas, S., 1997. Mechanical properties of pineapple leaf fiber-reinforced polyester composites. J. Appl. Polym. Sci. 64 (9), 1739–1748. Available from: https://doi.org/10.1002/(SICI)1097-4628(19970531)64:9 < 1739::AID-APP10 > 3.0.CO;2-T.

Devi, L.U., Bhagawan, S.S., Thomas, S., 2011. Dynamic mechanical properties of pineapple leaf fiber polyester composites. Polym. Compos. 32 (11), 1741–1750.

French, A.D., 2014. Idealized powder diffraction patterns for cellulose polymorphs. Cellulose 21 (2), 885–896.

George, J., Bhagawan, S.S., Prabhakaran, N., Thomas, S., 1995. Short pineapple-leaf-fiber-reinforced low-density polyethylene composites. J. Appl. Polym. Sci. 57 (7), 843–854.

George, J., Bhagawan, S.S., Thomas, S., 1998. Effects of environment on the properties of low-density polyethylene composites reinforced with pineapple-leaf fibre. Compos. Sci. Technol. 58 (9), 1471–1485.

Girish, M.N., Hari Rao, A.N., Veeramanju, K.T., 2018. Effect of erosive wear on pineapple leaf fiber reinforced vinylester based hybrid composite as acoustic barrier. In: Proceedings of TRIBOINDIA-2018 An International Conference on Tribology.

Hoang, H.A., Luu, Q.H., Le, P.T.K., Tran, V.T., 2019. Enhancement of pineapple residue composting by food waste addition. Chem. Eng. Trans. 72, 217–222.

Jain, J., Sinha, S., Jain, S., 2021. Compendious characterization of chemically treated natural fiber from pineapple leaves for reinforcement in polymer composites. J. Nat. Fibers 18 (6), 845–856.

Jamal, S.N., Abdul Rahman, W.A.W., Majid, N.A., Syed Ismail, S.N., 2018. PALF/skin extracted pineapple juice: New potential as pulping and bleaching agent in papermaking and its effect on mechanical and physical properties. AIP Conf. Proc. 2031 (1), 020023AIP Publishing LLC. Available from: https://doi.org/10.1063/1.5066979.

Jaramillo-Quiceno, N., Cadena Ch, E.M., Restrepo-Osorio, A., Felipe Santa, J., 2018. Improvement of mechanical properties of pineapple leaf fibers by mercerization process. Fibers Polym. 19 (12), 2604–2611. Available from: https://doi.org/10.1007/s12221-018-8522-3.

Kaewpirom, S., Worrarat, C., 2014. Preparation and properties of pineapple leaf fiber reinforced poly (lactic acid) green composites. Fibers Polym. 15 (7), 1469–1477.

Kassim, N.A., Mohamed, A.Z., Zainudin, E.S., Zakaria, S., Zakiah Azman, S.K., Abdullah, H. H., 2019. Isolation and characterization of macerated cellulose from pineapple leaf. BioResources 14 (1), 1198–1209.

Kengkhetkit, N., Amornsakchai, T., 2012. Utilisation of pineapple leaf waste for plastic reinforcement: 1. A novel extraction method for short pineapple leaf fiber. Ind. Crop. Prod. 40, 55–61. Available from: https://doi.org/10.1016/j.indcrop.2012.02.037.

Laftah, W.A., Rahaman, W.A.W.A., 2015. Chemical pulping of waste pineapple leaves fiber for kraft paper production. J. Mater. Res. Technol. 4 (3), 254–261. Available from: https://doi.org/10.1016/j.jmrt.2014.12.006.

Mahardika, M., Abral, H., Kasim, A., Arief, S., Asrofi, M., 2018. Production of nanocellulose from pineapple leaf fibers via high-shear homogenization and ultrasonication. Fibers 6 (2), 28.

Mangal, R., Saxena, N.S., Sreekala, M.S., Thomas, S., Singh, K., 2003. Thermal properties of pineapple leaf fiber reinforced composites. Mater. Sci. Eng.: A 339 (1-2), 281–285. Available from: https://doi.org/10.1016/S0921-5093(02)00166-1.

Ming, X., Li, Q., Feng, J., Jiang, W., 2021. Analysis and optimization of process parameters of the degradable fiber mulch paper made from pineapple leaf and rice straw by response surface method. BioResources 16 (2), 2021.

Mishra, S., Misra, M., Tripathy, S.S., Nayak, S.K., Mohanty, A.K., 2001. Potentiality of pineapple leaf fibre as reinforcement in PALF-polyester composite: surface modification and mechanical performance. J. Reinforced Plast. Compos. 20 (4), 321–334.

Moon, R.J., Martini, A., Nairn, J., Simonsen, J., Youngblood, J., 2011. Cellulose nanomaterials review: structure, properties and nanocomposites. Chem. Soc. Rev. 40 (7), 3941–3994.

Motaleb, K.Z.M., Islam, M.S., Hoque, M.B., 2018. Improvement of physicomechanical properties of pineapple leaf fiber reinforced composite. Int. J. Biomater. Available from: https://doi.org/10.1155/2018/73843602018.

Mtibe, A., Linganiso, L.Z., Mathew, A.P., Oksman, K., John, M.J., Anandjiwala, R.D., 2015. A comparative study on properties of micro and nanopapers produced from cellulose and cellulose nanofibres. Carbohydr. Polym. 118, 1–8.

Neto, A.R.S., Araujo, M.A.M., Barboza, R.M.P., Fonseca, A.S., Tonoli, G.H.D., Souza, F.V.D., et al., 2015. Comparative study of 12 pineapple leaf fiber varieties for use as mechanical reinforcement in polymer composites. Ind. Crop. Prod. 64, 68–78. Available from: https://doi.org/10.1016/j.indcrop.2014.10.042.

Nishiyama, Y., Langan, P., Chanzy, H., 2002. Crystal structure and hydrogen-bonding system in cellulose Iβ from synchrotron X-ray and neutron fiber diffraction. J. Am. Chem. Soc. 124 (31), 9074–9082.

Park, S., Baker, J.O., Himmel, M.E., Parilla, P.A., Johnson, D.K., 2010. Cellulose crystallinity index: measurement techniques and their impact on interpreting cellulase performance. Biotechnol. Biofuels 3 (1), 1–10.

Pavithran, C., Mukherjee, P.S., Brahmakumar, M., Damodaran, A.D., 1987. Impact properties of natural fibre composites. J. Mater. Sci. Lett. 6 (8), 882–884.

Rajeshkumar, G., Ramakrishnan, S., Pugalenthi, T., Ravikumar, P., 2020. Performance of surface modified pineapple leaf fiber and its applications. Pineapple Leaf Fibers. Springer, Singapore, pp. 309–321.

Rambabu, N., Panthapulakkal, S., Sain, M., Dalai, A.K., 2016. Production of nanocellulose fibers from pinecone biomass: evaluation and optimization of chemical and mechanical treatment conditions on mechanical properties of nanocellulose films. Ind. Crop. Prod. 83, 746–754.

Ramnath, B., Vijaya, C., Vinodh Krishna, S., Karthik, K., Saravanan, V.M., Manickavasagam, et al., 2014. Evaluation of the flexural properties of pineapple reinforced polymer composite for automotive and electrical applications. Adv. Mater. Res. 893, 271–274. Trans Tech Publications Ltd.

Salleh, S., Majid, R.A., Yahya, W.J., Abd Kadir, H., Ezzat Chan, A.F., Munthoub, D.I., et al., 2021. Composite paper from an agricultural waste of bagasse sugarcane and pineapple leaf fibre: a novel random and multilayer hybrid fibre reinforced composite paper. Nordic Pulp Pap. Res. J. Available from: https://doi.org/10.1515/npprj-2019-0081.

Samal, R.K., Mukul, C.R., 1997. Effect of chemical modifications on FTIR spectra. II. Physicochemical behavior of pineapple leaf fiber (PALF). J. Appl. Polym. Sci. 64 (11), 2119–2125.

Sasikala, M., Umapathy, M.J., 2018. Preparation and characterization of pineapple leaf cellulose nanocrystal reinforced gelatin bio-nanocomposite with antibacterial banana leaf extract for application in food packaging. N. J. Chem. 42 (24), 19979–19986.

Shih, Y.F., Chou, M.Y., Lian, H.Y., Hsu, L.R., Chen-Wei, S.M., 2018. Highly transparent and impact-resistant PMMA nanocomposites reinforced by cellulose nanofibers of pineapple leaves modified by eco-friendly methods. Exp. Polym. Lett. 12 (9), 844–854.

Stephen Pointing, B., Parungao, M.M., Hyde, K.D., 2003. Production of wood-decay enzymes, mass loss and lignin solubilization in wood by tropical Xylariaceae. Mycol. Res. 107 (2), 231–235.

Sufian, A.A.M., Othman, S.A., 2021. Preparation and characterization of irradiated magnetic pineapple leaves for paper production. AIP Conf. Proc. 2355 (1), 050002. AIP Publishing LLC.

Tanpichai, S., Witayakran, S., Boonmahitthisud, A., 2019. Study on structural and thermal properties of cellulose microfibers isolated from pineapple leaves using steam explosion. J. Environ. Chem. Eng. 7 (1), 102836.

Ter Teo, P., Zakaria, S.K., Taib, M.A.A., Budiman, F., Mohamed, M., Yusoff, A.H., et al., 2020. Recycling of pineapple (*Ananas comosus*) leaf agro-waste as one of the raw materials for production of eco-friendly new paper, IOP Conference Series: Earth and Environmental Science, vol. 596. IOP Publishing, p. 012018, no. 1.

Wada, M., Okano, T., Sugiyama, J., 1997. Synchrotron-radiated X-ray and neutron diffraction study of native cellulose. Cellulose 4 (3), 221–232.

Yue, Y., Han, G., Wu, Q., 2013. Transitional properties of cotton fibers from cellulose I to cellulose II structure. BioResources 8 (4), 6460–6471.

Yusof, Y., Ahmad, M.R., Saidin, W., Mustapa, M.S., Tahar, M.S., 2012. Producing paper using pineapple leaf fiber. Adv. Mater. Res. 383, 3382–3386. Trans Tech Publications Ltd.

Chapter 14

Pulping and papermaking of rice straw

Jannatun Nayeem, M. Mostafizur Rahman, M. Sarwar Jahan and Razia Sultana Popy
Pulp and Paper Research Division, Bangladesh Council of Scientific and Industrial Research (BCSIR), Dhanmondi, Dhaka, Bangladesh

14.1 Introduction

Pulp and paper consumption can be marked as one of the progress indices of any country. The consumption of paper increased manyfold with the increasing living standard of the people, population growth, urbanization as well as industrialization. An increased demand of paper has been considered a looming threat to our life-saving forests. A search for alternative and efficient raw materials was always on the table. But it was not always easy to find a suitable alternative because of the varied efficiency of raw materials. A plentiful accessibility to those raw materials is another important aspect to investigate. Asian countries like China, India, and Bangladesh, for example, produce a lot of paddies, and rice straw is abundant in this region.

The use of rice straw has low economical value as it is mainly used as domestic fuel and cattle feed. It is not a good fuel because of its low calorific value and needs additional processing for efficient energy use (Hossain and Narciso, 2004). Rice straw creates multifaceted problems, both for the people and environment. Farmers get a hard time to discard this huge stock of residue. They resort to burning those in open. As a result, thousands of organic acids are released in the air that poses health hazardous threats and are a serious reason behind global warming.

Rice straw use as a feedstock for the pulp and paper industry will offer double advantage to both farmers and paper industries.

14.1.1 Global demand of paper

The global demand for pulp, paper, and paperboard has reached around 402 MT/annum. This increment occurred within only 2 years (Kulkarni,

Pulping and Papermaking of Nonwood Plant Fibers. DOI: https://doi.org/10.1016/B978-0-323-91625-7.00001-1

2013). It has been projected that this demand will increase to 521 MT/year by 2021 (Kaur et al., 2017). Consumption of paper has seen a steady increase over the course from 1999 to 2005. In 1999 paper consumption was around 316 MT, and by the end of 2005, it reached 351 MT. This increment, however, is expected to cross 500 MT in 2025, and growth rate will be 1.6% per year (García et al., 2008).

The ever-increasing demand of paper has forced the scientific community to find economically feasible, yet environmentally friendly raw materials for paper production. It needs to be tested in terms of efficiency, that is, low cost and adequate desired properties. Thus the shift toward pulping of nonwood, for instance, agricultural wastes and annual plants, has evolved as an alternative and effective process in paper and related items. This is not only important as a suitable alternative but also imperative to save the existing reserve forests around the planet. This is a crucial step to curb the looming concern of climate change against the backdrop of widespread deforestation used in paper industries.

14.1.2 Paper production

Paper production is around 408 MT in the year 2018 with a little decrease over the year 2017 production (World Food and Agriculture – Statistical Yearbook 2020, 2020; Byrd and Hurter, 2005; Ferdous et al., 2020), half of which is supplied by China, USA, and Japan. Only 8% of the total production is using nonwood raw materials (Fahmy et al., 2017). It has been estimated that the market value of pulp and paper industry will experience an increase from 63.3 billion USD to 79.4 billion USD in the time span from 2019 to 2024 (Ferdous et al., 2020). Developing nations like China will be the deal breaker in this market. Per capita, paper consumption is 44 kg in Asia. It contributes 45% of world paper consumption (Sarkar et al., 2021a,b).

World-dominated pulp and paper producer countries are the USA, Canada, Northern European countries, East Siberian Russia, China, and Japan. Brazil and Australia also have many pulp and paper mills. Before 2009, the United States was leading the production (Kaur et al., 2017). China took over the leader board from then, and now it produces around 102,500 kilo tons followed by the United, respectively, to the forest-based pulp production.

Pulp and paper industries have been struggling with a number of serious issues regarding the lack of raw materials with a persistent increase of demand. Against the backdrop of raw material crisis, the utilization of agricultural residue can be an answer (Jahan et al., 2021). In rice-producing countries, fractionating agricultural residue can be attractive in terms of availability and affordability.

Due to a limited supply of wood-based raw materials, country like Bangladesh, which is a forest deficit country, continuously investigate non-wood fibers for pulp manufacturing. Despite wood being the suitable material for pulp production in the existing processes, nonwoods are considered the best alternative option for pulp without destroying the forest area due to its availability in many agricultural nations (Ferdous et al., 2020).

14.1.3 Alternative raw materials

Paper industry is highly dependent on materials derived from wood. Because of the high consumption of paper, the demand for woody raw materials may outstrip their production and supply (Kaur et al., 2017). This eventually prompts the destruction of natural forest areas. The need for alternate ligno-cellulosic raw materials for the pulp and paper industry is driven by population density and forest scarcity (Jahan and Jin, 2006). The growing interest in forest-based products as well as increased awareness about environmental security has forced pulp and paper industries to concentrate on raw material substitution (Sharma and Kumar, 1999). The growing concern of climate change due to greenhouse gas emissions through petroleum derivatives encouraged the conversion of lignocellulosic biomass into paper and other valuable chemicals. This has created opportunity for nonwood to be used as an input feedstock in pulp industry for cleaner production (Huang et al., 2008). Nonwood biomass has low lignin contrasted with wood and usually easily separated the fiber (Pande and Roy, 1996). Nonwoods are derived from different mono-or dicotyledonous plants and are botanically sorted as grass, bast, flax, and fruit fibers. The pulps derived from nonwood can be separated in two general classes, as indicated by their fiber length, like short fiber and long fiber (Jahan et al., 2013).

14.1.4 Opportunity of agricultural residues

After harvesting grain rice straw, wheat straw, cornstalks, and oil palm that are significant lignocellulosic nonwood materials remain as waste in the field (Jahan et al., 2021). These biomasses could be prominent raw materials for the pulp production. Almost all countries in the world generate significant amounts of agricultural residues. The utilization of these crop residues as feedstock for pulp mills can bring an extra advantage by staying away from disposal costs just as other undesirable consequences, like air pollution and pest control from residue burning (Kaur et al., 2017). Their use for producing pulp and paper items is continually engaging eco-friendly, financial, and natural benefits. Due to low lignin content pulping of nonwood is easier than wood and less energy is required to defiber due to lower fiber length (Jahan et al., 2013).

Agricultural residues are good quality pulping raw materials in terms of pulping abilities, bleachability, and fiber content to make good quality paper (Jiménez et al., 2008). The economic development has to be done with the least destruction of the ecological parts as well as saving, using, and reusing of the costly chemicals and raw materials for reasonable development. Many countries like China, India, Egypt, the USA, and European countries that are producing huge agricultural residues are using these valuable biomasses can be an important course-changing decision (Kaur et al., 2017).

In every agricultural country, huge amount of residues are created each year that would be an expected source of pulp fiber. In every year, around 2.5 billion tons of agricultural waste are generated, which can be used for papermaking and would deliver just about 10 MT of pulp (Sarkar et al., 2021a,b). Straw, bagasse, reed, and bamboo represent about 95% of the nonwood pulp and the rest from different sorts of biomass (Liu et al., 2018). China and India are the main contributors of nonwood pulp in the world market. China alone contributed about 53% of the global nonwood pulp (Sarkar et al., 2021a,b). Wheat straw, rice straw, and bagasse are the most investigated agrarian remaining parts (Xu et al., 2013).

14.1.5 Rice straw as pulping raw materials

Rice straw is one of the biomasses produced in large amounts worldwide and contains no economical value. It can be a significant lignocellulosic raw material for the pulp industries in Asia because of its huge production around here. Around 90% of the total produced rice is grown and consumed by Asian countries (Hossain and Narciso, 2004). In respect of Bangladesh, the accessible papermaking raw materials are agricultural wastes, industrial crops, and hardwoods. With the increasing demand for paper and the shortage of forest land, investigations for the suitability of nonwood fiber and fast-growing local plants for pulp have been strengthened (Jahan, 2009).

Rice straw is produced in large amounts worldwide consistently. Straw is viewed as agricultural waste in industrialized countries since it cannot be changed into significant products. As we endeavor to save forest lands, to diminish the antagonistic effect on the environment, and to utilize biomass resources to make up deficiencies of fiber in future, the improvement of appropriate technologies for use of crop residue appears to be both significant and huge. Countries that suffer from forest per capita, straw would be utilized broadly as the raw material for pulp and paper. For instance, China is producing 22 MT of paper using nonwood fibers, which account for 70% of its total production (Pan et al., 1999a,b).

Rice straw is an abundant and easily accessible nonwood raw material in the developing nations like China, India, Indonesia, and Bangladesh (Rodríguez et al., 2010; Siemers and Bioenergy, 2011). These countries produce huge rice straws and increase the production every year over the world, so the equivalent amount of straw is produced (Kaur et al., 2017). Thus the utilization of these important pulp raw materials is an eco-friendly approach for waste management. It will be an excellent choice to preserve our natural assets.

Rice straw is being used for paper grade pulp production in China, India, Egypt, Srilanka, and so on. Not only the rice straw but also other agricultural leftover biomasses are being used for producing various grades of paper like writing, printing, low-quality wrapping papers, and greaseproof papers. Since the world is facing the challenges of maintaining the nonrenewable fossil fuels, a worldwide consideration is needed on introducing sustainable second-generation fuels. Lignocellulosic biomass, particularly the crop residues, might be used for the fuel production. Jahan et al. (2015a,b) suggested that the pentose and hexose sugar extracted from pulping liquor can be used for biofuel manufacturing. Moreover, the isolated lignin and dissolved biomass from the liquor can be used as soil nutrients in the agro field that can be helpful for both soil fertility and fill in as a possible removal alternative (Xiao et al., 2007).

Rice, wheat, maize, and millet contribute 70% of the whole agricultural residues produced in the world. It has been projected that rice will be about 800 MT by 2030 to meet the increasing demand (Siemers and Bioenergy, 2011). Rice straw is a plentiful and minimal expense biomass for producing paper pulp for the Asian nations (Kaur et al., 2017). From the reports of the Ministry of Economic Affairs, Asian nations produced 684,595 kt of rice in 2013 and yielded 727,400 kt of straw. Lim et al. (2012) revealed that rice straw production varies from 41% to 396% of grain.

Rice is the main crop in Bangladesh, and it produces the highest amount of residue. Rice production was around 34.4 MT I nFY 2013–14 and gradually increased to 37.1 MT in FY 2019–20 (Jahan et al., 2015a,b). According to Jahan et al. (2015a,b), nearly 34–52 MT of rice straw was produced in Bangladesh excluding other crop residues. So, rice straw could be a potential source for biofuel and biochemicals in addition to pulp production.

14.2 Environmental and social aspects

The utilization of agricultural residues in potential purposes has social, environmental, and economic benefits. Burning is the most straightforward choice for the farmers to get rid of the crop waste. Farmers have little information about the industrial values of agro residues. Because of effective automation, especially the utilization of consolidated harvesting, declined quantities of livestock, extensive time consumption in composting, and inaccessibility of alternate financial arrangements, farmers are constrained to destroy their agro residues (Gupta et al., 2004). Furthermore, the narrow interval between two consecutive crop harvesting times encourages farmers to burn residues as an easy-going option. Large-scale burning of crop residues involves basic concern for greenhouse gases emission as well as issues of air pollution and loss of soil nutrients. Moreover, it creates different types of organic acids that result in upgraded respiratory illnesses like bronchitis, asthma, and eye disturbances (Rodríguez et al., 2010). It also affects the existence of animals, birds, and pests.

14.3 Chemical, physical, and morphological properties of rice straw

Despite the fact that wood and nonwood are composed of cellulose, hemicellulose, and lignin, the chemical composition and morphological properties are no longer the same. Harvesting time and geological areas influence the morphology and chemical composition of the nonwood. Rice straw contains α-cellulose varied from 28% to 36% and hemicellulose from 23% to 28% (Sridach, 2010). The physical properties and chemical composition of common nonwoods are shown in Table 14.1. It is seen that nonwoods

TABLE 14.1 Chemical and physical properties of non-wood fiber.

Properties, unit	Jute fiber	Wheat straw	Rice straw	Bamboo	Bagasse
Fiber width, μm	20	16	15	14	20
Fiber length, mm	2.5	0.82	1.4	2.7	1.7
Lumen width, μm	5	5.46	4.03	3.45	9.72
Aspect ratio (L/d)	139:1	110:1	175:1	135–175:1	85:1
α-Cellulose, %	45–63	38–39	28–36	26–43	32–44
Hemicellulose, %	15–26	26–32	23–28	15–26	27–32
Lignin, %	12–16	18	12–16	21–31	19–24
Pentosan, %	18–22	30	23–28	15–16	27–32
Alcohol benzene solubility, %	2.4	5.5	7.3	4.8	1.7
Hot water solubility, %	3.7	12.3	0.6	4.8	4.4
Cold water solubility, %	2.2	7.7	14.4	3.5	10.6
1% NaOH solubility, %	28.5	40.8	57.7	24.9	33.9
Ash, %	0.5–2	7	15–20	1.7–5	1.5–5
Silica, %	<1	5.4	9–14	1.5–3	0.7–3

Jahan et al. (2013), Kaur et al. (2017), Liu et al. (2018), Sadiku et al. (2016), Sridach (2010).

comparatively contain lower lignin and α-cellulose, however, contain higher hemicellulose and ash. Rice straws contain excessively high amounts of ash that makes it difficult to produce pulp in soda, kraft, and sulfite processes. The ash is mainly composed of silica that varied from 13.0% and 5.43%, respectively, in the rice straw (Tutus, 2003; Van Soest, 2006).

The amount of α-cellulose present in raw materials has emphatically impact on pulp yield during chemical pulping and furthermore profoundly affects paper strength. If the α-cellulose content is 34% or higher, that ligno-cellulosic materials can be considered appropriate for pulp manufacturing (Sarwar Jahan et al., 2016). Among nonwoods, bast fiber from jute has the most noteworthy sum of α-cellulose, despite the fact that the α-cellulose content of the majority of the nonwoods is lower than that of hardwoods (Table 14.1). As displayed in Table 14.1, wheat and rice straws contain almost similar amounts of α-cellulose, while it was high in bagasse, bamboo, and jute.

Although lignin is the targeted chemical constituent in chemical pulping process, hemicelluloses are also solubilized in the cooking liquor to some extent (Pejic et al., 2008). Hemicelluloses assume a significant part in fiber bonding that improves the tensile strength of paper; it diminishes the tearing resistance and fold endurance (Sitch and Marshall, 1950). Pentosan represents the hemicellulose content in lignocellulosic materials. The hemicellulose content in rice straw is comparable with white straw and bagasse which, however, seems higher than bamboo and jute (Table 14.1).

Lignin is removed from the lignocellulosic biomass in the pulping process and separated as a byproduct in the soda, kraft, and sulfite processes (Pan and Sano, 1999). Lower lignin content is appropriated for delignification, with lower energy utilization and lower chemical load, to get the desirable level of residual lignin in the pulp. Rice straw has lower lignin content than white straw, bagasse, and bamboo, however, higher than jute. Moreover, the composition of nonwood lignin is not similar to wood lignin. Nonwood lignins are composed of p-hydroxyphenyl (H), guaiacyl (G), and syringyl (S) monomeric units, while wood lignins are composed mainly of G and S units (Fidalgo et al., 1993).

There are some unwanted nonfibrous organic matters, such as tannins, sugars, color compounds, and inorganic matters, present in biomass, which are extractable with polar and nonpolar solvents. Rice straw has high alkali and water-soluble extractives (Kumar Sharma et al., 2011).

Fiber properties like length, diameter, and density pose significant effects on pulping and end application. The longer the length of fiber, the higher the strength of paper is expected (Ververis et al., 2004). Density of fiber is critically considered in surveying raw materials quality for pulping. Higher density is favorable bearing in mind the digester capacity. Nonwoods are bulky in nature in comparison to wood. The bulky nature of rice straw is one of the drawbacks as pulping raw materials (Sarkar et al., 2021a,b). However, the low density of rice straw permits simple penetration of chemicals, and pulp can be produced at atmospheric pressure (Kaur et al., 2017). Lower lignin content of rice straw compared to wood needs mild cooking conditions during pulping process. Thus rice straw pulping can end up being economical and eco-friendly.

14.4 Limitations of rice straw

The major challenges of utilizing rice straws as pulping raw materials are

1. High silica content is the primary obstacle to produce pulp from rice straw in the established pulping processes. It makes issues both in pulping process and pulp quality. In the time of black liquor recovery, the formation of scale in recovery chamber hampers heat transfer. Researchers are proposing alternative pulping processes that would be well fit for producing from rice straw. Among proposed processes, organic acid pulping is thought to be the best alternative for rice straw. In organic acid process, dissolved lignin and hemicellulose can be isolated effortlessly and utilized for bio-products. Although silica is not separated from the pulp fiber in organic acid process, it can effectively be dissolved by alkaline extraction (Sarwar Jahan et al., 2012; Quoc Lam et al., 2001; Rodríguez et al., 2008).
2. Cellulose fibers from rice straw are short. A large number of fines are generated from the shortest fibers. High amounts of fines lengthen the drainage time and hamper the paper machine's runability (Sarwar Jahan et al., 2012).
3. Nonwoods are naturally bulkier in nature. Its fibers are loosely connected, which eases the pulping. However, the low bulk density creates other problems. Because of bulkier nature, it is tedious, time-consuming, and costly to transport the raw materials in industrial zone. So, without developing suitable technology, businesses are not cordial ecologically and financially. Rice straw is utilized in a number of little pulp and paper industries but without a recovery system to eliminate silica (Julian, 2013).

14.5 Conventional pulping process

Pulp is produced from lignocellulosic material by means of mechanical, chemical, or semichemical treatment or a combination of two processes. Rice straw is mostly subjected to chemical or chemimechanical treatment for pulp production rather than mechanical process that renders better quality pulp. Pulp has been produced from rice straw in different chemical processes, including conventional, such as soda, soda anthraquinone (AQ), and kraft, as well as nonconventional processes like potassium hydroxide and organosolv (Jahan et al., 2021). Table 14.2 depicts comparison among conventional pulping of a few nonwoods.

14.5.1 Soda pulping

Soda pulping technique is popular for producing pulp from rice straw. Soda pulping of rice straw carried at high temperature (170°C) reduces the residual lignin very low and makes it easily bleachability pulp (Kaur et al., 2017). The soda pulping finished with oxygen (soda oxygen pulping) gives pulp with higher physical properties and higher yield. It is anticipated that soda oxygen pulping includes the oxidative elimination of lignin and less degrade the cellulose and hemicelluloses and, as a result, gives a higher yield. It is also proved that caustic soda pulping with mechanochemical technique is more viable than high-pressure pulping since it essentially brings down the ash content, which is a significant issue considering rice straw pulping (Rodríguez et al., 2010).

TABLE 14.2 Nonwood pulping in conventional process.

Pulping process	Raw materials	Pulp yield (%)	Kappa number
Kraft	Jute fiber	59.8	10.9
Soda		52.9	9.0
Soda–AQ		68	10
Kraft	Wheat straw	42.6	31
Prehydrolyzed kraft		30.2	13.8
Soda		48	16
Kraft	Bagasse	34	10
Soda		50.4	17.5
Soda–AQ		43	10.3
Kraft	Bamboo	43.4	11
Soda		44.1	30.1
Soda–AQ		41.2	18
Kraft	Rice straw	42	21
Prehydrolyzed kraft		30.8	12.0
Soda–AQ		37	15.9
Soda		35.3	15.8

AQ, Anthraquinone; Chen et al. (2019), Jahan et al. (2007), Rodríguez et al. (2008), Jahan et al. (2015, 2016a,b, 2021), Pydimalla et al. (2019).

14.5.2 Soda–anthraquinone pulping

The limitation of soda process is the requirement of higher alkali that impact negatively on environment as well as production cost. However, AQ with soda reduces the chemical consumption for delignification. But the presence of a high amount of ash in the black liquor still makes this process unfavorable for chemical recovery, and discharge to the environment is longer acceptable. The properties of soda and soda–AQ pulps have been shown in Table 14.3.

14.5.3 Kraft process

Kraft pulping is in minimal use taking everything into account. The chemical required in kraft process is costly and must need a recovery of chemicals. High ash content of rice straw makes the chemical recovery process unfavorable. Rather a neutral sulfite pulping is a more decent process for rice straw. Around 85% pulp yield is obtained with excellent papermaking properties (Kaur et al., 2017).

14.6 Nonconventional pulping process

Removal of lignin from the lignocellulosic material is the foremost target pulping process with a minimum damage of cellulose fiber (Jahan and Jin, 2006). In the conventional pulping process, the black liquor contains a complex mixture of mostly lignin, hemicellulose, and other sugar moieties. It is difficult to fractionate the individual chemical to utilize as basic raw material for conversion. Moreover, the processes are no longer environmentally friendly. So, interest grows to find an environmentally sound and economically feasible alternative pulping process specially to utilize the huge amount of agricultural residue. Among the alternative pulping chemicals, organic solvents and organic acids are the foremost considerations (Jahan and Jin, 2006; Young et al., 1986; Parajo et al., 1993). The objectives of these processes are making them environment friendly and utilizing all parts of the lignocellulosic biomass.

14.6.1 Organosolv pulping

Different organic chemicals, such as organic acids like acetic acid and formic acid, low boiling solvents like methanol, ethanol, and acetone, or high boiling solvents like ethylene glycol and ethanolamine, have been used to dissolve the lignin and hemicelluloses from the nonwood in organosolv pulping (Sridach, 2010). In organosolv pulping, acids show the most interesting process in terms of easy recovery, being environmentally friendly and compatible with the issue of silica content. Organic acids efficiently separated

TABLE 14.3 Soda/soda anthraquinone (AQ) pulping and pulp properties of rice straw.

References	Cooking liquor	Pulp yield (%)	Kappa number	Tear index [mN (m^2/g)]	Burst index (kN/g)	Tensile index (Nm/g)/ Breaking length (m)	Viscosity (mL/g)
Rodríguez et al. (2008)	Soda	33.8	24.4	0.31	1.2	2664	835
	Soda AQ	35.3	15.8	0.33	2.5	3494	732
	Soda-parabenzoquinone	32.33	18	0.36	1.3	2377	709
Rodríguez et al. (2010)	Soda AQ	40.4	9.5	0.44	1.63	2776	826
	Soda AQ (0.5)	40.4	9.4	0.45	1.65	2783	830
	Soda AQ (1)	40.1	9.3	0.46	1.68	2788	843
Sarwar Jahan et al. (2012)	Soda AQ	51.9	9.8	41.7	3.2	41.7	–
	Water pre-extraction + soda AQ	39.8	8.3	36.1	2.9	36.1	–
	Alkali pre-extraction + soda AQ	42.3	9.6	40.2	3.1	40.2	–

the pulp fibers and lignin from straw-type materials like rice straw and wheat straw and produced good quality pulp (Jimonez et al., 1998; Jahan and Jin, 2006; Quoc Lam et al., 2001). This pulping process is mainly performed at atmospheric pressure. Acetic acid–based pulping got researchers' attention due to its low environmental effect and lower energy utilization. Acetic acid effectively digested the rice straw into pulp with little quantity of mineral acid (H_2SO_4 or HCl) as a catalyst (Pan et al., 1999a,b). If the acetic acid–treated pulp is followed by extraction with alkaline provided a pulp with a higher kappa number and lower yield in comparison to followed by peroxyacetic acid treatment (Jahan and Jin, 2006) of rice straw. In addition, formic acid and mixed acid (acetic acid/formic acid/water mixture) (Jahan et al., 2015a,b) also have been used for rice straw pulping at their boiling point. The produced pulp and alkali-treated pulp shows similar physical properties. It is mentioned that organic acids do not remove silica from the pulp. Consequently, followed by alkali treatment is required to remove the silica (Jahan et al., 2015a,b).

The removed lignin and hemicellulose separations are easier from the organic acid black liquor than the conventional process. It has been reported that lignin from the organic acid has a higher phenolic group than the protolignin (Islam and Sarkanen, 1993). Lower molecular weight and higher reactivity of lignin will make it suitable for many lignin derivatives and other value-added products (Kubo et al., 1998; Nihat et al., 2002). The spent liquor lignin can be utilized for carbon fiber productions (Kadla et al., 2002). In this process, the maximum amount of hemicellulose goes in the black liquor. The hemicellulose recovered from the black liquor can be effectively converted to biochemicals and fuels. Table 14.4 addresses different pulp properties of rice straw from alternative pulping from various research studies, and Table 14.5 depicts the change of silica and ash percentage in organic pulp.

14.6.2 Nonsodium alkali pulping system

Rice straw has been pulped with different alkali systems other than soda to make the process more eco-friendly. The objective of this process is to omit the black liquor chemical recovery as well as use it as fertilizer. Potassium hydroxide and ammonium hydroxide are used alone or combined together. Huang et al. (2008) studied on the pulping with NH_4OH–KOH–AQ pulping process for rice straw. Jahan et al. (2016a,b) produced pulp with the KOH and subsequently separated lignin from the black liquor. Later the KOH lignin was used to prepare activated carbon. The KOH pulping is well fitted with the rice straw like organic acid process. Fig. 14.1 shows the flow diagram of pulping of integrated biorefinery of rice straw by KOH pulping (Jahan et al., 2016a,b).

TABLE 14.4 Pulp properties from alternative pulping of rice straw.

References	Cooking liquor	Pulp yield (%)	Kappa number	Tear factor	Burst factor	Breaking length (m)	Viscosity
Rodríguez et al. (2008)	KOH (170°C)	42.82	34.93	0.33[a]	1.45[a]	2551	888
	KOH (180°C)	35.81	19.29	0.35[a]	0.96[a]	1769	923
Jahan et al. (2016a)	KOH	40.2	6.3	8.7[a]	3.2[a]	37[a]	–
Mohammadi-Rovshandeh et al. (2005)	DEG	60.59	35.0	42.5	68	7220	–
	DEG/EG	62.5	34.0	42.0	62	7800	–
	DEG/EG/S	64.57	21.0	47.0	63	8140	–
Ho et al. (2009)	THFA(95%)	59.0	17.3	7.45[a]	2.3[a]	35.5[a]	–
	THFA(90%)	65	36	7.8[a]	2.6[a]	31[a]	
Quoc Lam et al. (2001)	FA(90%)	44.4	25.1	3.95[a]	2.39[a]	3919	–
	FA(70%)	46.6	39	4.78[a]	2.49[a]	4524	

[a]*Tear index [mN (m^2/g)], burst index (kP m^2/g) and tensile index (Nm/g).* KOH, *Potassium hydroxide;* DEG, *diethyleneglycol;* S, *soda,* EG, *ethyleneglycol;* FA, *formic acid;* THFA, *tetrahydrofurfuryl alcohol.*

TABLE 14.5 Change of silica and ash percentage in (1) formic acid pulping and (2) acetic acid pulping.

Properties	Rice straw	Quoc Lam et al. (2001)	Pan et al. (1999b)		
		Formic acid pulp	Acetic acid pulp	After O_2 bleaching	After bleaching
Yield (%)	100	44.4	51.8	42.5	39.3
Brightness (%)	–		27.8	–	88.4
Ash (% of Pulp)	14.6	16.4	18	18	15
Silica (% ratio to ash)	47.8	89.5	92	93	97
Ash yield (%)	100	49.9	57	47	37
Silica yield (%)	100	93.4	75	62	50
Burst index ($kPam^2/g$)	–	2.39	2.26	–	2.11
Tear index (mNm^2/g)	–	3.95	4.78	–	6.49
Breaking length (m)	–	3919	4010	–	3600

Conditions (1) Pulping liquor: 0.32% H_2SO_4 in 80% AcOH, Time: 3 h, L/W: 12. (2) 90% FA(v/v), Time: 1 hr, L/W: 12.

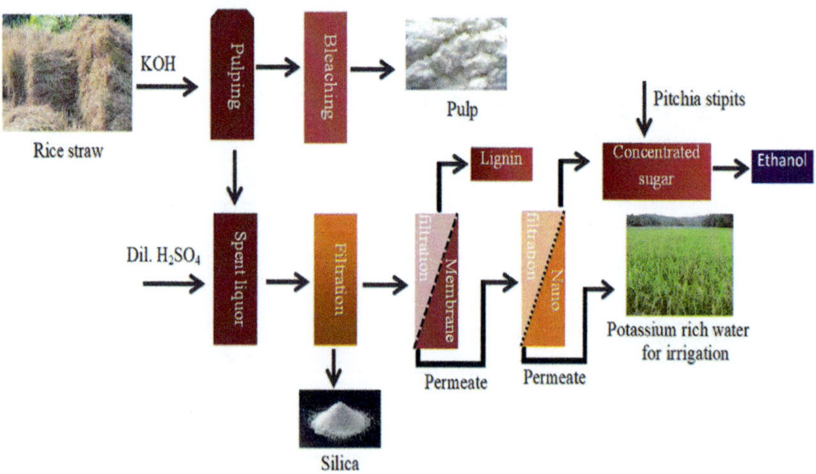

FIGURE 14.1 Integrated biorefinery of rice straw by Potassium hydroxide pulping (Jahan et al., 2016a).

14.6.3 Nitric acid and ammonium sulfite pulping

Nitric acid has also been studied to produce different grades of pulp depending on the severity of acid concentration, temperature, and time as well as followed by alkali treatment. The nitric chemimechanical pulp contains appreciable strength with higher opacity due to holding of silica. The nitric-treated pulp provided brightness of 48%–82% with the chlorination–alkali extraction hypochlorite (CEH) process (El-Taraboulsiq et al., 1983).

Ammonium sulfite (AS) is used as an eco-friendly pulping chemical for rice straw pulping. The black liquor from AS pulping contains inorganic nitrogen (ammonia and ammonium ion) and organically bonded nitrogen, along with lignosulfonate, all of which are further used as fertilizer to develop soil nutrients (Liu et al., 2018).

14.7 Other organosolv pulpings

Utilizations of various organic solvents with high boiling points like ethylene glycol, ethanolamine, diethanolamine, and diethylene glycol are used to solve the limitation of conventional nonwood pulping. Rice straw pulping using different mixtures of cooking liquor were examined by Mohammadi-Rovshandeh et al. (2004). They used diethylene glycol (DEG), the mixture of diethylene glycol (DEG) and ethylene glycol (EG), and the mixture of diethylene glycol (DEG) and ethylene glycol (EG) with 2% NaOH as cooking liquor. DEG + EG + 2%NaOH process papers were produced with higher mechanical properties. González et al. (2008) researched rice straw pulping utilizing three distinctive organic solvents: ethylene glycol 80 wt.%, diethylene glycol 80 wt.%, and ethanolamine 70 wt.% and compared pulp properties with the produced pulp in soda pulping of 15% NaOH.

Rice straw contains a large amount of ash that makes this material less reasonable for traditional pulping, whereas pulping with distinctive organic solvents presented that diethylene glycol pulp introduced the most reduced percentage of ash followed by ethanolamine pulp. However, pulp produced by ethylene glycol liquor contained a higher percentage of ash compared to soda pulp. Pulp treated with ethanolamine exhibited the best paper-production properties with the most holocellulose and α-cellulose contents.

Ho et al. studied on tetrahydrofurfuryl (THFA) pulping of rice straw at atmospheric pressure using HCl as a catalyst. Concerning pulp yields, the process showed a high rate of delignification. Pulp with a kappa value of 20 contained 60% yield that is higher than the conventional alkali pulping process. Expanding the concentration of THFA and the amount of catalyst also accelerates delignification. But the THFA pulp handsheets displayed lower properties compared to those produced by kraft pulping. Table 14.6 represents the pulp and paper properties of rice straw as starting materials by using organic solvent as cooking liquor studied by different scientists.

TABLE 14.6 Rice straw pulping using organic solvents.

Parameters	Pan et al. (1999b)	Quoc Lam et al. (2001)		Delmas et al. (2003)	Jahan et al. (2006)				Karim et al. (2010)	Ferdous et al. (2020)	Kumar et al. (2012)
Solvents	Formic acid: water (%) 80:20	Formic acid: water (%) 70:30	Formic acid: water (%) 90:10	Formic acid: acetic acid: water (%) 20:60:20	Acetic acid/ peroxy acetic acid 80% AA	Formic acid/ peroxy acetic acid 80% FA	Acetic acid/alkali extraction 80%AA	Formic acid/alkali extraction 80% FA	Acetic acid 300 mL	Formic acid:water (%)80:20	Acetic acid 76.7%
Catalyst (%)	0.3	–	–	–	0.3	–	0.5	–	7.5		0.67
Time (h)	3	1	1	3	2	2	2	2	0.5	4	4.5
Temperature (°C)	Boiling	100	100	107	80	80	80	80	116 ± 2	Boiling	79
Material: liquor	1:12	1:12	1:12	1:12	1:10	1:10	1:10	1:10	1:3	1:10	1:10
Yield (%)	51.8	46.6	44.4	53	52	49–51	51	45–46	51	55.6	56
Kappa number	9.2[a]	39	25	45	28	32	25	28	–	14	21
Tear index [mN (m^2/g)]	4.78	4.7	3.95	4.38	5–7	8–10	6.3	9.1	75[b]	2.6 ± 0.3	4.5–4.7
Tensile index (Nm/g)	4010[b]	–	–	–	36–44	48–53	40	48	2262[b]	45.0 ± 3.1	–
Burst index (kN/g)	2.26	2.49	2.39	2.52	2.4–3.4	5–7	2.9	3.3	13[b]	1.9 ± 0.4	0.6–0.65

[a]*Klason lignin.*
[b]*Tear factor, burst fector, breaking length, m.*

14.8 Pretreatment effect

Preextraction with water, acid (acetic acid), or alkali of rice straw at different temperatures further brings down the pulp yield and develops delignification rate. Preextractions contain little impact on pulp brightness and the percentage of cellulose. Working on pulp quality centrifugal type washings is truly useful for straw materials. According to the study of Kaur et al., pretreatments improve the effectiveness of pulping processes, and the utilization of different catalysts like AQ or its other derivatives upgrades the aftereffects of soda pulping on a more extensive way. Soda pulping with an AQ catalyst of nonwood is a reasonable process that can work with silica deposition on fiber as opposed to their disintegration in spent liquor. AQ catalysts slow down cellulose degradation and speed up the delignification rate. At lower cooking times, pulping process utilized by AQ helps additionally in lessening kappa value and expanding the yield of produced pulp (Issa et al., 2016).

14.9 Future imminent of nonconventional rice straw pulping

The improvement of normal air pressure technology of pulping can be an intrigue in future research. Agro residues (cornstalks, wheat straw, rice straw, bagasse, bamboo, etc.) contain loosely bonded fiber, so, defibering and delignification of these residues will require less concentrated chemical/mechanical conditions. These can be pulped at atmospheric conditions so no prerequisite of reactor with higher pressure in an agricultural residue–based pulp mill.

Organosolv pulping ends up being more eco-friendly than conventional processes yet is not viably utilized by the pulp mill of developing countries. It requires more developed procedures and different types of equipment to overcome the danger implied.

14.10 Conclusion

An equivalent amount of rice straw is produced in the time of rice production. Farmers normally burn the residues, but after characterizing these lignocellulosic crop residues, it represents itself as one of the promising pulp industry raw materials. High ash (silica) content of rice straw like the other nonwood raw materials limits its utilization in ordinary pulping processes due to numerous difficulties in the recovery of chemicals. Spread over a wide region and low density also limit their application to large-scale pulp production. Alternative pulping processes can be a good option for future pulping industries regarding rice straw. Compared with traditional pulping processes, organosolv, ammonium sulfite, and potassium hydroxide are promising processes, and pulp properties are also similar. Organosolv pulping solvents can easily be recovered by distillation and reused immediately.

Potassium hydroxide and ammonium sulfite processes black liquor can be used direct to field as fertilizer. So, there is no need for recovery plant. Potassium hydroxide and organic acid processes overcome many drawbacks of nonwood pulping using the idea of biorefinery concept.

Demonstrating an alternative raw material, more research should have been performed to calculate the cost-effectiveness and introduce environment-friendly technology of rice straw pulping. Study of pilot plant including equipment design ought to be considered for the alternative pulping processes.

References

Byrd, M.V., Hurter, R.W., 2005. A Simplified Pulping & Bleaching Process for Pith-Containing Nonwoods: Trials Whole Corn Stalk.

Chen, Z., Zhang, H., He, Z., Zhang, L., Yue, X., 2019. Bamboo as an emerging resource for worldwide pulping and papermaking. BioResources 14 (1), 3–5.

Delmas, M., Lam, H.Q., Le Bigot, Y., Avignon, G., 2003. A new non-wood pulping process for high silicon content raw materials. Application to rice straw. Appita J. 56 (2), 102–106. Available from: https://pascal-francis.inist.fr/vibad/index.php?action=getRecordDetail&idt=14728608.

El-Taraboulsiq, M.A., Nassafl, M.M., Abd El-Rehim, E.A., 1983. Rapid nitric acid cooking of rice straw. J. Chem. Tech. Biotechnol. 33.

Fahmy, Y., Fahmy, T.A., Mobarak, F., El-Sakhawy, M., Fadl, M., A Fahmy, T.Y., 2017. Overview and future prospects. Int. J. ChemTech Res. 10 (2). Available from: https://hal.archives-ouvertes.fr/hal-01613635.

Ferdous, T., Quaiyyum, M.A., Jahan, M.S., 2020. Chlorine dioxide bleaching of nineteen non-wood plant pulps. Nordic Pulp Pap. Res. J. 35 (4), 569–576. Available from: https://doi.org/10.1515/npprj-2020-0043.

Fidalgo, M.L., Terr6n, M.C., Martinez, A.T., Gonzdez,'j, A.E., Gonzdez-Vila, F.J., Gallettig, G. C., 1993. Comparative study of fractions from alkaline extraction of wheat straw through chemical degradation, analytical pyrolysis, and spectroscopic techniques (n.d.) J. Agric. Food Chem. 41.

García, M.M., López, F., Alfaro, A., Ariza, J., Tapias, R., 2008. The use of Tagasaste (*Chamaecytisus proliferus*) from different origins for biomass and paper production. Bioresour. Technol. 99 (9), 3451–3457. Available from: https://doi.org/10.1016/j.biortech.2007.08.004.

Gupta, P.K., Sahai, S., Singh, N., Dixit, C.K., Singh, D.P., Sharma, C., et al., 2004. Residue burning in rice-wheat cropping system: Causes and implications. Curr. Sci. 87 (12).

González, M., Cantón, L., Rodríguez, A., Labidi, J., 2008. Effect of organosolv and soda pulping processes on the metals content of non-woody pulps. Bioresour. Technol. 99 (14), 6621–6625. Available from: https://doi.org/10.1016/j.biortech.2007.12.038.

Ho, C.L., Wang, E.I.C., Su, Y.C., 2009. Tetrahydrofurfuryl alcohol (THFA) pulping of rice straw. J. Wood Chem. Technol. 29 (2), 101–118. Available from: https://doi.org/10.1080/02773810902796138.

Hossain, M., Narciso, J., 2004. Long-Term Prospects For the Global Rice Economy Global Rice Economy: Long-Term Perspectives 1.

Huang, G.L., Shi, J.X., Langrish, T.A.G., 2008. Environmentally friendly bagasse pulping with NH4OH-KOH-AQ. J. Clean. Prod. 16 (12), 1287–1293. Available from: https://doi.org/10.1016/j.jclepro.2007.06.011.

Islam, A., Sarkanen, K., 1993. The isolation and characterization of the lignins of jute (*Corchorus capsularis*). Holzforschung 47.

Issa, R.M., Abou-Sekkina, M.M., Khedr, A.M., Bastawisy, A.E.D.M., El-Helece, W.A., 2016. Trace the exploitation of Egyptian rice straw through spectral and thermal measurements. Arab. J. Chem. 9, S130–S137. Available from: https://doi.org/10.1016/j.arabjc.2011.02.022.

Jahan, M.S., 2009. Studies on the effect of prehydrolysis and amine in cooking liquor on producing dissolving pulp from jute (*Corchorus capsularis*). Wood Sci. Technol. 43, 213. Available from: https://doi.org/10.1007/s00226-008-0213-6.

Jahan, M.S., Jin, Y., 2006. Organic Acid Pulping of Rice Straw, Part-I: Cook. https://www.researchgate.net/publication/236296419

Jahan, M.S., Al-Maruf, A., Quaiyyum, M.A., 2007. Comparative studies of pulping of jute fiber, jute cutting and jute caddis. Bangladesh J. Sci. Ind. Res. 42 (4), 425–434.

Jahan, M.S., Rukhsana, B., Mojtaba Baktash, M., Ahsan, L., Fatehi, P., Ni, Y., 2013. Pulping of non-wood and its related biorefinery potential in Bangladesh: a review. Curr. Org. Chem. 17.

Jahan, M.S., Sarkar, M., Rahman, M.M., 2015a. Sodium carbonate pre-extraction of bamboo prior to soda-anthraquinone pulping. Biomass Conv. Bioref. 5, 417–423.

Jahan, M.S., Sutradhar, S., Rahman, M.M., Quaiyyum, M.A., 2015b. Fractionation of rice straw for producing dissolving pulp in biorefinery concept. Nordic Pulp Pap. Res. J. 30 (4).

Jahan, M.S., Haris, F., Rahman, M.M., Samaddar, P.R., Sutradhar, S., 2016a. Potassium hydroxide pulping of rice straw in biorefinery initiatives. Bioresour. Technol. 219, 445–450. Available from: https://doi.org/10.1016/j.biortech.2016.08.008.

Jahan, M.S., Rahman, J.N.M.M., Islam, M., Quaiyyum, M.A., 2016b. Chemical characteristics of ribbon retted jute and its effect on pulping and papermaking properties. Ind. Crop. Prod. 84, 116–120.

Jahan, M.S., Rahman, M.M., Ni, Y., 2021. Alternative initiatives for non-wood chemical pulping and integration with the biorefinery concept: a review, Biofuels, Bioproducts and Biorefining, 15. John Wiley and Sons Ltd., pp. 100–118No. 1. Available from: https://doi.org/10.1002/bbb.2143.

Jiménez, L., Rodríguez, A., Pérez, A., Moral, A., Serrano, L., 2008. Alternative raw materials and pulping process using clean technologies. Ind. Crop. Prod. 28 (1), 11–16. Available from: https://doi.org/10.1016/j.indcrop.2007.12.005.

Jimonez, L., de La Torre, M.J., Maestre, F., Ferrer, J.L., Porez, L., 1998. Delignification of wheat straw by organic acids. Holzforsch. 52.

Julian, J., 2013. Utilization of Soda Ash in Pulping and Bleaching Operation as a Substitution or Partial Replacement of sodium Related papers Pine Needle-An Evaluation of Pulp and Paper Making Potential Journal of Forest Product s and Industries JFPI Pretreatment Approaches in Non-wood Plant s for Pulp and Paper Production: A Review Journal of Forest Product s and Industries JFPI 7Pulping, Bleaching & Recovery Process. J. For. Prod. Ind. 2.

Karim, M.S., Seal, H.P., Rouf, M.A., Rahman, M.I., Talukder, M.H.R., Karmaker, P.G., 2010. Acetic acid pulp from jute stick, rice-straw and bagasse. J. Agrofor. Environ. 3 (2), 171–174. Available from: https://pas.cseas.kyoto-u.ac.jp/activity/HP_SPIRITS/Brahmaputra/data/JAE_ALL/ASFBpdf/6.AFSB3(2)pdf/40.%20S.%20Karim-2.pdf.

Kadla, J.F., Kubo, S., Venditti, R.A., Gilbert, R.D., Compere, A.L., Griffith, W., 2002. Lignin-based carbon fibers for composite fiber applications. Carbon 40.

Kaur, D., Bhardwaj, N.K., Lohchab, R.K., 2017. Prospects of rice straw as a raw material for paper making, Waste Management, 60. Elsevier Ltd, pp. 127–139. Available from: https://doi.org/10.1016/j.wasman.2016.08.001.

Kubo, S., Uraki, Y., Sano, Y., 1998. Preparation of carbon fibers from softwood lignin by atmospheric acetic acid pulping. Carbon 36 (8).

Kulkarni, H.D., 2013. Pulp and paper industry raw material scenario-ITC plantation a case study. IPPTA 25 (1), 79–90.

Kumar, P., Kumar, S., Bhardwaj, N.K., Choudhary, A.K., 2012. Optimization of process parameters for the photocatalytic treatment of paper mill waste water. Environ. Eng. Manage. J. 10 (5), 595–601. Available from: http://www.eemj.icpm.tuiasi.ro/pdfs/vol10/no5/2_226_Kumar_10.pdf.

Kumar Sharma, A., Dutt, D., Upadhyaya, J.S., Roy, T.K., 2011. *Bambusa* spp. & papermaking. Bioresour. 6 (4).

Lim, J.S., Abdul Manan, Z., Wan Alwi, S.R., Hashim, H., 2012. A review on utilization of biomass from rice industry as a source of renewable energy. Renew. Sustain. Energy Rev. 16 (5), 3084–3094. Available from: https://doi.org/10.1016/j.rser.2012.02.051.

Liu, Z., Wang, H., Hui, L., 2018. Pulping and papermaking of non-wood fibers. Pulp and Paper Processing. InTech. Available from: https://doi.org/10.5772/intechopen.79017.

Mohammadi-Rovshandeh, J., Talebizadeh, A., Rezayati-Charani, P., 2005. Pulping of Rice Straw by High Boiling Solvents in Atmospheric Pressure. http://www.SID.ir.

Pan, X.J., Sano, Y., 1999. Atmospheric Acetic Acid Pulping of Rice Straw IV: Physico-Chemical Characterization of Acetic Acid Lignins from Rice Straw and Woods. Part 1. Physical Characteristics.

Pan, X.-J., Sano, Y., Nakashima, H., Uraki, Y., 1999a. Atmospheric Acetic Acid Pulping of Rice Straw (1) Pulping Conditions and Properties of Pulp.

Pan, X.-J., Sano, Y., Nakashima, H., Uraki, Y., 1999b. Atmospheric Acetic Acid Pulping of Rice Straw (11): Behavior of Ash and Silica In Rice Straw during Atmospheric Acetic Acid Pulping and Bleaching.

Pande, H., Roy, D.N., 1996. Delignification kinetics of soda pulping of kenaf. J. Wood Chem. Technol. 16 (3), 311–325. Available from: https://doi.org/10.1080/02773819608545811.

Parajo, J.C., Alonso, J.L., Vazquez, D., Santos, V., 1993. Optimization of Catalysed Acetosolv Fractionation Holzforschung (47).

Pejic, B.M., Kostic, M.M., Skundric, P.D., Praskalo, J.Z., 2008. The effects of hemicelluloses and lignin removal on water uptake behavior of hemp fibers. Bioresour. Technol. 99 (15), 7152–7159. Available from: https://doi.org/10.1016/j.biortech.2007.12.073.

Pydimalla, M., Muthyala, B.R., Adusumalli, R.B., 2019. Influence of temperature on kraft pulping of whole bagasse and depithed bagasse. Sugar Tech. 21, 1003–1015. Available from: https://doi.org/10.1007/s12355-019-00719-8.

Quoc Lam, H., le Bigot, Y., Delmas, M., Avignon, G., 2001. Formic acid pulping of rice straw. Ind. Crop. Prod. 14. Available from: http://www.elsevier.com.

Rodríguez, A., Moral, A., Serrano, L., Labidi, J., Jiménez, L., 2008. Rice straw pulp obtained by using various methods. Bioresour. Technol. 99 (8), 2881–2886. Available from: https://doi.org/10.1016/j.biortech.2007.06.003.

Rodríguez, A., Sánchez, R., Requejo, A., Ferrer, A., 2010. Feasibility of rice straw as a raw material for the production of soda cellulose pulp. J. Clean. Prod. 18 (10–11), 1084–1091. Available from: https://doi.org/10.1016/j.jclepro.2010.03.011.

Sadiku, N.A., Oluyege, A.O., Sadiku, I.B., 2016. Analysis of the calorific and fuel value index of bamboo as a source of renewable biomass feedstock for energy generation in Nigeria. Lignocellulose 5 (1), 34–49.

Sarkar, A.M., Farzana, M., Rahman, M.M., Jin, Y., Jahan, M.S., 2021a. Future cellulose based industries in Bangladesh – a mini review. Cellulose Chem. Technol. 55 (5–6), 443–459. Available from: https://doi.org/10.35812/cellulosechemtechnol.2021.55.41.

Sarkar, A.M., Nayeem, J., Rahaman, M.M., Jahan, M.S., 2021b. Dissolving pulp from non-wood plants by prehydrolysis potassium hydroxide process. Cellulose Chem. Technol. 55 (1–2), 117–124. Available from: https://doi.org/10.35812/CelluloseChemTechnol.2021.55.12.

Sarwar Jahan, M., Nasir Uddin, M., Rahman, A., Rahman, M.M., Amin, M.N., 2016. Soda pulping of umbrella palm grass (*Cyperus flabelliformis*) (n.d.)J. Bioresour. Bioprod. (2), Available from: http://www.Bioresources-Bioproducts.com.

Sarwar Jahan, M., Shamsuzzaman, M., Rahman, M.M., Iqbal Moeiz, S.M., Ni, Y., 2012. Effect of pre-extraction on soda-anthraquinone (AQ) pulping of rice straw. Ind. Crop. Prod. 37 (1), 164–169. Available from: https://doi.org/10.1016/j.indcrop.2011.11.035.

Sharma, C., Kumar, S., 1999, n.d. Detection of Chlorophenolics in Effluent's from Bleaching Processes of Rice-Straw Pulp.

Siemers, Dipl-I.W., Bioenergy, S., 2011. Technical and Economic Prospects of Rice Residues for Energy in Asia.

Sitch, D.A., Marshall, H.B., 1950. The effect of hemicelluloses on the papermaking properties of white birch. Can. J. Res. 28 (10), 376–389.

Sridach, W., 2010. The environmentally benign pulping process of non-wood fibers. Suranaree J. Sci. Technol. 17 (2).

Tutus, A., 2003. A practical solution to the silica problem in straw pulping BAP PROJESİ View project Türkiye'de Kağıt ve Karton Sektörünün Durum Analizi View project. Article in Appita Journal . Available from: https://www.researchgate.net/publication/279572437.

Van Soest, P.J., 2006. Rice straw, the role of silica and treatments to improve quality. Anim. Feed. Sci. Technol. 130 (3–4), 137–171. Available from: https://doi.org/10.1016/j.anifeedsci.2006.01.023.

Ververis, C., Georghiou, K., Christodoulakis, N., Santas, P., Santas, R., 2004. Fiber dimensions, lignin and cellulose content of various plant materials and their suitability for paper production. Ind. Crop. Prod. 19 (3), 245–254. Available from: https://doi.org/10.1016/j.indcrop.2003.10.006.

World Food and Agriculture - Statistical Yearbook 2020. 2020. World Food and Agriculture – Statistical Yearbook 2020. FAO. https://doi.org/10.4060/cb1329en

Xiao, C., Bolton, R., Pan, W.L., 2007. Lignin from rice straw Kraft pulping: Effects on soil aggregation and chemical properties. Bioresour. Technol. 98 (7), 1482–1488. Available from: https://doi.org/10.1016/j.biortech.2005.11.014.

Xu, G., Wang, X., Hu, J., 2013, n.d. Biobleaching of Wheat Straw Pulp using Laccase and Xylanase.

Young, R.A., Davis, J.L., Wiesmann, E.-B., 1986. 40 (1986) H. 2 Organic Acid Pulping of Wood, Part II. Holzforsch. 99.

Chapter 15

Pulping and papermaking of aquatic macrophyte fibers

Z. Muta Harah[1,2], N.N.F. Syed[1], S.D. Ramaiya[3] and J.S. Bujang[4]
[1]Department of Aquaculture, Faculty of Agriculture, Universiti Putra Malaysia, Serdang, Selangor Darul Ehsan, Malaysia, [2]International Institute of Aquaculture and Aquatic Sciences (I-AQUAS), Universiti Putra Malaysia (UPM), Port Dickson, Negeri Sembilan, Malaysia, [3]Department of Crop Science, Faculty of Agricultural Science and Forestry, Universiti Putra Malaysia Bintulu Campus, Bintulu, Sarawak, Malaysia, [4]Department of Biology, Faculty of Science, Universiti Putra Malaysia, Serdang, Selangor Darul Ehsan, Malaysia

15.1 Introduction

Aquatic macrophytes are diverse groups of larger plants adapted to living entirely or partially in freshwater to marine water bodies. Four major categories of aquatic macrophytes are emergent, submerged, floating, and marginal vegetation (Muta Harah et al., 2014; Nather Khan, 1990). They occurred in different wetlands ranging from stagnant to flowing water bodies, including mudflats, swamps, peat swamps, ponds, lakes, streams, river systems, marshes, wet rice fields, and drainage canals (Muta Harah et al., 2014; Mashhor, 1996; Said, 1990). Nevertheless, these aquatic macrophytes, often described as a nuisance or noxious weeds due to their rapid growth ability and dense vegetation, created unfavorable conditions that blocked the water flow and affected the production of fish and crop plants in the water bodies (Mashhor, 1996; Nather Khan, 1990; Baki, 1982). Thus instead of discarding the weeds, they can be raw materials transformed into valuable items, such as a hat, dressing accessories, necklaces, and handbags (Abobi et al., 2015), as well as papermaking and paperboard production (Emmclan et al., 2018; Syed et al., 2016; Bidin et al., 2015; Poovarasi et al., 2014).

The pulp and paper industry has a significant impact on society, which is evident in almost every human daily activity. It is undeniably true that overconsumption of paper has significantly reduced the forest area in recent decades, deteriorating the environmental quality. Global paper consumption at 419.72 million metric tons in 2018 is expected to increase in the future. It has led to the search for alternative sustainable fiber sources that can meet demand when the

Pulping and Papermaking of Nonwood Plant Fibers. DOI: https://doi.org/10.1016/B978-0-323-91625-7.00014-X

supply of wood fibers is limited. Nonforest fibers cannot replace wood-based fibers production on a large scale; however, they can become an excellent commodity to support the wood pulp and paper industry and reduce forest fibers.

Pulping raw materials, for example, agricultural residues and nonwood plants, have been the choice, especially in countries with scarce forest resources, such as Spain, Italy, Greece, China, and India (Ververis et al., 2004; Saijonkari-Pahkala, 2008; Oinonen and Koskivirta, 1999). Although the primary source of fiber for paper production is from the wood plant, nonwood plants, including aquatic macrophytes, can be used as an alternative fiber source. Aquatic plants have significant differences in their physical and chemical characteristics (Hurter, 1997; Hurter and Riccio, 1998). According to Ververis et al. (2004), fiber dimension, lignin, and cellulose content of nonwood plants determine their suitability for paper production. Under certain conditions, the paper's tearing resistance is highly dependent on fiber dimension, that is, fiber length (Seth and Page, 1998). Nonwood aquatic plants have certain advantages, such as short growth cycles, low irrigation, fertilization requirements, and low lignin content, reducing energy and the need for fewer chemicals during pulping (Hurter and Riccio, 1998).

The handmade paper industry is environmentally friendly and very promising for local entrepreneurship. Papers made from aquatic macrophytes are multiuse: for writing, food wrapping, tissue paper, bookmark, and as value-added materials for handmade crafts. This chapter presents the evaluation of fiber characteristics, derived values, and chemical composition of aquatic macrophyte species for utilization as raw materials for pulp and papermaking. Their suitability is determined by comparing their fiber characteristics and the chemical composition with other tested nonwood plants for pulp production and papermaking. In addition, handmade paper sheets produced of selected aquatic macrophytes were tested for their quality for tensile strength, breaking length, and moisture content. Fig. 15.1 demonstrates the diagrammatic representation of fiber characterization and chemical composition of aquatic macrophyte materials, pulping and handmade papermaking, and testing mechanical properties of handmade paper.

All methods on determining fiber physical dimensions, biometric ratio, chemical composition, and mechanical properties evaluation are given briefly in Sections 15.2–15.5. Pulping and papermaking process are provided in detail in Section 15.4.

15.2 Fiber dimensions and biometric ratio

Table 15.1 shows the nonwood renewable materials from many freshwater and marine macrophytes as raw materials for pulp and papermaking. The parts used are the petioles, leaves, stems, stolons, rhizomes, and even plants. Fiber dimensions (fiber length, fiber diameter, fiber lumen diameter, and cell wall thickness) are determined according to Ogbonnaya et al. (1997). The

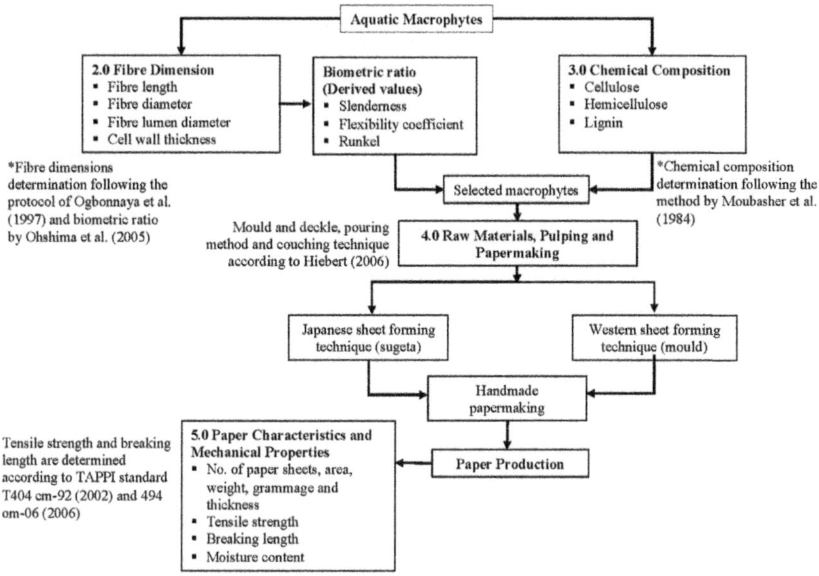

FIGURE 15.1 Diagrammatic representation of fiber characterization of aquatic macrophytes, chemical composition for pulping and handmade papermaking, and paper sheet characteristics and mechanical properties.

biometric ratio (derived values), Slenderness, Flexibility coefficient, and Runkels are derived based on Ohshima et al. (2005). The data in Table 15.2 shows aquatic macrophytes have significant differences in their fiber dimensions and chemical characteristics, similar to other nonwood plants, according to Hurter (1997) and Hurter and Riccio (1998). According to Ververis et al. (2004), fiber dimension, lignin, and cellulose content of nonwood plants determine their suitability for paper production. Fiber characteristics and chemical compositions are crucial because they influence the chemical processing of the material, pulp, and papermaking process. The systematic analysis of fiber and chemical composition in aquatic macrophyte is useful compared to those used in manufactured or tested materials for pulp production and papermaking. The principal component analysis (PCA) ordination is performed to assess the similarity or difference of the gathered data on fiber characteristics and the biometric ratio (Slenderness, Flexibility coefficient, and Runkel) with tested other nonwood plant material fibers (Tables 15.2 and 15.3). Fig. 15.2A shows PCA variables of fiber dimension and biometric ratio distribution. Fig. 15.2B comprises aquatic macrophytes, agricultural crop residues, and inland weeds clustered in four groups according to PCA variables and AHC (agglomerative hierarchical clustering) analysis.

Group 1 comprises aquatic macrophytes, *Oryza sativa*, *Monochoria hastata*, *Typha angustifolia*, *Pistia stratiotes*, *Ludwigia sedioides*, *Eichhornia*

TABLE 15.1 The freshwater and marine macrophytes, category, species, and parts as raw materials for pulp and papermaking.

Category	no	Family and species	Part used
Freshwater plants			
A. Emergent		Alismataceae	
	1	*Limnocharis flava* (L.) Buch. (Poovarasi et al., 2014)	Petiole, leaf
		Nymphaeaceae	
	2	*Nymphaea nouchali* Burm. *f.* (Poovarasi et al., 2014)	Petiole, leaf
	3	*Nymphaea pubescens* Willd. (Muta Harah and Japar Sidik, 2014–2018, unpublished)	Petiole, leaf
		Nelumbonaceae	
	4	*Nelumbo nucifera* Gaertn. (Muta Harah and Japar Sidik, 2014–2018, unpublished)	Petiole, leaf
		Poaceae	
	5	*Oryza sativa* L. (Muta Harah and Japar Sidik, 2014–2018, unpublished)	Leaf, stem
		Pontederiaceae	
	6	*Monochoria hastata* (L.) Solms (Muta Harah and Japar Sidik, 2014–2018, unpublished)	Petiole, leaf
		Typhaceae	
	7	*Typha angustifolia* Linn. (Bidin et al., 2015)	Petiole, leaf
B. Submerged		Hydrocharitaceae	
	8	*Blyxa aubertii* Rich. (Poovarasi et al., 2014)	Leaf
C. Floating		Araceae	
	9	*Pistia stratiotes* L. (Muta Harah and Japar Sidik, 2014–2018, unpublished)	Leaf
		Fabaceae (Leguminosae)	
	10	*Neptunia oleracea* Lour (Muta Harah and Japar Sidik, 2014–2018, unpublished)	Leaf
		Onagraceae	
	11	*Ludwigia adscendens* (L.) Hara (Muta Harah and Japar Sidik, 2014–2018, unpublished)	Leaf

(*Continued*)

TABLE 15.1 (Continued)

Category no		Family and species	Part used
	12	*Ludwigia sedioides* (Humb. & Bonpl.) H. Hara (Muta Harah and Japar Sidik, 2014–2018, unpublished)	Leaf
		Pontederiaceae	
	13	*Eichhornia crassipes* (Mart.) Solms. (Muta Harah and Japar Sidik, 2014–2018, unpublished)	Petiole, leaf
D. Marginal/ Semiaquatic		Alismataceae	
	14	*Alisma plantago-aquatica* L. (Muta Harah and Japar Sidik, 2014–2018, unpublished)	Petiole, leaf
		Cyperaceae	
	15	*Cyperus compressus* L. (Muta Harah and Japar Sidik, 2014–2018, unpublished)	Leaf
	16	*Cyperus digitatus* Roxb. (Emmclan et al., 2018; Bidin et al., 2015)	Stem, leaf
	17	*Cyperus halpan* L. (Bidin et al., 2015)	Stem, leaf
	18	*Cyperus iria* L. (Emmclan et al., 2018)	Stem, leaf
	19	*Cyperus malaccensis* Lam (Poovarasi et al., 2014)	Stem, leaf
	20	*Cyperus papyrus* L. (Muta Harah and Japar Sidik, 2014–2018, unpublished)	Leaf
	21	*Cyperus rotundus* L. (Bidin et al., 2015)	Stem, leaf
	22	*Eleocharis dulcis* (Burm. f.) Henschel (Poovarasi et al., 2014)	Leaf
	23	*Fimbristylis miliacea* (L.) Vahl (Muta Harah and Japar Sidik, 2014–2018, unpublished)	Leaf
	24	*Rhynchospora corymbosa* (L.) Britt (Muta Harah and Japar Sidik, 2014–2018, unpublished)	Leaf
	25	*Scirpus grossus* L.*f.* (Emmclan et al., 2018; Bidin et al., 2015)	Stem, leaf
	26	*Scirpus mucronatus* L. (Muta Harah and Japar Sidik, 2014–2018, unpublished)	Leaf
		Hanguanaceae	

(*Continued*)

TABLE 15.1 (Continued)

Category no		Family and species	Part used
	27	*Hanguana malayana* (Jack) Merr (Muta Harah and Japar Sidik, 2014–2018, unpublished)	Leaf
		Poaceae	
	28	*Brachiaria mutica* (Forssk.) Stapf (Muta Harah and Japar Sidik, 2014–2018, unpublished)	Leaf
	29	*Panicum repens* L. (Muta Harah and Japar Sidik, 2014–2018, unpublished)	Stolon, leaf
Marine plants			
Seagrass		Hydrocharitaceae	
	30	*Enhalus acoroides* (L.f.) Royle (Syed et al., 2016)	Leaf
	31	*Thalassia hemprichii* (Ehrenb.) Aschers. (Syed et al., 2016)	Leaf, rhizome
	32	*Halophila beccarii* Aschers. (Muta Harah and Japar Sidik, 2014–2018, unpublished)	Leaf, rhizome
	33	*Halophila ovalis* (R. Br.) Hook. *f.* (Syed et al., 2016)	Leaf, rhizome
	34	*Halophila spinulosa* (R. Br.) Aschers. (Syed et al., 2016)	Leaf, rhizome
		Cymodoceaceae	
	35	*Cymodocea serrulata* (R. Br.) Aschers. & Magnus (Syed et al., 2016)	Leaf, rhizome
	36	*Halodule pinifolia* (Miki) den Hartog (Muta Harah and Japar Sidik, 2014–2018, unpublished)	Leaf, rhizome
	37	*Halodule uninervis* (Forssk.) Aschers. (Muta Harah and Japar Sidik, 2014–2018, unpublished)	Leaf, rhizome
Mangrove		Arecaceae	
	38	*Nypa fruticans* Wurmb (Muta Harah and Japar Sidik, 2014–2018, unpublished)	Leaf
		Pandanaceae	
	39	*Pandanus odoratissimus* Linn. (Muta Harah and Japar Sidik, 2014–2018, unpublished)	Leaf

Source: The authors' database.

TABLE 15.2 Summary of fiber dimension and the biometric ratio of freshwater and marine macrophytes and other nonwood plant species.

		Fiber dimensions				Biometric ratio (derived values)		
Macrophytes species	Parts	Fiber length (mm)	Fiber diameter (μm)	Lumen diameter (μm)	Cell wall thickness (μm)	Slenderness (fiber length (μm)/fiber diameter (μm))	Flexibility coefficient (fiber lumen diameter (μm)/ fiber diameter (μm)) × 100	Runkel (cell wall thickness (μm)/fiber lumen diameter (μm)) × 2
Freshwater plants								
Limnocharis flava	Leaf	0.56 ± 0.05	23.26 ± 2.81	10.80 ± 1.32	7.23 ± 1.02	52.08 ± 7.78	55.36 ± 4.56	1.44 ± 0.18
Nymphaea nouchali	Petiole + leaf	0.57 ± 0.03	48.17 ± 2.23	38.46 ± 3.83	5.73 ± 0.87	11.88	80.06	0.30
Nymphaea pubescens	Petiole + leaf	0.40 ± 0.05	24.63 ± 1.98	8.96 ± 1.2	4.76 ± 0.57	18.92 ± 2.75	41.34 ± 5.73	2.23 ± 0.53
Nelumbo nucifera	Petiole + leaf	0.51 ± 0.11	21.21 ± 1.69	11.48 ± 0.92	5.24 ± 0.22	35.92 ± 8.22	55.12 ± 1.58	1.06 ± 0.07
Oryza sativa	Leaf + stem	0.77 ± 0.03	7.55 ± 0.34	2.4 ± 0.14	1.78 ± 0.06	106.07 ± 6.76	33.26 ± 2.20	1.55 ± 0.10

(*Continued*)

TABLE 15.2 (Continued)

		Fiber dimensions				Biometric ratio (derived values)		
Macrophytes species	Parts	Fiber length (mm)	Fiber diameter (μm)	Lumen diameter (μm)	Cell wall thickness (μm)	Slenderness (fiber length (μm)/fiber diameter (μm))	Flexibility coefficient (fiber lumen diameter (μm)/ fiber diameter (μm)) × 100	Runkel (cell wall thickness (μm)/fiber lumen diameter (μm)) × 2
Monochoria hastata	Petiole + leaf	1.15 ± 0.04	11.26 ± 0.44	5.25 ± 0.20	3.21 ± 0.13	111.36 ± 5.73	47.46 ± 1.08	1.26 ± 0.04
Typha angustifolia	Stem	0.83 ± 0.02	10.01 ± 0.66	4.35 ± 0.42	2.83 ± 0.18	89.34 ± 5.62	42.52 ± 2.19	1.52 ± 0.18
Blyxa aubertii	Leaf	0.26 ± 0.03	16.10 ± 1.71	9.83 ± 2.13	3.77 ± 1.12	16.40	61.78	0.80
Pistia stratiotes	Leaf	0.17 ± 0.03	12.70 ± 0.50	4.86 ± 0.60	3.75 ± 0.22	14.02 ± 2.66	38.11 ± 4.03	1.94 ± 0.18
Neptunia oleracea	Leaf	0.49 ± 0.04	36.67 ± 3.74	25.10 ± 2.60	5.37 ± 0.63	18.29 ± 2.80	70.85 ± 3.22	0.44 ± 0.04
Ludwigia adscendens	Leaf	0.62 ± 0.07	19.28 ± 0.79	14.25 ± 0.76	3.27 ± 0.14	35.07 ± 5.30	75.54 ± 4.33	0.48 ± 0.03
Ludwigia sedioides	Leaf	0.20 ± 0.04	5.03 ± 0.21	2.46 ± 0.19	2.14 ± 0.17	41.92 ± 8.40	49.45 ± 3.55	2.12 ± 0.30
Eichhornia crassipes	Petiole + leaf	0.31 ± 0.04	10.75 ± 0.69	5.38 ± 0.42	3.61 ± 0.24	34.7 ± 5.93	53.51 ± 3.76	1.53 ± 0.11

Alisma plantago-aquatica	Petiole + leaf	0.46 ± 0.62	21.15 ± 1.10	14.97 ± 0.76	4.46 ± 0.22	21.26 ± 2.76	74.68 ± 4.72	0.63 ± 0.03
Cyperus compressus	Leaf	1.16 ± 0.06	8.54 ± 0.27	3.77 ± 0.15	2.91 ± 0.11	141.71 ± 7.90	43.95 ± 0.90	1.60 ± 0.05
Cyperus digitatus	Stem	1.03 ± 0.06	9.31 ± 0.20	3.37 ± 0.15	2.97 ± 0.05	108.09 ± 4.87	35.20 ± 0.91	2.02 ± 0.08
Cyperus halpan	Stem + leaf	0.47 ± 0.07	24.84 ± 2.08	16.38 ± 1.86	7.01 ± 0.73	23.39 ± 3.97	68.90 ± 6.51	1.06 ± 0.16
Cyperus iria	Stem + leaf	1.11 ± 0.06	10.00 ± 0.27	4.54 ± 0.27	2.73 ± 0.09	113.08 ± 6.28	43.60 ± 1.72	1.65 ± 0.11
Cyperus malaccensis	Stem + leaf	0.09 ± 0.02	34.80 ± 2.01	18.63 ± 3.06	8.37 ± 2.04	2.70	53.8	0.91
Cyperus papyrus	Leaf	0.57 ± 0.04	9.26 ± 0.53	3.65 ± 0.24	1.75 ± 0.07	63.88 ± 4.32	40.91 ± 3.12	1.03 ± 0.07
Cyperus rotundus	Stem	0.71 ± 0.02	9.13 ± 0.47	4.32 ± 0.37	2.41 ± 016	81.57 ± 4.95	46.63 ± 2.56	1.28 ± 0.13
Eleocharis dulcis	Leaf	0.67 ± 0.10	6.31 ± 0.42	2.98 ± 0.21	3.04 ± 0.24	70.89 ± 15.07	48.47 ± 2.89	2.19 ± 0.20
Fimbristylis miliacea	Stem + leaf	0.99 ± 0.03	6.14 ± 0.26	2.44 ± 0.12	2.56 ± 0.11	177.85 ± 8.53	40.15 ± 1.25	2.20 ± 0.07
Rhynchospora corymbosa	Leaf	0.94 ± 0.05	8.02 ± 0.27	4.93 ± 0.21	2.83 ± 0.10	121.34 ± 6.80	61.36 ± 1.43	1.20 ± 0.04
Scirpus grossus	Stem	0.83 ± 0.02	12.11 ± 0.98	7.30 ± 0.89	2.41 ± 0.16	73.77 ± 7.32	58.08 ± 4.07	0.84 ± 0.17

(Continued)

TABLE 15.2 (Continued)

		Fiber dimensions				Biometric ratio (derived values)		
Macrophytes species	Parts	Fiber length (mm)	Fiber diameter (μm)	Lumen diameter (μm)	Cell wall thickness (μm)	Slenderness (fiber length (μm)/fiber diameter (μm))	Flexibility coefficient (fiber lumen diameter (μm)/ fiber diameter (μm)) × 100	Runkel (cell wall thickness (μm)/fiber lumen diameter (μm)) × 2
Scirpus mucronatus	Leaf	0.48 ± 0.06	24.91 ± 1.32	13.53 ± 0.52	17.68 ± 0.98	23.39 ± 3.98	48.64 ± 3.99	2.65 ± 0.14
Hanguana malayana	Leaf	0.61 ± 0.05	15.70 ± 0.70	7.63 ± 0.34	3.39 ± 0.19	37.83 ± 2.43	49.87 ± 2.08	0.88 ± 0.03
Brachiaria mutica	Leaf	0.86 ± 0.10	79.90 ± 2.40	44.68 ± 5.99	28.02 ± 1.46	11.12 ± 1.51	57.04 ± 7.93	1.74 ± 0.22
Panicum repens	Stolon + leaf	0.65 ± 0.02	65.57 ± 4.22	24.66 ± 1.19	23.29 ± 2.82	10.52 ± 0.62	40.04 ± 3.24	1.92 ± 0.22
Marine plants								
Seagrass								
Enhalus acoroides	Leaf	2.04 ± 0.13	15.96 ± 1.18	5.31 ± 0.57	5.43 ± 0.44	154.08 ± 12.82	31.79 ± 1.78	1.60 ± 0.19
Thalassia hemprichii	Leaf + rhizome	1.53 ± 0.05	16.13 ± 0.69	4.91 ± 0.24	6.31 ± 0.29	104.89 ± 5.59	30.07 ± 0.58	1.43 ± 0.10

Cymodocea serrulata	Leaf + rhizome	1.51 ± 0.07	16.82 ± 0.69	5.15 ± 0.23	6.53 ± 0.29	100.23 ± 7.06	30.44 ± 0.45	1.33 ± 0.05
Halodule pinifolia	Leaf + rhizome	0.34 ± 0.0	11.70 ± 0.51	6.03 ± 0.31	2.62 ± 0.15	30.20 ± 2.79	53.50 ± 3.08	0.89 ± 0.05
Halodule uninervis	Leaf + rhizome	1.22 ± 0.09	18.41 ± 0.97	6.45 ± 0.40	5.50 ± 0.32	75.19 ± 5.83	34.40 ± 0.67	0.98 ± 0.06
Halophila beccari	Leaf + rhizome	0.44 ± 0.03	12.14 ± 0.53	7.54 ± 0.35	3.28 ± 0.18	38.24 ± 2.75	64.67 ± 3.53	0.89 ± 0.04
Halophila ovalis	Leaf + rhizome	1.55 ± 0.07	20.87 ± 2.07	6.91 ± 0.50	7.81 ± 1.07	90.38 ± 6.09	35.18 ± 1.55	1.11 ± 0.06
Halophila spinulosa	Leaf + rhizome	1.54 ± 0.07	17.44 ± 0.84	5.50 ± 0.30	6.43 ± 0.32	98.12 ± 5.78	31.36 ± 0.59	1.26 ± 0.07
Mangrove								
Nypa fruticans	Leaf	1.31 ± 0.04	10.21 ± 0.28	4.16 ± 0.12	3.44 ± 0.10	133.31 ± 4.88	41.27 ± 1.01	1.70 ± 0.04
Pandanus odoratissimus	Leaf	1.49 ± 0.04	10.85 ± 0.39	5.52 ± 0.24	3.69 ± 0.17	154.47 ± 13.00	52.83 ± 2.16	1.47 ± 0.10
Other nonwood plants								
Eichhornia crassipes (Goswami and Saikia, 1994)	Leaf	1.60	5.50	9.00	2.50	290.90	163.64	0.56
Arundo donax (Ververis et al., 2004)	Internode	1.22	17.30	8.50	4.40	70.50	49.20	1.00

(Continued)

TABLE 15.2 (Continued)

		Fiber dimensions				Biometric ratio (derived values)		
Macrophytes species	Parts	Fiber length (mm)	Fiber diameter (μm)	Lumen diameter (μm)	Cell wall thickness (μm)	Slenderness (fiber length (μm)/fiber diameter (μm))	Flexibility coefficient (fiber lumen diameter (μm)/ fiber diameter (μm)) × 100	Runkel (cell wall thickness (μm)/fiber lumen diameter (μm)) × 2
Gossypium hirsutum (Ververis et al., 2004)	Stalk	0.84	18.07	12.32	3.18	44.6	65.7	0.52
Musa paradisiaca (Goswami et al., 2008)	Stem	1.55	22.00	14.20	5.50	70.50	64.55	0.77
Saccharum officinarum (Agnihotri et al., 2010)	Bagasse	1.51	21.40	6.27	7.74	70.56	29.29	2.46
Sorghum bicolor(Kiaei et al., 2011)	Stem	0.88	20.12	10.92	4.59	44.08	54.27	0.84

Oryza sativa (Kiaei et al., 2011)	Stem	0.83	10.89	4.57	3.16	76.58	41.96	1.38
Helianthus annuus (Kiaei et al., 2011)	Stem	0.96	22.84	11.12	5.85	42.03	48.68	1.05
Bambusa tulda (Sharma et al., 2013)	Stalk	1.89	17.00	3.45	6.78	111.20	20.29	3.93
Brassica napus (Tofanica et al., 2011)	Stalk	1.20	13.10	8.60	2.25	91.00	64.00	0.58
B. napus (Enayati et al., 2009)	Stalk	1.17	23.02	12.5	5.26	50.83	54.3	0.84
Nicotiana tabacum (Shakhes et al., 2011a)	Stalk	1.23	24.31	15.38	8.93	50.59	63.26	1.16
Hibiscus cannabinus (Shakhes et al., 2011b)	Stalk	1.31	24.12	16.4	4.25	54.32	67	0.52
Passiflora foetida (Brindha et al., 2012)	Stem	1.2	10	6	2	120	60	0.66

(Continued)

TABLE 15.2 (Continued)

		Fiber dimensions				Biometric ratio (derived values)		
Macrophytes species	**Parts**	**Fiber length (mm)**	**Fiber diameter (μm)**	**Lumen diameter (μm)**	**Cell wall thickness (μm)**	**Slenderness (fiber length (μm)/fiber diameter (μm))**	**Flexibility coefficient (fiber lumen diameter (μm)/ fiber diameter (μm)) × 100**	**Runkel (cell wall thickness (μm)/fiber lumen diameter (μm)) × 2**
Scoparia dulcis (Sharma et al., 2013)	Stem	0.93	27.14	16.81	5.17	34.27	61.89	0.62
Urena lobata (Sharma et al., 2013)	Stem	1.58	22.03	13.4	4.32	71.81	60.78	0.66
Vernonia cinerea (Sharma et al., 2013)	Stem	0.94	22.59	14.09	4.25	47.12	62.36	0.61
Sida cordifolia (Sharma et al., 2013)	Stem	1.3	22.74	14	4.37	58.22	61.83	0.69

Bidens pilosa (Sharma et al., 2013)	Stem	0.93	24.2	16.67	3.76	47.33	69.01	0.46
Solanum trilobatum (Thongpukdee et al., 2013)	Stem	4.1	24.48	17.04	3.44	167.48	54.15	0.81
Solanum diphyllum (Thongpukdee et al., 2013)	Stem	2.5	17.08	4.76	5.6	146.37	24.32	0.82
Solanum lasiocarpum (Thongpukdee et al., 2013)	Stem	1.61	27.04	14.48	6.08	59.54	52.79	0.91

TABLE 15.3 Types of paper produced from tested other nonwood plant species.

Species	Part	Parameter tested	Paper properties	Paper types	Reference(s)
Freshwater macrophytes					
1. *Eichhornia crassipes*	Leaf	Physical properties	Fiber dimension, biometric ratio (derived values)	Greaseproof paper	Goswami and Saikia (1994)
		Chemical properties	Cellulose, hemicelluloses, pentosan		
		Paper properties	Degree of freeness, burst index, tear index, tensile index		
Agriculture residues					
2. *Arundo donax*	Internode	Physical properties	Fiber dimension, biometric ratio	Printing and writing paper	Ververis et al. (2004)
		Chemical properties	Cellulose, lignin, ash		
		Paper properties	Slenderness ratio, tearing resistance		
3. *Gossypium hirsutum*	Stalk	Physical properties	Fiber dimension, biometric ratio	Writing, printing, mixed with hardwoods to produce various paper	Ververis et al. (2004)
		Chemical properties	α-Cellulose, lignin, ash		

4. *Musa paradisiaca*	Stem	Physical properties	Fiber dimension, biometric ratio	Greaseproof paper	Goswami et al. (2008)
		Chemical properties	Cellulose, hemicelluloses, lignin, ash, silica		
		Paper properties	Burst index, tear index, tensile index, blister, double-fold number, degree of freeness		
5. *Saccharum officinarum*	Bagasse	Physical properties	Fiber dimension, derived value	Writing and printing paper	Agnihotri et al. (2010)
		Chemical properties	Halocellulose, cellulose, hemicelluloses, ash, silica		
		Paper properties	Tensile index, tear index, burst index, double-fold number		
6. *Sorghum bicolor* 7. *Oryza sativa* 8. *Helianthus annuus*	Stem	Physical properties	Fiber dimension, derived value	Fiber plate, rigid cardboard and cardboard paper	Kiaei et al. (2011)
		Chemical properties	Lignin, cellulose, extractives, ash		
9. *Brassica napus*	Stem	Physical properties	Fiber dimension, derived value	Composites, paper and paperboard	Tofanica et al. (2011)
		Chemical properties	Cellulose, holocellulose, pentosan, lignin, ash, silica		

(*Continued*)

TABLE 15.3 (Continued)

Species	Part	Parameter tested	Paper properties	Paper types	Reference(s)
10. *B. napus*	Stalk	Physical properties	Fiber dimension, derived value	Not directly mentioned—raw materials-mixed with hardwood and softwood for papermaking	Enayati et al. (2009)
		Chemical properties	Holocellulose, α-cellulose, lignin, ash, extractives		
		Paper properties	NaOH charge, black liquor pH, screen yield, reject yield, kappa number, degree of polymerization		
11. *Nicotiana tabacum*	Stalk	Physical properties	Fiber dimension, derived value	Not directly mentioned—stated that suitable for papermaking	Shakhes et al. (2011a)
		Chemical properties	Holocellulose, α-cellulose, lignin, ash, extractives		
		Paper properties	NaOH charge, presence of AQ, kappa number, total yield, screen yield, rejects yield, ISO brightness		
12. *Hibiscus cannabinus*	Stalk	Physical properties	Fiber dimension, derived value	Not directly mentioned—alternative material for commercial long fiber pulp production	Shakhes et al. (2011b)
		Chemical properties	Holocellulose, α-cellulose, lignin, ash, alcohol benzene, 1% NaOH, hot water		
		Paper properties	NaOH charge, AQ, kappa number, total yield, rejects yield, freeness level (CSF), PFI revolution		

Commercial crop					
13. *Bambusa tulda*	Stalk	Physical properties	Fiber dimension, derived value	Writing and printing paper	Sharma et al. (2013)
		Chemical properties	Cellulose, holocellulose, lignin, ash		
Inland weeds					
14. *Passiflora foetida*	Stem	Physical properties	Fiber dimension, derived value	Suitable for various grades of paper production	Brindha et al. (2012)
		Chemical properties	Cellulose, hemicellulose, lignin		
15. *Scoparia dulcis* 16. *Urena lobata* 17. *Vernonia cinerea* 18. *Sida cordifolia* 19. *Bidens pilosa*	Stem	Physical properties	Fiber dimension, derived value	Not directly mentioned—all the fibers are flexible and fulfill the criteria for pulp and paper	Sharma et al. (2013)
20. *Solanum trilobatum* 21. *Solanum diphyllum* 22. *Solanum lasiocarpum*	Stem	Physical properties	Fiber dimension, derived value	Subsidiary materials or a potential substitute for hardwood and softwood in traditional forest product industries	Thongpukdee et al. (2013)

Freshwater macrophytes (1); agriculture residues (2–12); commercial crop (13); inland weeds (14–22), with reference to Table 15.2.
AQ, Anthraquinone.

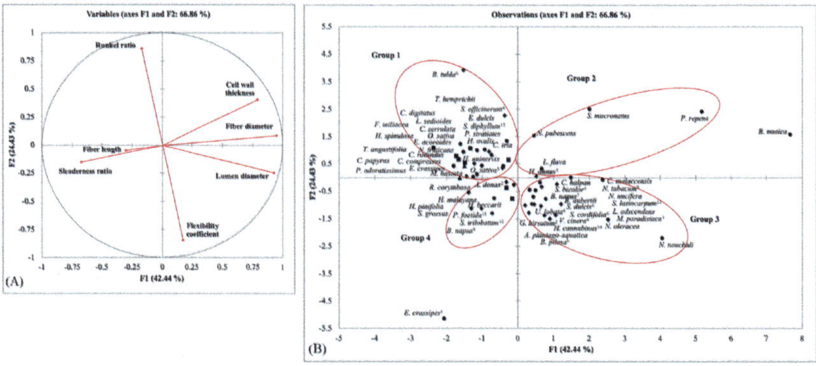

FIGURE 15.2 (A) PCA variables of fiber dimension (fiber length, fiber diameter, fiber lumen diameter, and cell wall thickness) and biometric ratio (Slenderness ratio, Flexibility coefficient, and Runkel ratio); (B) PCA score of fiber dimension and biometric ratio according to aquatic macrophyte and tested other nonwood species. ●—freshwater macrophytes, ■—seagrasses, and ▲—mangroves and tested nonwood materials and paper production,[1–12] with reference to Tables 15.2 and 15.3. *PCA*, Principal component analysis.

crassipes, *Cyperus compressus*, *Cyperus digitatus*, *Cyperus iria*, *Cyperus papyrus*, *Cyperus rotundus*, *Eleocharis dulcis*, *Fimbristylis miliacea*, *Enhalus acoroides*, *Thalassia hemprichii*, *Cymodocea serrulata*, *Halodule uninervis*, *Halophila ovalis*, *Halophila spinulosa*, *Nypa fruticans*, *Pandanus odoratissimus*, including the tested nonwood plants, *Saccharum officinarum*, *O. sativa*, *Bambusa tulda*, and *B. tulda* with Runkel ratio 1.00–2.00 and exceed 2.00 (Fig. 15.3).

The fibers are stiffer, less flexible, considered inferior for pulp and papermaking, and form bulkier paper of low bonded area (Ezeibekwe et al., 2009). Fibers having a Runkel ratio of less than 1 are considered very good for pulp and papermaking because they collapse and provide ample surface area for bonding. The pulp is of reasonable quality with a Runkel ratio range of 0.25–1.5 (Singh et al., 1991). Eroglu (1980) has classified the most suitable Runkel ratio as <1, where the fibers are with the thin cell wall, more flexible, and form a paper with extensive bonding. Runkel's ratio equal to 1 is also suitable for paper production, but less preferable. The fiber properties with this Runkel value are hard and stiff resulting in poor bonding ability and produce less paper quality (Wahab et al., 2006). For nonwood fibers, Runkel ratio >1 and exceeding 2.00 are still considered suitable for pulp and production of writing or printing paper (Agnihotri et al., 2010; Sharma et al., 2013) and fiber plate and rigid cardboard (Kiaei et al., 2011) (Fig. 15.2B; Table 15.3).

Group 2 are *Limnocharis flava*, *Nymphaea pubescens*, *Scirpus mucronatus*, *Panicum repens*, *Brachiaria mutica*, and macrophytes with similarity in cell wall thickness and fiber diameter. They are clustered with tested

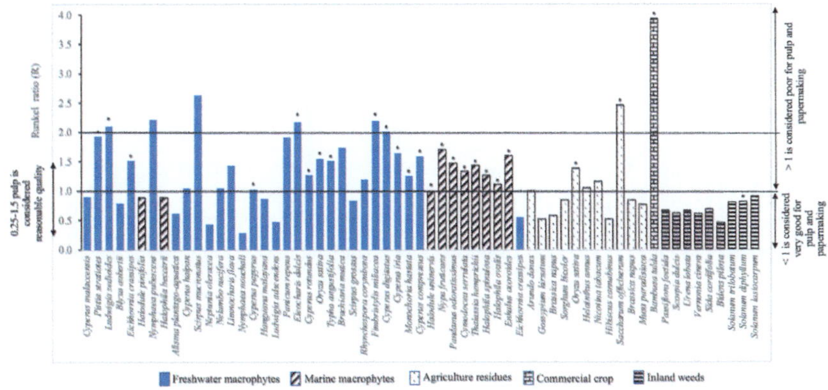

FIGURE 15.3 Runkel ratio (*R*) of fibers of freshwater and marine macrophytes, agriculture residue, commercial crop, and inland weeds. Plant fiber labeled with asterisks (*) are in PCA cluster Group 1 with *R* 1–2.00 and exceed *R* 2.00 (mediocre fiber). *R* < 0.25 (excellent fibers for papermaking), *R* 0.25–0.50 (very good fibers), *R* 0.50–1.00 (good fibers), *R* 1.00–2.00 (mediocre fibers), *R* > 2.0 (thick wall, fibers does not collapse and retain their tubular shape leading to poor adhesion). *PCA*, Principal component analysis. *Bar graph data from Table 15.2 is from our research and other references (Goswami and Saikia, 1994; Ververis et al., 2004; Goswami et al., 2008; Agnihotri et al., 2010; Kiaei et al., 2011; Sharma et al., 2013; Tofanica et al., 2011; Enayati et al., 2009; Shakhes et al., 2011a,b; Brindha et al., 2012; and Thongpukdee et al., 2013). The* R *values are shown in the figure from Area Maria and Popa Valentin (2014).*

nonwood *Helianthus annuus*, raw materials used by Kiaei et al. (2011) to produce fiber plate, rigid cardboard, and cardboard (Fig. 15.2B; Table 15.3).

Group 3 are *Nymphaea nouchali, Nelumbo nucifera, Blyxa aubertii, Alisma plantago-aquatica, Neptunia oleracea, Ludwigia adscendens, Cyperus halpan*, and *Cyperus malaccensis* with preferable Flexibility coefficient, *F* > 50–75, in the category of elastic fibers (Figs. 15.2B and 15.4; Bektaş et al., 1999). Based on Bektaş et al. (1999) and Hemmasi et al. (2011), fibers' classification are (1) high elastic fibers with *F* > 75, (2) elastic fibers with *F*, 50–75, (3) rigid fibers with *F*, 30–50, and (4) highly rigid *F* < 30. The reference value of the Flexibility coefficient of wood-based fibers, *F* > 50, produces good pulp strengths, mainly tensile and burst (Area Maria and Popa Valentin, 2014). Most compared tested nonwood fibers from agriculture residues and inland weeds; *Gossypium hirsutum, Musa paradisiaca, Sorghum bicolor, Brassica napus, Nicotiana tabacum, Hibiscus cannabinus, Scoparia dulcis, Urena lobata, Vernonia cinerea, Sida cordifolia*, and *Bidens pilosa* are in the same group with elastic fibers.

Fibers with such characteristics are for the paper production of writing and printing and are mixed with hardwoods to produce various paper (Ververis et al., 2004), greaseproof paper (Goswami et al., 2008), fiber plate, rigid cardboard and cardboard paper (Kiaei et al., 2011), and composites, paper, and paperboard (Tofanica et al., 2011).

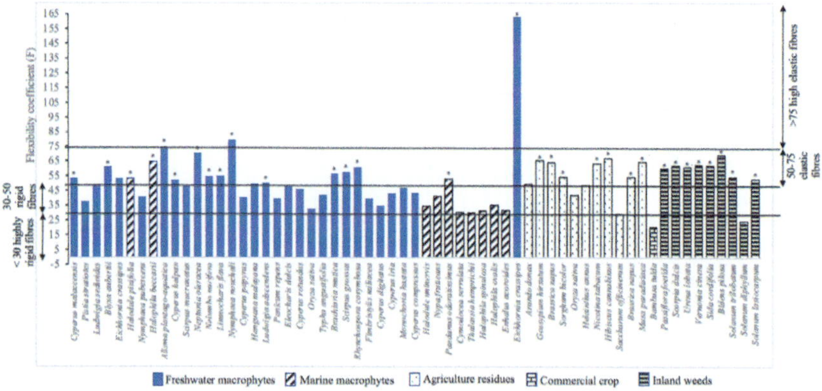

FIGURE 15.4 Flexibility coefficient (F) of fibers of freshwater and marine macrophyte species, agriculture residue, commercial crop, and inland weeds. Plant fibers labeled with asterisks (*) are in PCA cluster Group 3, with F 50–75 elastic fibers. Flexibility coefficient and fiber elasticity; (A) $F > 75$, high elastic fibers, (B) F 50–75, elastic fibers, (C) F 30–50, rigid fibers, and (D) $F < 30$ highly rigid. *PCA*, Principal component analysis. *Bar graph data from Table 15.2 from our research and other references (Goswami and Saikia, 1994; Ververis et al., 2004; Goswami et al., 2008; Agnihotri et al., 2010; Kiaei et al., 2011; Sharma et al., 2013; Tofanica et al., 2011; Enayati et al., 2009; Shakhes et al., 2011a,b; Brindha et al., 2012; and Thongpukdee et al., 2013).*

Group 4 comprises aquatic macrophytes, *Rhynchospora corymbosa*, *Scirpus grossus*, *Hanguana malayana*, *Halodule pinifolia*, and *Halophila beccarii*, in the same cluster with the tested other nonwood, *E. crassipes*, *Arundo donax*, *B. napus*, *Passiflora foetida*, and *Solanum trilobatum*. They are characterized by short-to-medium-sized fiber length (Fig. 15.5) and the Slenderness ratio (Fig. 15.6). In hard wood-based pulp production and papermaking, the fiber length was the influential feature on paper properties. For example, long fibers make a strong sheet because their length allows a better interweave, whereas this does not happen with very short or rigid fibers. However, certain paper qualities require short fibers as they fill the interstitial spaces making a sheet more "closed," which improves formation (Area Maria and Popa Valentin, 2014). The Slenderness ratio for most freshwater macrophytes and the tested nonwood plants in this group is between >33 and 70, while some exceed above 80. A Slenderness ratio of >33 is considered suitable for pulp and paper production (Sharma et al., 2013; Xu et al., 2006), and <70 for fibrous material is invaluable for pulp and paper quality (Young, 1981). A higher Slenderness ratio in fibers produces a higher rate of paper tear resistance (Akpakpan et al., 2012). Freshwater and marine macrophytes with short-to-medium-sized fibers and Slenderness ratio even above 80, which deviates from the norm, are suitable for the production of grease-proof paper (Goswami and Saikia, 1994), printing and writing paper (Ververis et al., 2004), various grades of paper production (Brindha et al., 2012),

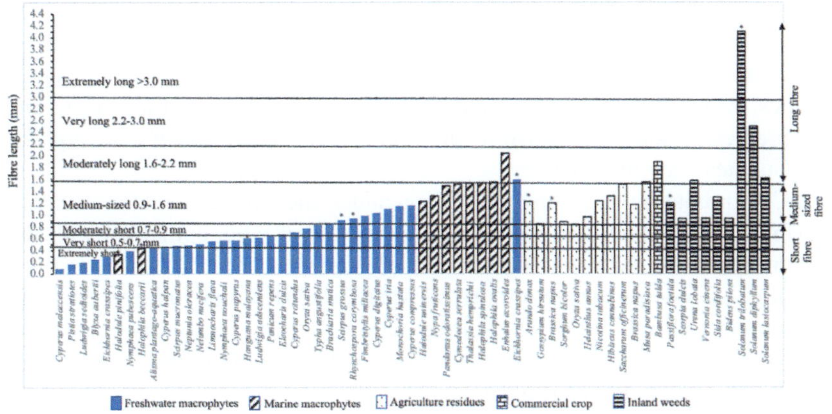

FIGURE 15.5 Fiber length of freshwater and marine macrophytes. Plant fibers labeled with asterisks (*) are in PCA cluster Group 4, with a medium-sized fiber length from 0.9 to 1.6 mm and a moderately long fiber length from 1.6 to 2.2 mm. The classification of fiber type is based on the Standard Classification published by the International Association of Wood Anatomist (IAWA, 1937) and Metcalfe and Chalk (1983). *PCA*, Principal component analysis.

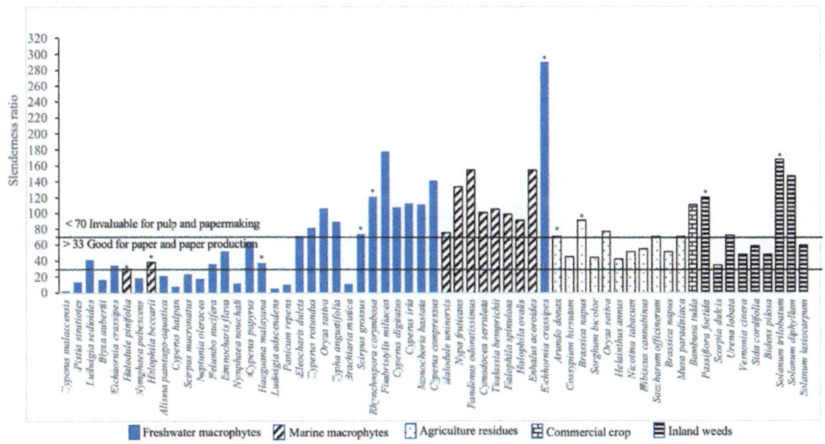

FIGURE 15.6 Slenderness ratio of fibers of freshwater and marine macrophyte species, agriculture residue, commercial crop, and inland weeds. Plant fibers assigned with asterisks (*) in the same cluster of Slenderness ratio of >33, <70 and >70. Slenderness ratio or Felting Index 70–80 is satisfactory pulp strengths, especially tear. *Bar graph data from Table 15.2 from our research and other references (Goswami and Saikia, 1994; Ververis et al., 2004; Goswami et al., 2008; Agnihotri et al., 2010; Kiaei et al., 2011; Sharma et al., 2013; Tofanica et al., 2011; Enayati et al., 2009; Shakhes et al., 2011a,b; Brindha et al., 2012; and Thongpukdee et al., 2013. Asterisks (*) referring to the source by Area Maria and Popa Valentin (2014).*

subsidiary materials (Enayati et al., 2009), or a potential substitute for hardwood and softwood in traditional forest product industries (Thongpukdee et al., 2013).

Aquatic macrophytes have significant differences in their physical characteristics, which influences the biometric ratio, as demonstrated above. A single predictor does not govern the pulp and papermaking for aquatic macrophytes' raw materials.

PCA ordination has separated them under the biometric ratio, Runkel, Flexibility coefficient, and Slenderness. The Runkel ratio, combined with the Flexibility ratio, also known as Felting Index (Area Maria and Popa Valentin, 2014), remains the most relevant in aquatic macrophytes and the tested nonwoods materials. The Runkel ratio value and Flexibility coefficient need to be reviewed or redefined for nonwood materials rather than following wood-based materials standards.

15.3 Chemical composition

The suitability of fibers for papermaking depends on a plant's cellulose, hemicellulose, and lignin contents. Differences in the composition may require differences in treatment and processing for the preparation of pulp. Pan et al. (2006) regarded lignin, cellulose, and hemicellulose contents of plants as essential aspects of evaluating and modifying the fiber properties. Screening plants for lignin, cellulose, and hemicellulose content can help identify the raw materials meeting the applicable requirements for pulping and papermaking. The chemical components of fibers from aquatic macrophytes are extracted and determined by the estimation method of Moubasher et al. (1984).

Table 15.4 shows the summary and comparison of aquatic macrophytes' chemical composition with other nonwood resources from various studies that have been tested suitable for paper production. Together with other nonwood resources, the aquatic macrophytes were ordinated with the PCA to assess their similarity or difference in their chemical composition. Fig. 15.7 is PCA based on chemical composition variables and the grouping of aquatic macrophytes, agricultural crop residues, and inland weeds in Fig. 15.7B.

Group 1 is represented by freshwater macrophytes, *M. hastata* and *H. malayana*, marine macrophytes, *E. acoroides*, *H. spinulosa*, *H. beccarii*, *H. ovalis*, *H. pinifolia*, and *C. serrulata*, and mangroves, *P. odoratissimus* and *N. fruticans*, including nonwood resources from other studies, *Cleome grandiflora*, *Cinnamomum burmanni*, *Phragmites* sp., *Musa* sp., *S. officinarum*, *Cymbopogon citratus*, *Cymbopogon martini*, and *Chamaecytisus* sp. They have similarity in a relatively high cellulose content ranging from 46.67% to 77.18%, comparable and higher than 38%–49% hardwood cellulose (Hurter and Eng, 2001). *C. grandiflora* and *C. burmanni* in the same group and with high cellulose (Brindha et al., 2013) are suitable for producing filter paper

TABLE 15.4 Chemical composition of aquatic macrophytes.

Aquatic macrophytes	Parts	Chemical composition (%)		
		Cellulose	Hemicellulose	Lignin
Freshwater plants				
1—Alismataceae				
Alisma plantago-aquatica	Petiole + leaf	21.91 ± 2.44	53.72 ± 5.46	24.37 ± 3.54
Limnocharis flava	Leaf	23.97 ± 1.69	63.66 ± 3.20	12.37 ± 2.10
2—Cyperaceae				
Cyperus compressus	Leaf	45.21 ± 1.56	41.67 ± 2.83	13.12 ± 1.52
Cyperus digitatus	Stem	44.82 ± 0.29	42.78 0.29	11.81 ± 0.12
Cyperus iria	Stem + leaf	44.60 ± 0.38	43.44 ± 0.60	10.63 ± 0.28
Cyperus malaccensis	Stem + leaf	28.9 ± 1.73		10.4 ± 1.24
Cyperus papyrus	Leaf	28.66 ± 3.04	49.70 ± 6.71	28.37 ± 0.99
Cyperus rotundus	Stem	42.58 ± 1.32	45.64 ± 1.12	9.54 ± 1.08
Eleocharis dulcis	Leaf	21.14 ± 0.72	51.08 ± 0.12	27.78 ± 0.68
Fimbristylis miliacea	Leaf	42.91 ± 8.16	47.53 ± 7.05	9.57 ± 5.39
Rhynchospora corymbosa	Leaf	39.34 ± 4.80	39.36 ± 7.43	21.30 ± 3.33
Scirpus grossus	Stem	42.13 ± 3.56	45.57 ± 0.51	11.44 ± 0.28

(Continued)

TABLE 15.4 (Continued)

Aquatic macrophytes	Parts	Chemical composition (%)		
		Cellulose	Hemicellulose	Lignin
Freshwater plants				
Scirpus mucronatus	Leaf	1.74 ± 0.54	2.53 ± 0.07	1.62 ± 0.15
3—Fabaceae (Leguminosea)				
Neptunia oleracea	Leaf	14.77 ± 1.58	64.47 ± 3.52	20.76 ± 2.25
4—Hanguanaceae				
Hanguana malayana	Leaf	46.67 ± 2.62	38.99 ± 2.23	14.33 ± 0.64
5—Hydrocharitaceae				
Blyxa aubertii	Leaf	7.81 ± 1.64		6.53 ± 1.46
6—Nelumbonaceae				
Nelumbo nucifera	Petiole + leaf	41.12 ± 2.79	34.97 ± 3.16	23.90 ± 2.34
7—Nymphaeaceae				
Nymphaea nouchali	Petiole + leaf	9.45 ± 2.74		6.98 ± 3.04
8—Onagraceae				
Ludwigia sedioides	Leaf	37.48 ± 9.79	53.58 ± 7.88	8.94 ± 2.08
9—Poaceae				
Oryza sativa	Leaf + stem	37.50 ± 2.22	51.91 ± 0.70	10.59 ± 2.70

10—Pontederiaceae				
Eichhornia crassipes	Petiole + leaf	53.27 ± 2.24	43.29 ± 2.33	3.44 ± 0.20
Monochoria hastata	Petiole + leaf	72.54 ± 10.41	4.82 ± 3.29	22.63 ± 7.42
11—Typhaceae				
Typha angustifolia	Stem	44.05 ± 0.49	54.84 ± 4.27	20.04 ± 3.37
Marine plants				
Seagrass				
5—Hydrocharitaceae				
Enhalus acoroides	Leaf	77.18 ± 3.65	28.31 ± 0.70	6.46 ± 1.32
Thalassia hemprichii	Leaf + rhizome	40.30 ± 1.15	14.33 ± 0.70	5.02 ± 0.08
Halophila beccarii	Leaf + rhizome	55.77 ± 2.58	38.63 ± 2.34	5.63 ± 0.32
Halophila ovalis	Leaf + rhizome	53.18 ± 4.58	27.62 ± 0.45	9.05 ± 0.55
Halophila spinulosa	Leaf + rhizome	54.77 ± 3.30	22.89 ± 1.58	10.83 ± 0.29
12—Cymodoceaceae				
Cymodocea serrulata	Leaf + rhizome	47.33 ± 2.83	25.92 ± 0.62	11.20 ± 0.36
Halodule pinifolia	Leaf + rhizome	57.07 ± 3.00	26.83 ± 3.32	16.06 ± 0.38
Halodule uninervis	Leaf + rhizome	36.53 ± 1.59	18.37 ± 0.56	6.84 ± 0.56

(Continued)

TABLE 15.4 (Continued)

Aquatic macrophytes	Parts	Chemical composition (%)		
		Cellulose	Hemicellulose	Lignin
Freshwater plants				
Mangrove				
13—Arecaceae				
Nypa fruticans	Leaf	39.96 ± 9.83	35.76 ± 3.86	24.28 ± 6.05
14—Pandanaceae				
Pandanus odoratissimus	Leaf	63.73 ± 4.77	14.18 ± 4.34	22.09 ± 0.97
Nonwood plant (previous studies)				
Arundo donax (Shatalov et al., 2001)	Stem	29.2	32.1	20.9
Musa sp. (Silveira et al., 2008)	Straw	53.45	28.56	15.42
Zea mays (Daud et al., 2014)	Stem	39	42	7.3
Saccharum officinarum (Agnihotri et al., 2010)	Bagasse	42.34	28.6	21.7
Panicum virgatum (Radiotis et al., 1999)	Stem	43.4	35.9	21.8
Passiflora foetida (Brindha et al., 2013)	Stem	40	36	24
Cleome grandiflora (Brindha et al., 2013)	Stem	56.53	21.74	21.73
Cinnamomum burmanni (Brindha et al., 2013)	Stem	49.25	31.88	18.84

Cymbopogon citratus (Kaur and Dutt, 2013)	Leaf	44.16	29.07	17.39
Cymbopogon martini (Kaur and Dutt, 2013)	Leaf	45.55	28.12	17.04
Phalaris arundinacea (Hattaka et al., 1996)	Stem	35	35	9
Festuca arundinacea (Janson et al., 1996)	Leaf, stem	34	29	19
Chamaecytisus sp. (Jimenez et al., 2008)	Leaf	44	31.7	14.8
Phragmites sp. (Jimenez et al. 2008)	Leaf	40	24.4	23.6
Retama monosperma (Jimenez et al., 2008)	Leaf	43	29	21.5

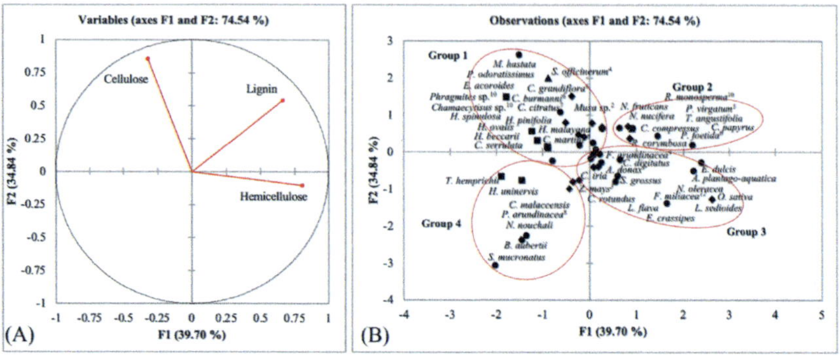

FIGURE 15.7 (A) Plot of chemical composition (cellulose, hemicellulose, and lignin); (B) position of principle component (PC) score of aquatic macrophytes species, ●—freshwater macrophytes, ■—seagrasses, and ▲—mangroves. Nonwood resources and tested nonwood fiber materials and paper production,[1−10] with reference to Table 15.4.

and boards and composite materials (Cordeiro et al., 2004). The cellulose content of more than 34%, for example, in wood-based resources, is an important component in papermaking as it gives high influence to the pulp produced. In chemical pulping, the pulp yield was positively correlated with cellulose contents as the paper strength and stability properties depended on raw materials' cellulose content (Amidon, 1981). The higher the cellulose content, the longer the pulping time needed to degrade the cellulose (Akpakpan et al., 2012). In Group 1, *M. hastata* and *E. acoroides* with a higher cellulose content of >70% potentially produce higher pulp yield than other species.

Group 2 is represented by *N. nucifera*, *T. angustifolia*, *C. compressus*, *C. papyrus*, and *R. corymbosa* and mangrove, *N. fruticans*, clustered with non-wood resources of other studies: *Retama monosperma*, *Panicum virgatum*, and *P. foetida*. Aquatic macrophyte possesses substantial lignin content of 13.12%–28.37%, while other nonwoods range from 29% (Jimenez et al., 2008) to 36% (Brindha et al., 2013). Lignin provides plant tissue and individual fiber with compressive strength and stiffens the fiber's cell wall to protect the carbohydrates from chemical and physical damages (Saheb and Jog, 1999). However, in papermaking, lignin is an unwanted polymer that needs to be removed during pulping using high energy and chemicals (Madakadze et al., 1999). Elimination of lignin from fiber during paper production is a major cost consumption in the paper industry and an important environmental polluting process (Ralph et al., 1997). Hence, lower lignin of raw materials is preferable in papermaking. Preferable lignin content in paper production should not exceed 30% (Nieschlag et al., 1960) to prevent the fiber from getting tougher and stiffer (Khalil et al., 2011). Lignin is undesirable as it causes unstable color, relatively dark, and hydrophobic surface, not favorable for forming the interfiber bonds of hemicellulose and cellulose (Hubbe and Bowden, 2009). Chemical removal of lignin or bleaching will affect the tensile strength rather than the tensile stiffness of fiber attributed to the low OH-group in lignin that prevents water to bond to the lignin molecules. Despite the substantial lignin, aquatic macrophyte species showed a considerable amount of cellulose ranging from 28.66% to 45.21%, similar to *Zea mays* (47.33%), *B. tulda* (47.00%), and *B. napus* (34.50%), are still utilizable for production of fiber plate, rigid cardboard and cardboard, paper printing and writing paper, composites, and paper and paperboard (Kiaei et al., 2011; Sharma et al., 2013; Tofanica et al., 2011).

Group 3 comprises mainly the freshwater macrophytes, *A. plantago-aquatica*, *L. flava*, and the majority from the Cyperaceae family: *C. digitatus*, *C. iria*, *C. rotundus*, *E. dulcis*, *F. miliacea*, *S. grossus*, *N. oleracea*, *L. sedioides*, *O. sativa*, *E. crassipes*, and including nonwood materials *A. donax* and *Festuca arundinacea* for the production of printing and writing paper by Shatalov et al. (2001) and by Janson et al. (1996). The aquatic macrophytes and tested other nonwood materials shared similarity in hemicellulose content from 29.00% to 64.47% and covered the hemicellulose range of nonwood materials 31.2% Flax fiber to 42% Canola straw for papermaking (Hosseinpour et al., 2010; Han and Rowell, 1997). Higher hemicellulose values result in a higher yield of pulp and paper strength (tensile, burst, and fold) as it reduced the internal fiber stress to improve the fiber bonding (Biermann, 1996).

Group 4 is a cluster of freshwater, *C. malaccensis*, *B. aubertii*, *N. nouchali*, *S. mucronatus*, and marine macrophytes *T. hemprichii* and *H. uninervis* and nonwood reed canary grass, *Phalaris arundinacea*, used by Hattaka et al. (1996). Their similarity is low in chemical components, cellulose or hemicellulose or a combination of both holocellulose and lignin. The freshwater macrophytes, *N. nouchali*, *B. aubertii*, and *S. mucronatus*, although with low lignin (1.62%–6.98%) but with low cellulose content, 1.74%–9.45%, make them not preferred materials in papermaking. Seagrasses in the same group with low lignin, substantial cellulose content, 36.53%–40.30% and low hemicellulose, 14.33%–18.37%, are still acceptable for papermaking. Furthermore, lower hemicellulose may also benefit during pulping, decreasing the water absorption capacity (Bakker and Jenkins, 2003).

15.4 Raw material, pulping and handmade papermaking

The pulp is the fibrous cellulosic material used in the production of paper. In papermaking, wood is the primary raw material for the production of pulp. Depending on the countries, the pulp is from available native plants. For examples, in China and Japan, mulberry is a common fiber that made a thin, soft, absorbent paper suitable for calligraphy and woodblock printing.

In hand papermaking, wood as raw material is rarely used. Noncellulose materials, for example, lignin, are not eliminated in the processing of wood pulp and causes the resulting paper to decay over time. Pulp from wood that will deteriorate over time is not a choice, and acceptable medium Hand papermakers generally produce paper for fine books and artwork (Hiebert, 2000). Malaysia is endowed with rich nonwood fiber resources in respect of diversity as well as abundance. The hot, wet climate with abundant rainfall nurture plant life, the aquatic macrophytes in freshwater and marine alike.

15.4.1 Raw materials from aquatic macrophytes

The selection of raw materials for pulping and papermaking is based on the fiber characteristics (Fig. 15.2B), chemical composition (Fig. 15.7B) evaluation, and the plants' availability. Eight freshwater macrophyte species, *C. digitatus*, *C. iria*, *C. rotundus*, *F. miliacea*, *R. corymbosa*, *S. grossus*, *M. hastata*, and *T. angustifolia*, five marine macrophytes, *E. acoroides*, *T. hemprichii*, *C. serrulata*, *H. ovalis*, *H. spinulosa*, and two mangrove *P. odoratissimus* and *N. fruticans*, are selected for pulping and papermaking production. The harvested plant parts, a combination of petiole and leaves, stem and leaves, stolon and leaves, and rhizome and leaves were cleaned (Table 15.1) and dried and fragmented before pulping. Fig. 15.8 shows the overview of raw material preparation, pulping process, and papermaking. The harvested plants' preparation, pulp production and the process of handmade papermaking are outlined based on Hiebert (2006).

15.4.2 Pulping process of raw material

Pulping is the process by which the bonds within the plant structure are ruptured chemically. Cooking is necessary to remove the binding agents from the raw material. Cooking removes noncellulose materials, such as lignins, pectins, waxes, and gums, which, if not removed, interfere with the bonding process and affect the paper's archival qualities. Chemicals and heat are used to dissolve the lignin and the binding agent that holds the raw material fibers together. Several alkalis can break down plant fibers and aids in removing noncellulose materials from plant fibers. This is achieved by the cooking of 1 kilogram of raw material with soda. For freshwater macrophytes, 100 g of soda ash (Na_2CO_3) in 8 L of aged tap water is added to the processed raw material in a stainless steel pot at 160°C for 1.5 hours (Poovarasi et al., 2014) or 20 g sodium carbonate (Na_2CO_3) in 2 L of water using induction cooker at 140 °C for 2 hours (Bidin et al., 2015) or 6 L of 9.0 g sodium hydroxide (NaOH) dissolved in 1 L of water cook in an ELBA induction cooker stainless steel pot at 100°C for 2 hours (Emmclan et al., 2018). Marine macrophytes' raw materials from *P. odoratissimus* and *N. fruticans*

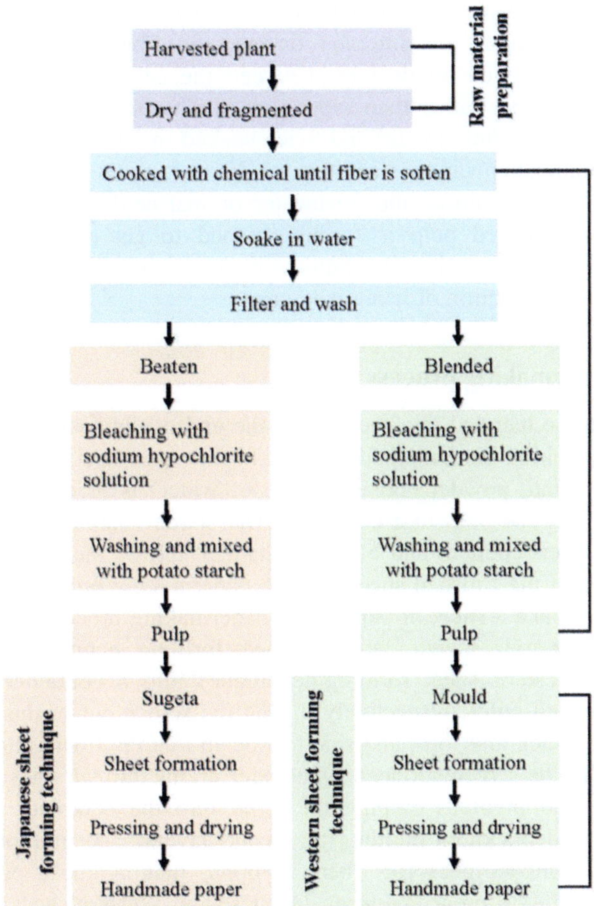

FIGURE 15.8 Pulp and papermaking process of aquatic macrophytes.

are cooked with 60 g of sodium hydroxide (NaOH) in 6 L of water using an ELBA induction cooker at 160°C for 8 hours (Muta Harah and Japar Sidik, unpublished). The raw material of seagrasses is cooked in a 10 L solution of 40 g sodium hydroxide (NaOH) in an ELBA induction cooker at a temperature of 160°C for 2–6 hours. The cooking duration is longer for the tougher *E. acoroides* and with additional water and NaOH solution to soften the pulp (Syed et al., 2016). The obtained chemical pulp after cooking is soaked in the water for 1–3 days. Pulp containing fibers are filtered using 250 μm sieved and washed with water repeatedly to remove excess chemical and residue. Pulp was further subjected to either hand-beating with a wooden mallet for 15–30 minutes until the fiber spread out and became soft or blend in a standard kitchen blender for 8–10 minutes. Beating also shortens the fibers, so they are evenly distributed during sheet forming. As water is removed and evaporated throughout pressing and drying, the fibers are pulled closer together, and hydrogen bonds form between the cellulose molecules. The blender cuts the fiber rather than separating it. The pulp stock is washed in water to dissolve organic matter and twice soaked in 6 L water with 120 mL of 0.02% sodium hypochlorite solution or 300 mL of 0.05% sodium hypochlorite solution delignified the freshwater or marine-derived pulp, respectively. The delignified pulp is further washed to get rid of the sodium hypochlorite from the pulp. The pulp is made up of plant fibers, the raw material in the production of handmade paper.

15.4.3 Papermaking process

Additives can be blended into the pulp at the end of the beating or blending cycle before making the handmade sheets. In this case, one L of starch solution (1 g of potato powder cooked in 1 L of water) is blended in the 1 of water containing the pulp in a container. The starch coats fibers and makes them more water-repellent and bleed-resistant. It allows the use of watercolors and inks on the finished sheets. In the papermaking process, there is no single way to make a sheet of paper. The papermaking process involves dipping a Japanese-style sugeta (Japanese sheet forming technique) or a mold and deckle (Western sheet forming technique) into a container of hand-beaten or blended pulps, respectively, lifting the screen out of the container, and shaking it back and forth and side to side so that the fibers interlock and bond on top of the screen surface as the water drains through the screen. The Japanese method involves dipping the mold into the container repeatedly until the desired thickness building up fibers' layers. The multiple laminations create more surfaces for fibers to bond, making these papers very strong. This is the difference between Japanese and Western sheet formation. The freshly produced Japanese or Western sheet formation is then transferred onto a flat surface and is slowly pressed with a sponge. Pressing removes excess water from the sheet of paper and compacts the fibers tightly together as they bond. After pressing, the sheet still contains moisture. It is air-dried at room temperature, 25°C–27°C, and placed in a herbarium press. Multiple sheets can be stacked evenly placed on top of the first sheet, each sheet separated by a similar size cloth. The herbarium press is placed in an air-circulating oven at 40°C–50°C, for 1–3 days, depending on the sheet thickness.

15.5 Paper characteristics and mechanical properties

Table 15.5 shows the paper characteristics from selected aquatic macrophytes from hand-beating or blended pulp for the Japanese and Western sheet formation, respectively. Dry paper sheets produced are of different size

TABLE 15.5 Paper sheet production from aquatic macrophytes using Japanese and Western methods.

Methods	Japanese method					Western method				
Paper production	No. of paper sheet	Total paper weight (g)	Area per sheet (cm^2)	Grammage per sheet (g/m^2)	Thickness per sheet (mm)	No. of paper sheet	Total paper weight (g)	Area per sheet (cm^2)	Grammage per sheet (g/m^2)	Thickness per sheet (mm)
Freshwater macrophytes										
Cyperus digitatus	4	63	619.5	237.15	0.92					
Cyperus iria	4	63.25	619.50	239.17	1.06					
Cyperus rotundus						2	26.97	630	234.07	0.81
Fimbristylis miliacea						4	63.43	619.50	325.77	0.43
Rhynchospora corymbosa		53.84	619.50	267.2	0.58	4	70.41	619.50	332.95	0.34
Scirpus grossus	4	65.13	619.50	254.34	1.18	2	27.28	630	252.47	0.75
Monochoria hastata	4	28.73	619.50	202.95	0.12	4	67.65	619.50	130.06	0.08
Typha angustifolia						3	39.72	630	232.33	0.96

(Continued)

TABLE 15.5 (Continued)

Methods	Japanese method					Western method				
Paper production	No. of paper sheet	Total paper weight (g)	Area per sheet (cm^2)	Grammage per sheet (g/m^2)	Thickness per sheet (mm)	No. of paper sheet	Total paper weight (g)	Area per sheet (cm^2)	Grammage per sheet (g/m^2)	Thickness per sheet (mm)
Marine macrophytes										
Enhalus acoroides	10	12.82	75	170.93	0.64	13	15.6	75	160.00	0.34
	8	25.12	132	237.9	0.60	11	26.36	132	181.51	0.44
	1	4.03	195	4.03	0.77	3	10.25	195	175.20	0.45
	2	39.84	496	401.6	1.12	3	26.89	496	180.71	0.55
Thalassia hemprichii	6	6.98	75	155.14	0.50	3	3.98	75	132.58	0.52
	1	3.07	195	157.61	0.62	2	3.12	195	80.10	0.39
Cymodocea serrulata	2	1.72	75	114.67	0.46	2	1.48	75	98.67	0.30
Halophila ovalis	7	11.54	75	219.79	0.69	5	5.297	75	141.09	0.59
Halophila spinulosa	3	2.10	75	93.33	0.63	4	2.25	75	75.00	0.48
	1	1.81	195	92.57	0.55					
Pandanus odoratissimus						4	55.67	619.50	258.91	0.573
Nypa fruticans						4	59.90	619.50	304.42	0.505

and number due to the different size of wooden sugeta or mold and deckle used during the papermaking process. Sheet area per sheet (cm^2), grammage per sheet (g/m^2), and thickness per sheet (mm) vary depending on the aquatic macrophyte species. The Japanese sheet formation techniques produce a thicker sheet. The paper sheet thickness is important depending on the paper's purpose, such as filling cards, printing papers, condenser paper, and decoration paper. The pulp-to-water ratio in the container is the main determinant of paper sheets' thickness (Hiebert, 1998). The paper characteristics in Table 15.5 possessed the corresponding mechanical properties of handmade paper sheet produced with the Japanese and Western techniques (Table 15.6). Moisture is present in paper sheets (moisture content of the produced paper sheet is determined by placing 1 g of paper sheet on AD-4715 Infrared Moisture Determination Balance) vary from 9% to 19%, depending on the relative humidity of the environment.

Higher moisture content in the paper will make the paper less tear resistant, whereas lesser moisture caused the paper to be brittle and easy to break down. According to TAPPI T 412 om-06 (2011), there are several standards for typical moisture values based on paper types that are newsprint paper, 7.5%–9.5%; office or business paper, 4%–4.5%; marketing wood pulp, 10%, printing paper, 6%–7%; and tissue, 2%–7%. *M. hastata* and *N. fruticans* fall between the standard range of printing paper, whereas the rest exceed the standard value.

Tensile strength and breaking length of each paper strip of 17.8 ± 0.1 cm $\times$ 2.5 ± 0.1 cm dimensions according to the TAPPI T404 cm-92 standard (1992) as in Table 15.6 is measured following TAPPI Method T-404 (1992) and modified TAPPI (Technical Association of the Pulp and Paper Industry, 494 om-06 (2006)). PCA ordination in Fig. 15.9A showed that variables' tensile strength (kN/m) and breaking length (km) and the corresponding paper from aquatic macrophytes compared to the TAPPI standard commercial paper, offset paper (MD), Rag bond paper (MD), and newsprint (MD) (Table 15.6; Fig. 15.9B).

Group 1 consists of paper sheets with great tensile strength includes *F. miliacea* (WM), *R. corymbosa* (JM), *M. hastata* (WM), *E. acoroides* (JM and WM), *T. hemprichii* (JM and WM), and *H. ovalis* (JM) clustered together with commercial offset paper type (MD) with standard tensile strength (5.5 kN/m). Group 2 paper consists of with a higher breaking length with *H. spinulosa* (JM), clustered together with the seaweed paper produced by Seo et al. (2010) and TAPPI offset paper (CD), Rag bond paper (MD and CD), and Newsprint paper (MD) with standard breaking length ranged 3.06–4.9 km. The rest of the paper sheets are clustered in Group 3 with low mechanical strength similar to Newsprint (CD) with standard tensile strength and breaking length of 0.9 kN/m and 1.84 km, respectively. Hence, all paper sheets produced (Fig. 15.10) are with tensile strength (kN/m) and breaking length (km) equivalent to commercial papers' TAPPI standard.

TABLE 15.6 Mechanical properties of handmade paper from aquatic macrophytes using Japanese and Western methods.

Methods	Japanese method			Western method		
Mechanical properties	Moisture (%)	Tensile strength (kN/m)	Breaking length (km)	Moisture (%)	Tensile strength (kN/m)	Breaking length (km)
Freshwater macrophytes						
Cyperus digitatus	13.00	2.61	1.2			
Cyperus iria	14.6	1.82	0.62			
Cyperus rotundus				10.11	1.69	0.7
Fimbristylis miliacea				11.58	3.68	1.2
Rhynchospora corymbosa	10.75	3.78	1.4	13.07	2.14	0.7
Scirpus grossus	13.80	1.41	0.65	13.08	1.52	0.6
Monochoria hastata	10.25	2.5	1.2	6.47	3.03	2.4
Typha angustifollia				13.13	0.94	0.4
Marine macrophytes						
Enhalus acoroides	18.68	3.46	2.49	15.12	4.10	2.12
Thalassia hemprichii	13.84	4.00	3.43	12.82	3.24	2.64
Cymodocea serrulata	12.00	1.24	2.24	12.15	2.24	1.11
Halophila ovalis	12.35	3.36	1.83	12.13	2.44	2.07
Halophila spinulosa	14.12	2.72	2.43	14.21	0.80	1.09

Pandanus odoratissimus				10.09	3.16	1.2
Nypa fruticans				9.33	1.62	0.5
Previous study (seaweeds)						
TAPPI test method (T 205 sp-02, 2006)						
Gelidium corneum (Seo et al., 2010)		2.29	3.84			
Gelidium amansii (Seo et al., 2010)		1.86	3.01			
TAPPI standard method (beating) (T 227 om-94, 1994)						
Rhizoclonium sp. (Chao et al., 2000)		2.96	3.6			
Commercial paper (TAPPI standard)		MD			CD	
Offset paper (MD)		5.55	5.3		3.21	3.06
Rag bond paper (MD)		3.6	4.9		2.55	3.47
Newsprint (MD)		1.79	3.65		0.9	1.84

CD, Cross direction; *MD*, machine direction.

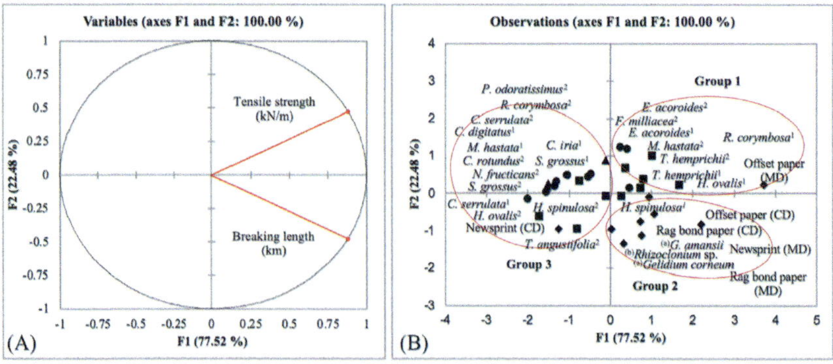

FIGURE 15.9 (A) Plot of tensile strength and breaking length; (B) position of PC score of aquatic macrophyte species, ●—freshwater macrophytes, ■—mangroves, and ▲—seagrass with comparison from previous studies: [(a)]Seo et al. (2010), [(b)]Chao et al. (2000), Commercial paper (TAPPI standard) MD, CD using [1]Japanese method and [2]Western methods. *CD*, Cross direction; *MD*, machine direction.

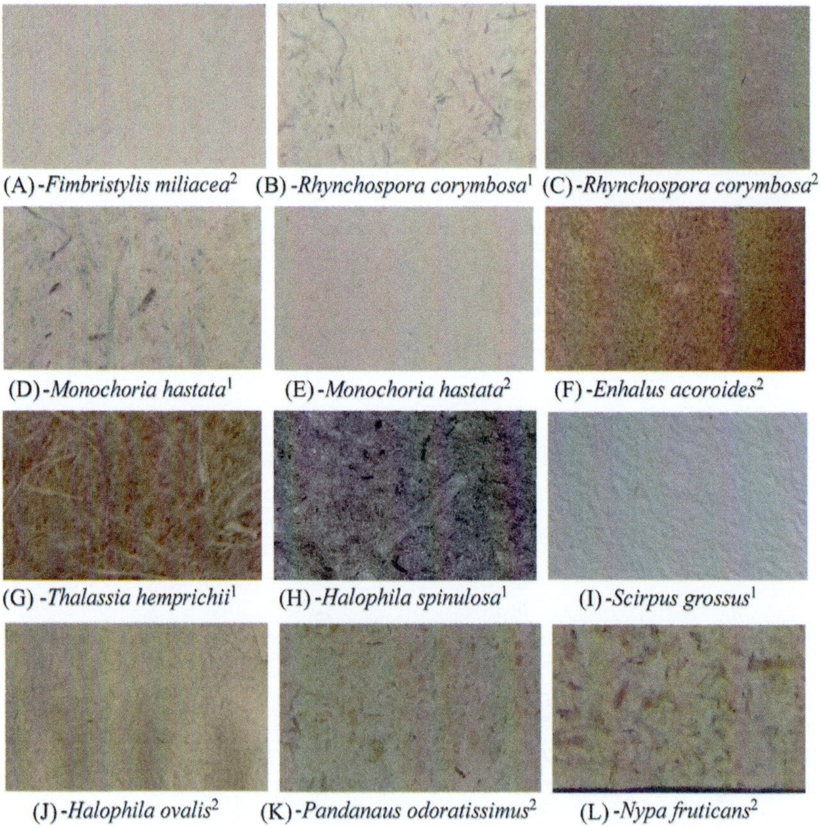

FIGURE 15.10 Examples of aquatic macrophyte paper sheet with equivalent TAPPI commercial papers' tensile strength (kN/m) and breaking length (mm). (A–F) Paper sheet produced by [1]Japanese sheet forming technique (hand-beaten fiber) and [2]Western sheet forming technique (blended fiber) tensile strength (kN/m) equivalent to offset paper type (MD) with standard tensile strength, 5.5 kN/m, (G–L) paper sheet produced by [1]Japanese sheet forming technique (hand-beaten fibers) and [2]Western sheet forming technique (blended fibers) equivalent to Offset paper (CD), Rag bond paper (MD and CD), and Newsprint paper (MD) with standard breaking length ranged 3.06 km–4.9 km. *CD*, Cross direction; *MD*, machine direction.

15.6 Conclusion

The comparative analysis of fiber dimension and chemical composition of aquatic macrophytes with those manufactured from traditional or other nonwood species tested for papermaking and production is very useful to obtain paper or paper products with predetermined characteristics.

Aquatic macrophytes, mentioned in this chapter, are with differences in fiber dimension, and chemical compositions (including poor chemical composition, e.g., low cellulose, hemicellulose, or a combination of both) are potentially utilizable for the production of printing and writing paper, fiber plate, rigid cardboard and cardboard, composites, paper and paperboard filter paper and boards, and composite materials.

The aquatic macrophyte handmade paper sheets are with properties as good as the TAPPI offset paper (MD) standard tensile strength (5.5 kN/m), offset paper (CD), Rag bond paper (MD and CD), and Newsprint paper (MD) standard breaking length ranged 3.06–4.9 km, and Newsprint (CD) standard tensile strength and breaking length of 0.9 kN/m and 1.84 km, respectively.

The effective use of nonwood fiber resources, for example, from aquatic macrophytes, available in Malaysia and many other countries as pulp for papermaking, promotes the concept of waste to valued products of a range of paper types or their derivatives. They can be diversified as subsidiary raw material fibers to softwood pulp and paper-related industry. The papermaking method uses readily available renewable resources, minimal chemicals and technology and is applicable to local community practice.

Acknowledgments

This research is funded by the Ministry of Science and Technology and Innovation Malaysia (MOSTI) under the Science Fund entitled Utilization of Aquatic Macrophytes for Papermaking (04-01-04-SF1184) and with the collaboration of the Japan Society for the Promotion of Science (JSPS) Asian CORE programme and the Establishment of Research Education Network on Coastal Marine Science in Southeast Asia (COMSEA).

References

Abobi, S.M., Ampofo-Yehoah, A., Kpodonu, T.A., Alhassan, E.H., Abarike, E.D., Atindaana, S.A., et al., 2015. Socio-ecological importance of aquatic macrophytes to some fishing communities in the Northern Region of Ghana. Elixir Biodivers 79, 30432–30437.

Agnihotri, S., Dutt, D., Tyagi, C.H., 2010. Complete characterization of bagasse of early species of *Saccharum officinarum*-Co 89003 for pulp and paper making. BioResources 5 (2), 1197–1214.

Akpakpan, A.E., Ukana, D.A., Obot, I.B., 2012. Evaluation of physicochemical properties and soda pulping of *Nypa fruticans* frond and petiole. Elixir Appl. Chem. 45, 7664–7668.

Amidon, T.S., 1981. Effect of the wood properties of hardwood on kraft paper properties. Tappi 64 (3), 123–126.

Area Maria, C., Popa Valentin, I., 2014. Wood Fibers for Papermaking. Smithers Rapra Technology Ltd.

Baki, B.B., 1982. Aquatic Weeds in Major Drainage and Irrigation Canals with Special Reference to Water Hyacinth in Malaysia. 1st Rice Advisory committee meeting.

Bakker, R.R., Jenkins, B.M., 2003. Feasibility of collecting naturally leached rice straw for thermal conversion. Biomass Bioenergy 25 (6), 597–614. Available from: https://doi.org/10.1016/S0961-9534(03)00053-9.

Bektaş, İ., Tutuş, A., Eroğlu, H., 1999. A study of the suitability of Calabrian pine (*Pinus brutia* Ten.) for pulp and paper manufacture. Turkish J Agric. For. 23 (EK3), 589–598.

Bidin, N., Muta Harah, Z., Japar Sidik, B., Nur Aznadia, A.A., 2015. Suitability of aquatic plant fibers for handmade papermaking. Int. J. Polym. Sci. 2015, 1–9. Available from: https://doi.org/10.1155/2015/165868.

Biermann, C.J., 1996. Handbook of Pulping and Papermaking. Elsevier, Amsterdam.

Brindha, D., Vinodhini, S., Alarmelumangai, K., 2012. Fiber dimension and chemical contents of fiber from *Passiflora foetida* L. and their suitability in paper production. J. Sci. Res. Rep. 2 (3), 210–219.

Brindha, D., Vinodhini, S., Alarmelumangai, K., 2013. Extraction and characterization of fiber from three plant species of the genus *Cleome*, L. Asian J. Exp. Biol. Sci. 4 (1), 69–73.

Chao, K.P., Su, Y.C., Chen, C.S., 2000. Feasibility of utilizing *Rhizoclonium* in pulping and papermaking. J. Appl. Technol. 12, 53–62.

Cordeiro, N., Belgacem, M.N., Torres, I.C., Moura, J., 2004. Chemical composition and pulping of banana pseudo-stems. Ind. Crop Prod. 19 (2), 147–154. Available from: https://doi.org/10.1016/J.INDCROP.2003.09.001.

Daud, Z., Mohd Zainuri, M.H., Angzzas, S.M.K., Halizah, A., Aripin, A.M., 2014. Exploring of agro waste (pineapple leaf, corn stalk, and Napier grass) by chemical composition and morphological study. BioResources 9 (1), 872–880.

Emmclan, L.S.H., Muta Harah, Z., Japar Sidik, B., 2018. Utilization of aquatic weeds fibers for handmade papermaking. BioResources 13 (3), 5684–5701.

Enayati, A.A., Hamzeh, Y., Mirshokraiei, S.A., Molaii, M., 2009. Papermaking potential of canola stalks. BioResources 4 (1), 245–256.

Eroglu, H., 1980. Investigating Possibilities of Obtaining Wood Pulp from Wheat Straw by O_2^- NaOH Method (PhD dissertation). Karadeniz Technical University.

Ezeibekwe, L.O., Ezenwaka, C.L., Mbagwu, F.N., Unamba, C.L.N., 2009. Effects of combination of different levels of auxin (NAA) and cytokinin (BAP) on in vino propagation of *Dioscorea rotundata* L. (white yam). J. Mol. Genet. 1 (2–4), 18–22.

Goswami, T., Saikia, C.N., 1994. Water hyacinth—a potential source of raw material for greaseproof paper. Bioresour. Technol. 50 (3), 235–238. Available from: https://doi.org/10.1016/0960-8524(94)90095-7.

Goswami, T., Kalita, D., Rao, P.G., 2008. Greaseproof paper from banana (*Musa paradisiaca* L.) pulp fibre. Indian J. Chem. Technol. 15, 457–461.

Han, J.S., Rowell, J., 1997. Chemical composition of agro-based fibers. In: Rowell, R.M., Young, R.A., Rowell, J.K. (Eds.), Paper and Composites from Agrobased Resources. CRC Lewis Publishers, New York, pp. 83–176.

Hattaka, A., Mettala, A., Harkonen, T., Paavilainen, L., 1996. Biopulping of Gramineous Plants by White-rot Fungi. In: International Conference; 6th, Biotechnology in the Pulp and Paper Industry, Vienna, Austria, pp. 229–232.

Hemmasi, A.H., Samariha, A., Tabei, A., Nemati, M., Khakifirooz, A., 2011. Study of morphological and chemical composition of fibers from Iranian sugarcane bagasse. Am.-Eurasian J. Agric. Environ. Sci. 11 (4), 478–481.

Hiebert, H., 1998. Papermaking with Plants: Creative Recipes and Projects Using Herbs, Flowers, Grasses, and Leaves. Storey Publishing LLC, North Adams.

Hiebert, H., 2000. The Papermaker's Companion: The Ultimate Guide to Making and Using Handmade Paper. Storey Publishing LLC, United States.

Hiebert, H., 2006. Papermaking with Garden Plants and Common Weeds. Storey Publishing LLC, North Adams.

Hosseinpour, R., Fatehi, P., Latibari, A.J., Ni, Y., Sepiddehdam, S.J., 2010. Canola straw chemimechanical pulping for pulp and paper production. Bioresour. Technol. 101 (11), 4193–4197. Available from: https://doi.org/10.1016/j.biortech.2010.01.055.

Hubbe, M.A., Bowden, C., 2009. Handmade paper: a review of its history, craft, and science. BioResources 4 (4), 1–2. Available from: https://doi.org/10.15376/biores.4.4.1736-1792.

Hurter, R.W., 1997. Nonwood plant fiber characteristics. Agric. Residues 1, 1–4.

Hurter, R.W., Eng, P., 2001. Sisal fibre: market opportunities in the pulp & paper industry. In: FAO/CFC Technical Paper 14 (2001): 61–74.

Hurter, R.W., Riccio, F.A., 1998. Why CEOS don't want to hear about nonwood or should they? In: TAPPI Proceedings, TAPPI Press, Atlanta, GA, pp. 1–11.

IAWA, 1937. Standard terms of length of vessel membranes of wood fiber. Tropical Woods 51 (21), 26–62.

Janson, J., Jousimaa, T., Huppa, M., Backman, R., 1996. Phospate-based pulping of agrofibre including papermaking and spent liquor recovery. Nordic Pulp Pap. Res. J. 111996, 1–6.

Jimenez, L., Rodriguez, A., Perez, A., Moral, A., Serrano, L., 2008. Alternative raw materials and pulping process using clean technologies. Ind. Crop Prod. 28, 11–16.

Kaur, H., Dutt, D., 2013. Anatomical, morphological and chemical characterization of lignocellulosic y-products of lemon and sofia grasses obtained after recuperation of essential oils by steam distillation. Cellulose Chem. Technol. 47 (1–2), 83–94.

Khalil, H.P., Abdul, S., Alwani, M.S., Mohd Omar, A.K., 2011. Chemical composition, anatomy, lignin distribution, and cell wall structure of Malaysian plant waste fibers. BioResources 1 (2), 220–232. Available from: https://doi.org/10.15376/biores.1.2.220-232.

Kiaei, M., Samariha, A., Kasmani, J.E., 2011. Characterization of biometry and the chemical and morphological properties of fibers from bagasse, corn, sunflower, rice and rapeseed residues in Iran. Afr. J. Agric. Res. 6 (16), 3762–3767. Available from: https://doi.org/10.5897/AJAR10.752.

Madakadze, I.C., Radiotis, T., Li, J., Goel, K., Smith, D.L., 1999. Kraft pulping characteristics and pulp properties of warm season grasses. Bioresour. Technol. 69 (1), 75–85. Available from: https://doi.org/10.1016/s0960-8524(98)00131-x.

Mashhor, M., 1996. Noxious floating weeds of Malaysia. Hydrobiologia 340 (1), 121–125. Available from: https://doi.org/10.1007/bf00012743.

Metcalfe, C.R., Chalk, L., 1983. Anatomy of the Dicotyledons. II. Leaves, Stem, and Wood in Relation to Taxonomy with Notes on Economic Uses. Clarendon Press, Oxford.

Moubasher, A.H., Abdel-Hafez, S.I.I., Abdel-Fattah, H.M., Moharram, A.M., 1984. Fungi of wheat and broad-bean straw composts II. Thermophilic fungi. Mycopathologia 84, 65–71. Available from: https://doi.org/10.1007/bf00436514.

Muta Harah, Japar Sidik, 2014–2018 (Unpublished). Utilization of Aquatic Macrophytes for Papermaking (04-01-04-SF1184). Science Fund. Ministry of Science and Technology and Innovation Malaysia (MOSTI).

Muta Harah, Z., Japar Sidik, B., Suzalina Akma, A., 2014. Aquatic macrophytes and their values. In: Amin, S.M.N., Kamaruddin, M.S., Arshad, A., Romano, N. (Eds.), Perspective of

Fisheries and Aquaculture in Malaysia. Universiti Putra Malaysia Press, Serdang, Selangor, pp. 84–104.

Nather Khan, I.S.A., 1990. Socio-Economic Values of Aquatic Plants (Freshwater Macrophytes) of Peninsular Malaysia. Asian Wetland Bureau, Kuala Lumpur.

Nieschlag, N.J., Nelson, G.H., Wolf, I.A., Perdue, R.E., 1960. A search for new fiber crops. Tappi 43 (3), 193–201.

Ogbonnaya, C.I., Roy-Macauley, H., Nwalozie, M.C., Annerose, D.J.M., 1997. Physical and histochemical properties of kenaf (*Hibiscus cannabinus* L.) grown under water deficit on a sandy soil. Ind. Crop Prod. 7 (1), 9–18. Available from: https://doi.org/10.1016/S0926-6690(97)00034-4.

Ohshima, J., Yokota, S., Yoshizawa, N., Ona, T., 2005. Examination of within-tree variations and the heights representing whole-tree values of derived wood properties for quasi-non-destructive breeding of *Eucalyptus camaldulensis* and *Eucalyptus globulus* as quality pulpwood. J. Wood Sci. 51 (2), 102–111. Available from: https://doi.org/10.1007/s10086-004-0625-3.

Oinonen, H., Koskivirta, M., 1999. Special challenges of pulp and paper industry in Asian populated countries, like Indian Sub-Continent and China. In: Proceedings of the Paperex: 49–68. Proceedings of the Paperex 99-4th: International Conference on Pulp and Paper Industry: Emerging Technologies in the Pulp and Paper Industry, Delhi, India, December 14–16, 1999.

Pan, X., Gilkes, N., Kadla, J., Pye, K., Saka, S., Gregg, D., et al., 2006. Bioconversion of hybrid poplar to ethanol and co-products using an organic solvent fractionation process: optimization of process yields. Biotechnol. Bioeng. 94, 851–861. Available from: https://doi.org/10.1002/bit.20905.

Poovarasi, G., Muta Harah, Z., Japar Sidik, B., 2014. Fiber characteristics in tropical aquatic plants for handmade papermaking. Bull Pure Appl. Sci.-Bot. 33 (1&2), 1–12. Available from: https://doi.org/10.5958/2320-3196.2014.00001.9.

Radiotis, T., Jian, L., Goel, K., Eisner, R., 1999. Fiber characteristics, pulpability, and bleachability of switchgrass. Tappi J. 82 (7), 100–105.

Ralph, J., MacKay, J.J., Hatfield, R.D., O'Malley, D.O., Whetten, R.W., Sederof, R.R., 1997. Abnormal lignin in a loblolly pine mutant. Science 277 (5323), 235–239. Available from: https://doi.org/10.1126/science.277.5323.235.

Saheb, D.N., Jog, J.P., 1999. Natural fiber polymer composites: a review. advances in polymer technology. J. Polym. Process Inst. 18 (4), 351–363.

Said, I.M., 1990. A Preliminary List of Wetland Plant Species of Peninsula Malaysia, with Particular Reference to Those Having Socio-Economic Value. Asian Wetland Bureau, Kuala Lumpur.

Saijonkari-Pahkala, K., 2008. Non-wood plants as raw material for pulp and paper. Agric. Food Sci. Finl. 10, 11–101. Available from: https://doi.org/10.23986/afsci.5707.

Seo, Y.B., Lee, Y.W., Lee, C.H., You, H.C., 2010. Red algae and their use in papermaking. Bioresour. Technol. 101, 2549–2553. Available from: https://doi.org/10.1016/j.biortech.2009.11.088.

Seth, R.S., Page, D.H., 1998. Fiber properties and tearing resistance. Tappi J. 71 (2), 103–107.

Shakhes, J., Marandi, M.A., Zeinaly, F., Saraian, A., Saghafi, T., 2011a. Tobacco residuals as promising lignocellulosic materials for pulp and paper industry. BioResources 6 (4), 4481–4493.

Shakhes, J., Zeinaly, F., Marandi, M.A., Saghafi, T., 2011b. The effects of processing variables on the soda and soda-AQ pulping of kenaf bast fiber. BioResources 6 (4), 4626–4639.

Sharma, M., Sharma, C.L., Kumar, Y.B., 2013. Evaluation of fiber characteristics in some seed of Arunachal Pradesh, India for pulp and paper making. Res. J. Agric. For. Sci. 1 (3), 15–21.

Shatalov, A.A., Quilho, T., Pereira, H., 2001. *Arundo donax* L. Reed. New perspectives for pulping and papermaking: I raw materials characterization. Tappi J. 84 (1), 45–61.

Silveira, M.L., Furlan, S.A., Ninow, J.L., 2008. Development of an alternative technology for the oyster mushroom production /using liquid inoculum. Food Sci. Technol. (Camp.) 28 (4), 858–862. Available from: https://doi.org/10.1590/s0101-20612008000400014.

Singh, S.V., Rai, A.K., Dhawan, R., 1991. Advances in Pulp and Paper Research in India. SRO, FRI, Dehradun.

Syed, N.N.F., Muta Harah, Z., Japar Sidik, B., 2016. Fiber characteristics and papermaking of seagrass using hand-beaten and blended pulp. BioResources 11 (2), 5358–5380. Available from: https://doi.org/10.15376/biores.11.2.5358-5380.

T 205 sp-02. 2006. Forming Handsheets for Physical Test of Pulp. Atlanta, GA: TAPPI Press, reaffirmed July 11, 2006.

T 227 om-94. 1994. Freeness of Pulp (Canadian Standard Method). Atlanta, GA: TAPPI Press, revised in 1994.

T 412 om-06. Moisture in Pulp, Paper and Paperboard. Atlanta, GA: TAPPI Press, revised January 1, 2011.

T 404 cm-92. 1992. Tensile Breaking Length and Elongation of Paper and Paperboard (Using Pendulum-Type Tester). Atlanta, GA: TAPPI Press, revised in 1992.

T 494 om-01. 2006. Tensile Properties of Paper and Paperboard (Using Rate of Elongation Apparatus). Atlanta, GA: TAPPI Press, approved June 13, 2006.

Thongpukdee, A., Thepsithar, C., Timchookul, S., 2013. Fiber microstructure in *Solanum* found in Thailand. Int. J. Bioeng. Life Sci. 7 (8), 793–796.

Tofanica, B.M., Cappelletto, E., Gavrilescu, D., Mueller, K., 2011. Properties of rapeseed (*Brassica napus*) stalks fibers. J. Nat. Fibers 8 (4), 241–262. Available from: https://doi.org/10.1080/15440478.2011.626189.

Ververis, C., Georhiou, K., Christodoulakis, N., Santas, P., Santas, R., 2004. Fiber dimensions, lignin and cellulose content of various plant materials and their suitability for paper production. J. Ind. Crop Prod. 19, 245–254. Available from: https://doi.org/10.1016/j.indcrop.2003.10.006.

Wahab, R., Mustafa, M.T., Rahman, S., Salam, S., Sulaiman, M.A., Sudin, O.M., et al., 2006. Relationship between physical, anatomical and strength properties of 3-year-old cultivated tropical bamboo *Gigantochloa scortechinii*. J. Agric. Biol. Sci. 7 (10), 782–791. Available from: https://doi.org/10.36347/sajb.2020.v08i02.003.

Xu, F., Zhonga, X.C., Sunb, R.C., Lu, Q., 2006. Anatomy, ultrastructure and lignin distribution in cell wall of *Caragana korshinkii*. Ind. Crop Prod. 24, 186–193. Available from: https://doi.org/10.1016/j.indcrop.2006.04.002.

Young, J.H., 1981. Fiber preparation and approach flow (papermaking process, pulp). Pulp Pap. Chem. Technol. 2, 821–913.

Chapter 16

Blending and beating of nonwood fibers

Z.M.A. Ainun[1], J.G. Boon[2], S. Amran[3], S. Zakiah[1] and R.A. Ilyas[4]
[1]Program of Pulp and Paper & Pollution Control, Institute of Tropical Forestry and Forest Products (INTROP), Universiti Putra Malaysia, Serdang, Selangor, Malaysia, [2]Department of Bio and Natural Resources Technology, Faculty of Bioengineering and Technology, Universiti Malaysia Kelantan (UMK) Kampus Jeli, Jeli, Kelantan, Malaysia, [3]Department of Wood Industry, Faculty of Applied Sciences, Universiti Teknologi MARA, Shah Alam, Selangor, Malaysia, [4]Department of Chemical Engineering, Faculty of Chemical and Energy Engineering, Universiti Teknologi Malaysia, Johor Bahru, Malaysia

16.1 Introduction

Strength of paper is very crucial in the papermaking industry because strength affects the value and price of a paper product. Hence, the correlation of factors that contributes to paper strength is really taken care of (Małachowska et al., 2021), such as beating and blending. In other words, the purpose of blending and beating in producing papers either uses woody or nonwoody materials remains the same, to increase the paper strength. In principle, blending means the mixing of at least two or more types of raw materials to increase the strength of initial fibers. Typically, the initial fibers are categorized as having shorter fibers than the fibers to be mixed. With this, it is expected that the longer fibers will increase the mechanical strength of the resulting paper. In pulp and paper technology the terms "beating" and "refining" refer to the same process. The term "beating" is used since the early days of papermaking where the raw material is beaten with a stick to flatten and separate the fibers manually (Lumiainen, 2000). Beating or refining is one of the mandatory processes in the mill, which involves mechanical action against the surface of fibers using a rotor in presence of water. Beating is carried out at stock preparation stage in the mill. The flow of papermaking process is shown in Fig. 16.1.

Pulping and Papermaking of Nonwood Plant Fibers. DOI: https://doi.org/10.1016/B978-0-323-91625-7.00004-7

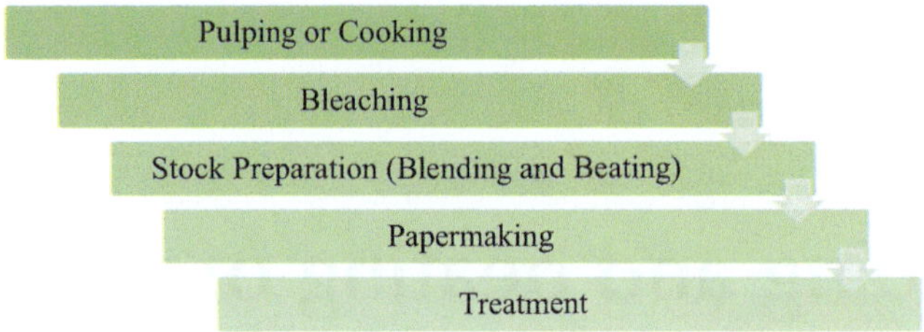

FIGURE 16.1 The beating or refining process appears after blending process in pulp and papermaking.

16.2 Blending process of nonwoody plant fibers

The main objective of blending is to obtain better properties of produced paper. Pulp and papermakers will apply blending process by adding other types of pulps into their main pulp slurries in the factory. Blending process is also known as the mixing process in stock preparation. The blending depends on pulp slurry condition; sometimes nonwood fibers become the strengthening fibers, but in some cases nonwood fibers need other fibers to reach certain strength of produced paper. But there are also cases where nonwoody fibers act as strengthening fibers to the other nonwoody fibers. To relate, a study conducted by Megra et al. (2022), who blended *Melia azedarach* with 1%–4% cotton waste (nonwoody fiber), obtained results that showed better physical strength of papers such as tear, tensile, and burst indices. In addition, Bhardwaj et al. (2019) mixed hardwood and agro pulp (nonwood pulp) to enhance the properties of paper with a proportion of 15% and 85% of hardwood to agro pulp achieving good-quality paper. In another study, Shaikh (1990) found out a significant improvement in paper properties by adding an amount of cotton-stalk pulp (nonwoody fiber) into bagasse pulp (nonwoody fiber). The breaking length, burst, tear, and folding properties are higher with the increase of cotton-stalk pulp. At the same time, the bulk density and porosity get lower. Hence, blending is a process to obtain better properties of paper; however, it is not mended to modify the flexibility of fibers like beating process (Syed et al., 2016a,b). Nasroun (2019) prepared a blend of three nonwoody fibers of kenaf bast, kenaf core, and bagasse fibers at 60:20:20, but the results showed decrement of tensile index from 79.5 to 58.8 Nm/g; however, after beating process, the strength increased to 94.1 Nm/g. Besides targeting for better properties, blending process decreases the factory cost (Pati and Vrat, 2010).

16.3 Beating process of nonwoody plant fibers

Beating is a mechanical action enforced on the outer surface of fibers. By this enforcement the fiber surface areas are open with more areas acceptable for bonding opportunities. Refining increases pulp specific surface area and fiber surface charge that provide good fiber bonding. Why does a beaten fiber surface have high chances and spacious for bonding? The outer surface of the beaten fiber experiences fibrillations whereby these new surfaces provide a new bonding space. Beaten pulp undergoes a variety of changes in terms of physical and electrokinetic properties. The physical properties of pulps are fiber length, fiber coarseness, fiber curl, fibrillations, fine formation, crystallinity index while the electrokinetic properties consist of surface and total charges of pulp. Increasing the degree of beating or refining can increase the surface charge of fibers (Bhardwaj et al., 2004) but not the total charge which remained constant. Among all pulp properties, pulp freeness is truly synonymous with beating. Bhagwat et al. (2020) conducted a comprehensive study on pulp properties of reinforced cotton rag upon beating time that linear relationship is obtained of decreasing trend for Canadian Standard Freeness versus beating level.

Is it true that beating is essential for all types of fibers in the processing of papermaking? A study has found that, in some instances of fiber condition, refining has no significant impact on pulp and paper properties as well. Goel et al. (2000) investigated those unbeaten chemical bleached pulps of wheat straw for both soda-anthraquinone (soda-AQ) and organosolv pulps response moderately to the beating process compared to woody materials of softwoods and hardwoods in which no external fibrillation is observed. In addition, at the same freeness readings, the organosolv and soda–AQ pulps have equivalent tensile strength as hardwood and softwood pulps accordingly.

The relationship between beatings, fiber properties, and paper properties are often discussed by the researchers. Typically, bleaching is carried out before beating. Is nonwoody material easier to beat than woody material? There are several studies that found out nonwood materials are more beatenable compared to woody materials. A study by Guo et al. (2009) obtained that wheat straw (nonwoody fibers) is more responsive to the mechanical action of beating compared to eucalyptus pulps. This is supported by the findings of morphological and ultrastructural differences of pulps as well as fines and xylan content. Fines content between 35% and 47% leads to the improvement of pulp drainability. The fines are observed by the presence of vessels and epidermal cells, while eucalyptus that produced 8%–10% fines only consists of broken fiber parts from the cell wall.

Processing nonwood materials is very different from woody materials due to their physical and chemical compositions. The difference starts from the process of preparing raw materials until the beating or refining process. Refining affects most properties of pulp or fiber, such as freeness, specific surface area, specific volume, surface charge, total charge, and even the properties of paper. Freeness readings acceptable in papermaking are at 225–430 mL of the CSF. This is because, at this range, the pulp approach flow, and the wet end paper manufacturing operation, is excellent (Naik et al., 2009). The equipment used to determine the properties of pulp before and after refining are shown in Table 16.1.

16.4 Importance of blending and beating process in papermaking industry of nonwood fibers and recent technologies

Paper made from nonwood fiber are available in small portions in the world market. According to Abd El-Sayed et al. (2020), 92% of world production still depends on wood. However, nonwood fiber has become more important, and in some countries, like India and China, nonwood pulp is the leading primary pulp source (Qian, 2021). Nonwoody fiber is used in China for last 2000 years for papermaking before the Industrial Revolution started to use woody material (Azeez, 2018). The common nonwoody fiber exclusively for paper manufacturing includes abaca (Waller and Wilsby, 2019), bagasse (Ativo and Pintor, 2022; Haile et al., 2021), bamboo (Liu et al., 2018), cotton, flax (Haile et al., 2021), hemp (Ahmed et al., 2022), jute (Akter et al., 2020), reed (Haile et al., 2021), sisal (Junle, 2021), and straw (Malik et al., 2020). The product range made from paper blended with nonwood fiber can be the printing and writing paper (Mari et al., 2019), molded pulp packaging (Rattanawongkun et al., 2020), tissue (An et al., 2020), paperboard (Laukala et al., 2019), and specialty papers.

Blending and beating are essential to accommodate nonwood fiber for the papermaking process especially at an industrial scale that can alter and

TABLE 16.1 List of pulp and paper properties with their tested equipment.

Pulp properties	Equipment
Freeness	Canadian Standard Freeness
Specific surface area	Permeability tester
Specific volume	Permeability tester
WRV	Centrifuge method
Surface charge	Conductometric titration/Particle Charge Detector using polyDADMAC (polydiallyldimethylammonium chloride) (polyelectrolyte titration)
Total fiber charge (not affected by refining)	Conductometric titration

WRV, Water retention value.

manipulate the physical properties of fiber in the pulp stock prior to papermaking. The differences in fiber's physical properties or fiber dimension, such as fiber length, cell wall thickness, and diameter, provide different properties to the paper. Unlike wood fiber, nonwood plant has a wider range of dimensions. Fiber properties have great relationship with pulp properties (Ferdous et al., 2021). According to the data published by Isenberg (1967), the fiber length of some common nonwood fiber can be as short as 0.2 mm and as long as 250 mm. The diameter of the nonwood fiber is also found in a wide range among which rice straw fiber is one of the thin fiber with the reported fiber diameter at an average 8 μm, while kenaf exhibits a large diameter at an average of 22 μm (Ververis et al., 2004). From the fiber length aspect, either long or short, both fibers are essential in paper manufacturing with their own roles. The presence of longer fiber may aid the strength of shorter fiber papermaking, depending on optimum proportion between two types of fibers. In other words, pulp blending of different fiber lengths can provide paper with better properties. Short fiber can provide a better fitment among fiber and thus provides better fiber–fiber bonding. The good formation will attribute to the printability and opacity of paper as well. Meanwhile, longer fiber works as the core of the mechanical properties and contributes strength and strong matrix to the paper. Besides the paper property aspects, too much short fiber in pulp furnish will create low freeness and unsuitable for high-speed paper manufacturing. The failure of removing water during forming will cause the defect during pressing as the paper web contains higher moisture content than the acceptable range.

The paper manufacturing using nonwood fiber can be solely made from nonwood fiber but foremost are blended with wood fiber. The portion of nonwood fiber will depend on the required quality and type of paper. Depending on the requirement, the manufacturer can customize the mixing ratio of the nonwood plant fiber according to the fiber length. The common long fibers from nonwood plant used in paper manufacturing include flax, hemp, and kenaf, while the short fiber from nonwood plant includes straw, bagasse, and oil palm empty fruit bunch (EFB) as shown in Table 16.2.

The central reason for blending nonwood fiber with wood fiber rather than solely using woody fiber is the continually increasing wood price due to the shortage of wood as the feedstock (Gajdačová et al., 2018). However, still no doubt wood is the best feedstock for papermaking (Deykun et al., 2018). On the other hand, the comparable properties of nonwood fiber in terms of physical, mechanical, and optical properties made them a perfect alternative to the wood fiber. The substitution of woody fiber with 10%–15% of nonwood fiber will not significantly affect paper strength, optical, and surface properties.

For recycled paper manufacturing, nonwood fiber can blend with secondary fiber, to restore the loss of fiber during the secondary fiber recovery. Blending the secondary fiber with nonwood fiber certainly will improve the

TABLE 16.2 Fiber length of some nonwood fibers.

Common name	Scientific name	Fiber length (mm)	References
Abaca	*Musa textilis*	3.31–6.92	Mari et al. (2019)
Bamboo	*Gigantochloa apus*	30	Dwiky et al. (2019)
Corn stalk	*Zea mays* L.	0.95	Rodriguez et al. (2010)
Cotton lint	*Gossypium* spp.	250–300	Morais et al. (2020)
Esparto	*Stipa tenacissima*	400–600	El-Abbassi et al. (2019)
Flax	*Linum usitatissimum*	25	Goudenhooft et al. (2019)
Hemp	*Cannabis sativa*	100–1100	Musio et al. (2018)
Jute	*Corchorus capsularis*	10–20	Bheel et al. (2021)
Kapok	*Ceiba pentandra*	16.3	Purnawati et al. (2018)
Kenaf	*Hibiscus cannabinus*	150–400	Noor Abbas et al. (2022)
Mulberry	*Broussonetia papyrifera*	10–40	Patxullayev et al. (2020)
Oil palm empty fruit bunch		1.27–1.17	Sumardi et al. (2020)
Papyrus	*Cyperus papyrus*	0.41–0.93	Zahran and El-Amier (2014)
Pineapple leave		914.4	Todkar and Patil (2019)
Ramie	*Boehmeria nivea*	20	Yu et al. (2019)
Raphia	*Raphia hookeri*	300–500	Oliveira Filho et al. (2020)
Reed bagasse		15	Shon et al. (2018)
Rice straw		0.843	Kiaei (2014)
Rosselle		2400	Tamta and Kalita (2020)
Sarkanda		0.939	Sharma et al. (2020)
Straw		30–50	Bouasker et al. (2014)

(*Continued*)

TABLE 16.2 (Continued)

Common name	Scientific name	Fiber length (mm)	References
Sugarcane bagasse	*Saccharum officinarum* L.	1.86	Mamaye et al. (2020)
Wheat straw		0.818	Tripathi et al. (2013)

strength properties of paper products. It helps in regaining the bonding properties of the secondary fiber by mixing it with nonwood fiber. The secondary fiber strength is relatively weak compared to virgin fiber, due to the various degradation on the paper before the recycling and the disintegration force that further lead to the damage on the fiber. Adding the nonwood virgin fiber to the recycled paper helps in good papermaking as it essentially increases the amount of strong fiber in the paper web and thus provides the paper with better mechanical properties.

Besides fiber length, as mentioned earlier, cell wall thickness is also one of the crucial factors in paper properties. Fiber before refining exhibits a smooth surface and high elasticity, and together with the crowdedness of long fiber will cause the paper very bulky and poor in strength. Refining is a necessary treatment to alter the fiber cell wall that is too stiff and smooth for papermaking. Refining involves two actions on fiber: fiber cutting and fiber fibrillation. Fiber cutting aims to trim down the long fiber and provides better formation during papermaking. However, the cutting mechanism should be controlled at minimum to preserve the paper's tensile strength and tear-resistant resistance. Meanwhile, fibrillation will rupture the fiber and improve the flexibility. Besides, the external fibrillation and cutting action during refining process will create active fine on the fiber surface. The active fine consists of short and detached fragments of primary and secondary walls of fiber. The active fine formation and the larger surface area of fiber after treatment can improve the fiber–fiber bonding. Compared to wood fibers, fibers of nonwood plant are more responsive to refining treatment due to the differences in their morphological properties and ultrastructure (Guo et al., 2009). Hence, the refining on nonwood fiber should be milder compared to wood fiber.

The basis of beating is to allow fibers to be treated between stator and rotor plate of refiner. The compression and shearing forces occur on the fiber that are trapped in between the bars and are responsible for changing the fiber structure and morphology. Compared to the laboratory refiner, the industrial refiner can treat the pulp in larger quantities and continually. It usually comes with certain flexibility in changing the stator plate with different patterns to provide different grooves and bar dimensions. The stator plate with a small angle, small bar width, and sharp bar edges cuts the fiber. In contrast, stator bars with large angles tend to offer fibrillation (Holik, 2006).

The industrial refiner commonly found today comes in different geometries such as disk, conical, and cylindrical. The disk refiners can be found in single disk, double disk, and multidisk types. They are favorable with high consistency operation that tend to fiber fibrillation over fiber cutting. The standard filling on the refiner stator plate usually is comprehensive to handle all ranges of fiber. In some cases, all range fillings may not be the best option to optimize the treatment. Some manufacturers such as Voith offered the Pluralis Line fillings specifically tailored to the different range of fiber, including super short fiber, short fiber, intermediate fiber, long fiber, OCC secondary fiber, and coarse fiber, respectively.

For conical refiners, Jordan refiner and Claflin refiner are the common types that can be found in industries. Jordan type is equipped with low-angle stator (11–16 degrees) and is tailored for low-consistency refining. Meanwhile, Claflin is equipped with high-angle stator, the taper degree offered by the modern Claflin refiners, such as Bematec and ZDJ, is commonly found in 60 degrees. Besides Jordan refiner and Claflin refiner, there are also other modern conical refiners. Valmet has claimed that the OptiFiner Conflo conical refiners made by it are more energy-saving and provide precise gap control. They also have a wide range of fillings to allow paper manufacturers to optimize the fiber properties. With the optional addition of turbine housing, Valmet claimed this to be upgradeable and helps improve the production by 20%–40% and allows the refiner to install fillings with the denser pattern that can refine more gently and extend the service life fillings up to 25%. The cylindrical refiner is the latest modern refiner developed by Andritz. The cylindrical refining zone offers gentle and stable fiber treatment. In addition, the small diameter of the refiner reduces the net refining energy and no-load energy.

Due to the energy cost becoming the primary concern in manufacturing industries, energy consumption has become a crucial element in developing industrial refiner. The refining process, especially at the industrial scale, is a highly energy-consuming process to drive the rotor moving against the friction among the rotor, stator, and fiber. The no-load energy and net refining energy have become the competition in developing new refiners. Another crucial element in developing an industrial refiner is the homogeneity of treatment (Bajpai, 2010). It is almost impossible to exteriorly control the movement in the refiner during the refining process. The conventional method and refiner produced highly untreated fiber. To reduce the untreated fiber, several concepts, including multiple disk refining and low-consistency refining with stable gap refining, are introduced to improve the situation. The improvement in fillings and the plates' design development offers a better adjustable operation gap, more extended filling zone, and longer service life.

16.5 Future technologies in blending and beating process of nonwood fibers

In papermaking industries, paper strength is the most crucial property to be taken care of due to its capability of disturbing the smoothness of mill operation and affecting the mill operation cost. Hence, pulp and paper millers will do their best to make sure that the strength of paper is controlled in many ways, including by applying beating and blending processes. Beating is a process that is normally carried out after pulping process prior to papermaking process.

The aim of beating process is to improve paper strength by obtaining more refined fibers that cause decrement in the porosity and roughness of produced paper (CNBM Paper Pulp, 2020). As we know, pulp beating is the mechanical process of pulp between two surfaces of constant loading that moves inversely to separate the fibers to achieve the desired requirements of quality paper (TAPPI, 2000). Different beating revolutions can affect the mechanical properties of paper such as tensile, tear, and burst indices. Beating revolutions up to 7000 showed that tensile, tear, and burst indices of sugarcane bagasse paper increase; however, the strength started to decrease after the beating process of more than 7000 revolutions. Another important process to enhance pulp and paper strength is blending process. Pulp blending is also one of the procedures to improve the quality of paper by mixing more than one type of fibers. Pulp blending can be divided into three methods: chips blending, pulp blending before beating that would slightly improve physical strength of paper, and pulp blending after beating that results in excellent paper strength properties (Nandkumar, 2009). Studies conducted by Rafidah et al. (2017) and Nur Aqeela et al. (2018) applied blending process of kenaf–oil palm EFB and citronella–oil palm EFB beaten pulps accordingly. They proved that blending the pulps resulted in good mechanical paper properties such as tensile and tearing indices for either chemical or mechanical pulps.

In pulp and paper industries, high production of paper has led to high order of pulp sources like wood. The slow growth of woody plants has made it difficult for the industries to keep up with current demands. Currently, many countries have been utilizing nonwood plant fibers, such as agricultural residues, for paper production particularly countries that lack forest area. In addition, increasing price of woods turns out to be a problem for paper industries. China, India, and Pakistan are the top three countries that use the most nonwood plant fibers with more than 77,000, 2000, and 400 thousand metric tonnes, respectively (Abd El-Sayed et al., 2020). Besides massive production, nonwood fibers have been proved to be great alternatives, like banana, kenaf, bamboo, and kapok, which are stronger than wood fibers and have longer fibers for effective fiber binding. Blended softwood with kapok pulps exhibits higher tensile and burst strength than kapok with hardwood pulps due to kapok's longer fibers (Chaiarrekij et al., 2012). Since 2000, National Kenaf and Tobacco Board (NKTB) has dedicatedly led the development of kenaf industry in Malaysia for products like paper, composites, and textiles (Saleh et al., 2012). High felting power and ease-deformation have placed kenaf fibers as a good raw material to produce quality printing paper as it has more interfiber contacts during papermaking.

Blending that involves wood and nonwood fibers is the mostly used approach that has been applied in papermaking production focally to enhance the physical and mechanical properties of paper. By looking at the demand for fibers greatly increasing, it is necessary to reduce the dependency on wood; hence, blending of different types of abundantly nonwood fibers is practiced. Study conducted by Rafidah et al. (2017) showed that the whole stem of kenaf blended with oil palm EFB has improved mechanical properties of paper and meets the standard requirement grade of paper for writing and printing. They used different pulping methods that are kraft pulping for kenaf and soda–AQ pulping for oil palm EFB. The pulps are disintegrated for 20 minutes and beaten up to 4000 revolutions separately. The pulp weight of kenaf to EFB is measured to redisintegrate for pulp blending, and pulp freeness has shown that more water is held by blended pulp.

In another case, nanofibrillated cellulose (NFC) has been used as an additive to increase strength properties of paper. In the study conducted by Kumar et al. (2016), three types of fibers are used to obtain blended NFC that must be added to the beaten and unbeaten pulps. The fibers were beaten at two levels of freeness: 32 and 90°SR. The study conducted showed that more revolutions are needed for softwood; 10,000 and 125,000 revolutions than hardwood and bagasse that only needs 3500 and 100,000 revolutions, respectively, because of the lower hemicellulose content in softwood pulp. Before NFC was added, the beaten pulps have shown higher mechanical properties with increasing beating intensity as compared to unbeaten pulps. Increasing beating revolutions have resulted in enhancing the hydrogen bonding capacity in the pulps that further causes an increase in tensile and burst indices, while the addition of NFC in beaten pulps has reduced more of the paper porousness as compared to unbeaten pulps.

Meanwhile, Syed et al. (2016a,b) used various types of seagrass plants in the study and have processed them by using two methods that are hand-beating and mechanical blending. The hand-beating is conducted by using wooden mallet for 20 minutes to spread out the fibers until they become soft. The study showed that hand-beating method has improved tensile and breaking length of the paper due to more flexible fiber caused by fibrillation, fiber shortening, and fines formation, while mechanical blending has cut the fiber into small fragments and results in increasing the bonding strength of the paper. Either hand-beating or mechanical blending has improved mechanical paper properties except longer time is required to obtain better fiber formation and distribution to produced paper.

The most common beating method used is by grooving the metallic plates of disk refiner with pulp fibers in the presence of water, which allows the fiber movement through the beating machine. There are two types of beating machines, laboratory Valley Beater and PFI Mill (Dolle and Bajrami, 2021). Valley Beater machine is filled with diluted 1.57% fiber consistency and runs for 3 minutes without load. The 5.5 kg of weight is added to the lever arm after the pulp is properly disintegrated to initiate beating process. Five samples of total 1.2 L are withdrawn from the beater at equal freeness intervals for pulp freeness and papermaking. Meanwhile for PFI Mill machine, the process involves fiber soaking and disintegration before 2 L of distilled water is added to obtain 1.2% pulp consistency. The pulp suspension is dewatered and diluted until the pulps weight of 240 g is used for 10% stock consistency. This beating process involves a temperature of 23°C. Both of the beating machines have shown similar trends of mechanical paper properties; only Valley Beater is suitable for beating large amounts of fibers compared to PFI Mill.

Two parts of kenaf fibers, bast and core, and sugarcane bagasse are blended with different ratios and beating time. The mixing of short and long fibers has been studied before by Nandkumar (2009) to overcome poor drainage and low wet strength of softwood fibers, and the study by Faiza (2019) has shown mixing of shorter fibers of kenaf core and bagasse with longer fibers of kenaf bast has improved mechanical properties of paper. Blended kenaf bast and kenaf core paper have shown better tensile and tear indices when kenaf bast proportion ratio is increased from 50 to 70. Blended kenaf bast and bagasse also showed enhanced tensile and tear indices paper properties and, however, decreased as proportion ratio of bagasse is increased to 70, and yet the situation is overcome as the beating time increases to 3 minutes. Similar situation is applied to blended kenaf bast–core–bagasse paper that earlier has shown good tensile index at 79.5 Nm/g is decreased to 58.8 Nm/g at 1 minute of beating but drastically increased to 94.1 Nm/g while 4.5 kPa (m/g) of burst index decreased more than half and risen up at 2 minutes of beating to 3.9 kPa (m/g).

Acknowledgments

The authors would like to thank the Ministry of Higher Education for the grant of Vot Nos. 5524855 and 6369111 under Fundamental Research Grant Scheme (FRGS) and Higher Education Center of Excellence (HICoE), respectively. The authors also would like to thank all laboratory technical team members from Institute of Tropical Forestry and Forest Products (INTROP), Universiti Putra Malaysia, Faculty of Bioengineering and Technology, Universiti Malaysia Kelantan, Faculty of Applied Sciences, Universiti Teknologi MARA (Jengka), and Faculty of Chemical and Energy Engineering & Centre for Advanced Composite Materials, Universiti Teknologi Malaysia for their endless support.

References

Abd El-Sayed, E.S., El-Sakhawy, M., Sakhawy, M., 2020. Non-wood fibers as raw material for pulp and paper industry. Nordic Pulp Pap. Res. J. 35 (2), 215–230. Available from: https://doi.org/10.1515/npprj-2019-0064.

Ahmed, A.F., Islam, M.Z., Mahmud, M.S., Sarker, M.E., Islam, M.R., 2022. Hemp as a potential raw material toward a sustainable world: a review. Heliyon 08753.

Akter, S., Sadekin, M.N., Islam, N., 2020. Jute and jute products of Bangladesh: contributions and challenges. Asian Bus. Rev. 10 (3), 143–152.

An, X., Liu, J., Liu, L., Zhang, H., Nie, S., Cao, H., et al., 2020. Improving the flexibility of bamboo mechanical pulp fibers for production of high soft tissue handsheets. Ind. Crop. Prod. 150, 112410.

Ativo, A.O., Pintor, L.L., 2022. Biomass from agricultural wastes for renewable energy in the Philippines. Biomethane Through Resource Circularity. CRC Press, pp. 151–156.

Azeez, M.A., 2018. Pulping of non-woody biomass. Pulp Pap. Process. 55–86.

Bajpai, P., 2010. Environmentally Friendly Production of Pulp and Paper. John Wiley & Sons, Inc, Hoboken, NJ.

Bhagwat, S., Karlekar, O., Padalkar, S., Chaudhari, S., Kulkarni, K., 2020. Nature of CSF based on beating time in fiber reinforced cotton rag. In: 2020 IEEE Pune Section International Conference (PuneCon), pp. 88–91.

Bhardwaj, N.K., Duong, T.D., Nguyen, K.L., 2004. Pulp charge determination by different methods: effect of beating/refining. Colloids Surf. A: Physicochem. Eng. Asp. 236 (1–3), 39–44.

Bhardwaj, S., Bhardwaj, N.K., Negi, Y.S., 2019. Cleaner approach for improving the papermaking from agro and hardwood blended pulps using biopolymers. J. Clean. Prod. 213, 134–142.

Bheel, N., Tafsirojjaman, T., Liu, Y., Awoyera, P., Kumar, A., Keerio, M.A., 2021. Experimental study on engineering properties of cement concrete reinforced with nylon and jute fibers. Buildings 11, 454.

Bouasker, M., Belayachi, N., Hoxha, D., Al-Mukhtar, M., 2014. Physical characterization of natural straw fibers as aggregates for construction materials applications. Materials 7, 3034–3048.

Chaiarrekij, S., Apirakchaiskul, A., Suvarnakich, K., Kiatkamjornwong, S., 2012. Kapok I: Characteristics of kapok fiber as a potential pulp source for papermaking. BioResources 7 (1), 475–488.

CNBM Paper Pulp, 2020. The Truth about Pulp Beating. Retrieved from https://www.paperpulpingmachine.com/truth-about-pulp-beating/.

Deykun, I.R.Y.N.A., Halysh, V., Barbash, V., 2018. Rapeseed straw as an alternative for pulping and papermaking. Cellulose Chem. Technol. 52 (9–10), 833–839.

Dolle, K., Bajrami, B., 2021. In situ precipitated calcium carbonate in the presence of pulp fibers—a beating study. J. Eng. Res. Rep. 20 (8), 1–17.

Dwiky, M.I., Kumala, S.N., Fety, I.R., Khaliq, F.N., 2019. The effect of NaOH concentration variation in the process of paper making from bamboo fiber. IOP Conf. Series: Mater. Sci. Eng. 535 (1), 012008.

El-Abbassi, F.E., Assarar, M., Ayad, R., Bourmaud, A., Baley, C., 2019. Compos. Part. A 128, 105677.

Faiza, E.A.E.A., 2019. Evaluation of Paper Properties Made from Bagasse and Kenaf Fiber, and Their Blends (Unpublished Master's thesis). Sudan University of Science and Technology, Khartoum, Sudan.

Ferdous, T., Ni, Y., Quaiyyum, M.A., Uddin, M.N., Jahan, M.S., 2021. Non-wood fibers: relationships of fiber properties with pulp properties. ACS Omega 6 (33), 21613–21622.
Gajdačová, P., Hýsek, Š., Jarský, V., 2018. Utilisation of winter rapeseed in wood-based materials as a solution of wood shortage and forest protection. BioResources 13 (2), 2546–2561.
Goel, A., Jewell, M., Tschirner, U., Ramaswamy, S., 2000. Comparison of refining response of Soda AQ and Organosolov Wheat Straw pulps with traditional wood pulps. In: TAPPI Pulping/Process and Product Quality Conference, Proceedings of the 2000 TAPPI Pulping/Process and Product Quality Process, Boston, MA, United States, 11/5/00, pp. 1071–1081.
Goudenhooft, C., Bournaud, A., Baley, C., 2019. Flax (*Linum usitatissimum* L.) fibers for composite reinforcement: exploring the link between plant growth, cell walls development, and fiber properties. Front. Plant. Sci. 10, 411.
Guo, S., Huaiyu, Z., Chengfeng, Z., Shinyu, F., Heijnesson-Hultern, A., Basta, J., et al., 2009. Pulp and fiber characterization of wheat straw and eucalyptus pulps—a comparison. Bioresources 4 (3), 1006–1016.
Haile, A., Gebino, G., Tesfaye, T., Mengie, W., Ayele, M., Abuhay, A., et al., 2021. Utilization of non-wood biomass for pulp manufacturing in paper industry: case of Ethiopia. Biomass Convers. Biorefinery 1–19.
Holik, H., 2006. Handbook of Paper and Board. WILEY, Weinheim.
Isenberg, H.I., 1967. Pulp and Paper Microscopy, third ed. The Institute of Paper Chemistry, Wisconsin.
Junle, M.A., (2021). The operating logic of Chinese capital in Africa from historical perspective: a case study of sisal industry in Tanzania. Retrieved from https://commons.ln.edu.hk/videos/895.
Kiaei, M., 2014. Investigation on biometrical properties and mineral content of rice residues and its application in pulp and paper production. Adv. Environ. Biol. 8 (13), 952–959.
Kumar, A., Singh, A.P., Singh, A.K., 2016. Comparative study of cellulose nanofiber blending effect on properties of paper made from bleached bagasse, hardwood and softwood pulps. Cellulose 23, 2663–2675.
Laukala, T., Ovaska, S.S., Tanninen, P., Pesonen, A., Jordan, J., Backfolk, K., 2019. Influence of pulp type on the three-dimensional thermomechanical convertibility of paperboard. Cellulose 26 (5), 3455–3471.
Liu, W., Hui, C., Wang, F., Wang, M., Liu, G., 2018. Review of the resources and utilization of bamboo in China. Bamboo-Curr. Fut. Prospect. 133–142.
Lumiainen, J., 2000. Refining of chemical pulp. Papermaking Part. 1, 86–122.
Małachowska, E., Lipkiewicz, A., Niemczyk, M., Dubowik, M., Boruszewski, P., Przybysz, P., 2021. Influences of fiber and pulp properties on papermaking ability of cellulosic pulps produced from alternative fibrous raw materials. J. Nat. Fibers 18 (11), 1751–1761.
Malik, S., Rana, V., Joshi, G., Gupta, P.K., Sharma, A., 2020. Valorization of wheat straw for the paper industry: pre-extraction of reducing sugars and its effect on pulping and papermaking properties. ACS Omega 5 (47), 30704–30715.
Mamaye, M., Kiflie, Z., Feleke, S., Yimam, A., 2020. Evaluation of soda delignification and single-stage hydrogen peroxide bleaching for the Ethiopian sugarcane bagasse for paper production. Sugar Tech. 22 (4), 706–717.
Mari, E.L., Austria, C.O., Torres, A.S., Domingo, E.P., 2019. Residual grade and waste abaca fibers as reinforcement for packaging and printing/writing papers from recycled fiber. Philippine J. Sci. 148 (2), 349–358.
Megra, M.B., Bachheti, R.K., Tadesse, M.G., Worku, L.A., 2022. Evaluation of pulp and papermaking properties of *Melia azedarach*. Forests 13 (2), 263.

Morais, J., James, J., Hinds, Z., Smith, W., Kelly, B., Hequet, E., 2020. A method to improve cotton fiber length measurement for laboratory analysis. MethodsX 7, 100859.

Musio, S., Mussig, J., Amaducci, S., 2018. Optimizing hemp fiber production for high performance composite applications. Fornt. Plant. Sci. 9, 1702.

Naik, B.H., Bhardwaj Nishi, K., Ray, A.K., 2009. Statistical Evaluation of the Effect of Refining on Fiber Charge of Bagasse Pulp.

Nandkumar, P., 2009. Effects of pulp blending on strength properties of *Ipomoea carnea* Jacq. Asian J. Chem. 21 (6), 4571–4576.

Nasroun, T.H., 2019. Effect of cooking mixtures of kenaf and bagasse and blending their pulps on paper properties. Eff. Cook. Mixtures Kenaf Bagasse Blending Their Pulps Pap. Prop. 37 (2), 11.

Noor Abbas, A.G., Abdul Aziz, F.N.A., Abdan, K., Mohd Nasir, N.A., Norizan, M.N., 2022. Kenaf fiber reinforced cementitious composites. Fibers 10 (3), 1–24.

Nur Aqeela, M.Y., Ainun, Z.M.A., Jawaid, M., 2018. Effect of pretreatment concentration on pulp blending between oil palm empty fruit bunch and *citronella* leaf fibers in terms of pulp and paper properties. IOP Conf. Series: Mater. Sci. Eng. 368, 1–8.

Oliveira Filho, E.G., Luz, F.S., Fujiyama, R.T., Silva, A.C., Candido, V.S., Monteiro, S.N., 2020. Effect of chemical treatment and length of raffia fiber (*Raphia vinifera*) on mechanical stiffening of polyester composites. Polymers 12, 2899.

Pati, R.K., Vrat, P., 2010. Economic paper blending optimization model with competing materials. Manage. Environ. Quality. 21 (5), 602–617. Available from: https://doi.org/10.1108/14777831011067917.

Patxullayev, S.U., Islambekova, N.M., Kulmetov, M., Ochilov, T.A., Valieva, Z.F., 2020. Determination of the quality characteristics of fibers obtained from mulberry bark. Int. J. Recent Technol. Eng. 8 (5), 5647–5651.

Purnawati, R., Febrianto, F., Wistara, I.N.J., Nikmatin, S., Hidayat, W., Lee, S.H., et al., 2018. Physical and chemical properties of kapok (*Ceiba pentandra*) and balsa (*Ochroma pyramidale*) fibers. J. Korean Wood Sci. Technol. 46 (4), 393–401.

Qian, K., 2021. Environmental concerns of the pulp and paper industry: focusing on household and sanitary paper products. Decision Science for Future Earth. Springer, Singapore, pp. 181–195.

Rafidah, D., Ainun, Z.M.A., Hazwani, H.A., Rushdan, I., Luqman, C.A., Sharmiza, A., et al., 2017. Characterisation of pulp and paper manufactured from oil palm empty fruit bunches and kenaf fibers. Pertanika J. Tropikal Agric. Sci. 40 (3), 449–458.

Rattanawongkun, P., Kerddonfag, N., Tawichai, N., Intatha, U., Soykeabkaew, N., 2020. Improving agricultural waste pulps via self-blending concept with potential use in moulded pulp packaging. J. Environ. Chem. Eng. 8 (5), 104320.

Rodriguez, M., Rodriguez, A., Bayer, R.J., Vilaseca, F., Girones, J., Mutje, P., 2010. Determination of corn stalk fibers' strength through modeling of the mechanical properties of its composites. BioResources 5 (4), 2535–2536.

Saleh, M., Asa'ari, A.Z.M., Jasmani, L., Othman, M.S.H., Tahir, P.M., Harun, J., et al., 2012. Kenaf – A Potential Fiber for Pulp and Paper Manufacture. Malaysian Timber Industry Board (MTIB), Kuala Lumpur, Malaysia.

Shaikh, A.J., 1990. Blending of cotton stalk pulp with bagasse pulp for paper making. Biol. Wastes 31 (1), 37–43.

Sharma, N., Tripathi, S.K., Bhardwaj, N.K., 2020. Utilization of sarkanda for making pulp and paper using elemental chlorine free and total chlorine free bleaching processes. Ind. Crop. Prod. 149, 112316.

Shon, C.S., Mukashev, T., Lee, D., Zhang, D., Kim, J.R., 2018. Can common reed fiber become an effective construction material? Physical and thermal properties of mortar mixture containing common reed fiber. Sustainability 11, 903.

Sumardi, I., Hadiyane, A., Rumidatul, A., Melani, L., 2020. Characteristics of empty palm bunch fibers as alternative pulp material. Am. J. Appl. Sci. 17, 129–134.

Syed, N.N.F., Zakaria, M.H., Bujang, J.S., 2016a. Paper from seagrass fiber. BioResources 11 (2), 5358–5380.

Syed, N.F.N., Zakaria, M.H., Bujang, J.S., 2016b. Fiber characteristics and papermaking of seagrass using hand-beaten and blended pulp. BioResources 11 (2), 5358–5380.

Tamta, K., Kalita, B.B., 2020. Properties of roselle and its blends. Int. J. Curr. Microbiol. Appl. Sci. 9 (10), 3616–3621.

TAPPI, 2000. Laboratory beating of pulp (PFI Mill method). TAPPI T248 SP-00, pp. 1–6.

Todkar, S.S., Patil, S.A., 2019. Review on mechanical properties evaluation of pineapple leaf fiber (PALF) reinforced polymer composites. Compos. Part B 1–43.

Tripathi, S., Singh, S., Gangwar, A., Mishra, O.P., Chakrabarti, S.K., Bhardwaj, N.K., et al., 2013. Blending of banana stem with wheat straw and bagasse to enhance physical strength properties of paper. Cellulose 2300, 4.

Ververis, C., Georghiou, K., Christodoulakis, N., Santas, P., Santas, R., 2004. Fiber dimensions, lignin and cellulose content of various plant materials and their suitability for paper production. Ind. Crop. Prod. 19 (3), 245–254. Available from: https://doi.org/10.1016/j.indcrop.2003.10.006.

Waller, V., Wilsby, A., 2019. Abaca in the Philippines, an Overview of a Potential Important Resource for the Country: Relating the Tensile Strength of the Single Fiber to the Microfibrilar Angle.

Yu, M., Zhang, H., Liu, Z., Ge, Z., Kong, F., Shao, H., et al., 2019. Effects of fiber dimension and its distribution on the properties of lyocell and ramie fibers reinforced polylactide composites. Fibers Polym. 20 (8), 1726–1732.

Zahran, M.A., El-Amier, 2014. Ecology and establishment of fiber producing taxa naturally growing in the Egyptian deserts. Egypt. J. Basic. Appl. Sci. I 144–150.

Index

Note: Page numbers followed by "*f*" and "*t*" refer to figures and tables, respectively.

Q

R

S

T

CPI Antony Rowe
Eastbourne, UK
July 08, 2023